SURFACE MECHANICS

SURFACE MECHANICS

FREDERICK FONGSUN LING

Rensselaer Polytechnic Institute

A Wiley Interscience Publication

JOHN WILEY & SONS, New York • London • Sydney • Toronto

Library of Congress Cataloging in Publication Data:

Ling, Frederick Fongsun.
   Surface mechanics.

   Includes bibliographical references.
   1. Surfaces (Technology).   2. Continuum mechanics.
3. Tribology.   I. Title.

TA418.7.L55   531   72-10012
ISBN 0-471-53905-8

Printed in the United States of America

10 9 8 7 6 5 4 3 2 1

To

HELEN AND FRANK

I often say that when you can measure what you are speaking about, and express it in numbers, you know something about it, when you cannot measure it in numbers, your knowledge is of a meager and unsatisfying kind; it may be the beginning of knowledge, but you have scarcely, in your thoughts, advanced to the stage of science whatever the matter may be.

LORD KELVIN

PREFACE

Surface phenomena have in recent years been receiving increasing attention in the study of materials science and engineering, notably the field of activity that has acquired a new designation—tribology. Surface behavior of all types is becoming of interest, and the awareness of the fact that bulk behavior of materials is often critically affected by surface conditions is increasing. This field of endeavor, so designated, covers several traditional academic disciplines. The name "materials" not only underscores the interdependence of various disciplines for fruitful pursuits but also brings to the fore the interrelatedness of several professional fields.

This book was written to satisfy a twofold purpose. The first is to set down concrete examples dealing with one facet of the body of interrelated knowledge, that of surface mechanics, which seems appropriate in filling a void within the larger context of surface physics. The second is to provide a collection of basic tools relevant to quantitative studies of problems involving surfaces.

Classical continuum mechanics are used throughout, that is, the field theory which assumes that the material body is indefinitely divisible while retaining its defining properties or continuum physics. In choosing the phenomenological method for describing materials, its preference over electron, molecular, or atomic theories of material is not inferred. In fact, newer continuum theories to be described briefly include first approximation of microscopic phenomena. Moreover, while the bulk of the text is analytical and theoretical; much of the theories are buttressed by decades of experience accrued through experimental mechanics.

By limiting the scope to surface mechanics, the second purpose is better illuminated, that is, to offer a set of critical ingredients for solving problems encountered in various professions while not committing space to professional problems. Not the least important is that quantitative tools are often catalytic to new discoveries in surface chemistry and surface physics.

Surface mechanics pertains to surfaces, to be sure, but it was coined also to characterize the notion that information on the surface may be obtained analytically and rigorously without the encumbrance of the entire solution. This may be achieved by employing various transform methods. It now embraces continuum treatment of surface, surface layer, and interface phenomena.

There are ten chapters and an appendix in the book. The first chapter provides a cursory review of irreversible thermodynamics which serves as a hub for interacting sets of field and constitutive equations. The appendix surveys singular integral equations as they may impinge on the subject matter. The remaining chapters are devoted to selected models and solutions which are considered basic to aspects of surface studies and descriptions of key experimental results. These fall, respectively, under categories on heat conduction in solids, interface temperatures, classical elasticity, thermoelasticity, viscoelasticity, perfect plasticity, rough surfaces, chemical effects, and applications. Smooth surfaces are used where permissible and nonsmooth surfaces used where necessary. Emphasis has been given to surfaces of bodies which involve motion and entail interaction of field, for example, thermoelastic.

While simplicity is sought, mathematical complexity has not been avoided when the situation so demands. I have used the material in a series of lectures bearing the same name given as a course on special topics in mechanics at Rensselaer Polytechnic Institute.

I am greatly indebted to former students who attended the lectures and made useful comments on the text material during its development.

Also, particular thanks are due to those who have graciously read the manuscript and have offered many suggestions: Professor H. Blok, Director, Institute of tribomechanics, University of Technology, Delft, Holland; Professor R. Courtel, Directeur de Recherches au Centre National de la Recherche Scientifique and Directeur, Laboratoire Mécanique des Surfaces, l'Institut Supérior des Matériaux et de la Construction Mécanique, France; Professor D. Dowson, Director, Institute of Tribology and Co-Director, Bioengineering Group for the Study of Human Joints, University of Leeds, England; Professor M. D. Hersey, Brown University; Professor W. F. Hughes, Carnegie-Mellon University; P. M. Ku. Director, Department of Fluids and Lubrication Technology, Southwest Research Institute; Dr. W. R. Osgood, Washington, D. C.; Professor E. A. Saibel, Carnegie-Mellon University; Professor M. C. Shaw, Head, Department of Mechanical Engineering and Director, Processing Research Institute, Carnegie-Mellon University; and Dr. D. Tabor, Head, Department of Surface Physics, Cavendish Laboratory, University of Cambridge, England.

During the last year of the preparation of this book, I was a 1970 National Science Foundation Senior Postdoctoral Fellow.

To Miss J. E. Doocey, I wish to express my special thanks for an excellent job of technical typing. Thanks are due also to Mr. P. S. Kounas for a careful job reading the proof.

FREDERICK FONGSUN LING

Troy, New York
April 1972

CREDITS FOR ILLUSTRATIONS AND TABLES

The author wishes to thank the following publishers for permission to use these illustrations.

Figures 3.4 (p. 67), 3.5 (p. 67), and 3.6 (p. 68): F. F. Ling and E. Saibel, "Thermal Aspects of Galling of Dry Metallic Surfaces in Sliding Contact," *Wear*, **1**, 80–91 (1957).

Figures 8.2 (p. 174) and 8.8 (p. 181): I. V. Kragelsky and N. B. Demkin, "Contact Area of Rough Surfaces," *Wear*, **3**, 170–187 (1960).

Figure 4.7 (p. 94): *Recent Advances in Engineering Science, II*, A. C. Eringen, ed., Gordon and Breach Science Publisher (1965).

Figures 6.3 (p. 150) and 6.4 (p. 155): *American Society of Lubrication Engineers Transactions*, **10**, 294–301 (1967).

Figures 8.9 (p. 184), 8.10 (p. 186), 8.11 (p. 187), 8.12 (p. 189), 8.13 (p. 190), and 8.14 (p. 191): F. F. Ling, *Journal of Applied Physics*, **29**, 1168–1174 (1958).

Figure 9.1 (p. 198): V. I. Likhtman, E. D. Shchukin, and P. A. Rehbinder, *Physicochemical Mechanics of Metals*, Academy of Sciences of the U. S. S. R. (English translation, 1962).

Figures 10.4 (p. 209), 10.5 (p. 212), and 10.6 (p. 213): H. Blok, "The Dissipation of Frictional Heat," *Applied Scientific Research*, **A5**, 151–181 (1954).

Figures 10.10 (p. 216), 10.11 (p. 217), 10.12 (p. 217), and 10.13 (p. 217): B. Jacobson, "On the Lubrication of Heavily Loaded Spherical Surfaces Considering Surface Deformations and Solidification of the Lubricant," *Acta Polytechnica Scandinavica*, **Me 54** (1970).

Figures 10.16 (p. 221), 10.17 (p. 222), 10.18 (p. 223), 10.19 (p. 224), and 10.20 (p. 225): F. W. Carter, "On the Action of a Locomotive Driving Wheel," *Proceedings of the Royal Society*, **A112,** 151–157 (1926).

Figures 10.36 (p. 239), 10.37 (p. 239), 10.38 (p. 240), 10.39 (p. 240), 10.40 (p. 241), 10.41 (p. 242), 10.42 (p. 243), 10.43 (p. 243), and 10.44 (p. 244): W. T. Chen, "Stresses in Some Anisotropic Materials Due to Indentation and Sliding," *International Journal of Solids and Structures*, **5,** 191–214 (1969).

Figures 10.41 (p. 242), 10.42 (p. 242), 10.43 (p. 243), and 10.44 (p. 243), Tables 10 (p. 244), 11 (p. 245), 12 (p. 245), and 13 (p. 246), and Figures 10.45 (p. 246), 10.46 (p. 247), and 10.47 (p. 244): W. T. Chen and P. A. Engel, "Impact and Contact Stress Analysis in Multilayer Media," *International Journal of Solids and Structures*, **8,** 1257–1281 (1972).

Figures 10.56 (p. 255), 10.57 (p. 256), 10.58 (p. 257), 10.59 (p. 257), and 10.60 (p. 258): *International Journal of Heat and Mass Transfer*, **14,** 199–206 (1970).

Figures 10.61 (p. 259), 10.62 (p. 264), 10.63 (p. 265), and 10.64 (p. 265): J. J. Wu and F. F. Ling, "A Method for Micro-hardness Analysis of an Elastoplastic Material," *Developments in Mechanics*, **6,** 359–372 (1971).

Figure 10.65 (p. 266): A. W. Crook, "Simulated Gear-Tooth Contacts: Some Experiments upon Their Lubrication and Subsurface Deformation," *Proceedings of the Institute of Mechanical Engineers*, **171,** 187 (1957).

Figures 10.74 (p. 275), 10.75 (p. 275), 10.76 (p. 276), and 10.77 (p. 277): D. Tabor, "The Hardness of Solids," *Review of Physics in Technology*, **1,** 145–179 (1970).

Figures 10.1 (p. 203) and 10.2 (p. 204): F. F. Ling, "A Quasi-Iterative Method for Computing Interface Temperature Distributions," *Zeitschrift für Angewandte Mathematik und Physik*, **X,** 461–474 (1959).

Figures 10.81 (p. 285), 10.82 (p. 286), 10.83 (p. 287), 10.84 (p. 288), and 10.85 (p. 289): M. C. Mow and F. F. Ling, "On Weeping Lubrication Theory," *Zeitschrift für Angewandte Mathematik und Physik*, **XX**, 156–166 (1969).

Acknowledgment is also made to the American Society of Mechanical Engineers for the use of the following figures.

Figures 2.2 (p. 23) and 2.3 (p. 27): F. F. Ling and C. W. Ng, *Proceedings of the 4th U.S. National Congress of Applied Mechanics*, 1343–1349 (1962).

Figure 10.66 (p. 267): K. L. Johnson, *Proceedings of the 4th U.S. National Congress of Applied Mechanics*, 971–975 (1962).

Figure 3.1 (p. 63): E. W. Gaylord, W. F. Hughes, F. C. Appl, and F. F. Ling, *Trans. ASME*, **80,** 307–310 (1958).

Figures 3.2 (p. 65) and 3.3 (p. 66): H. H. H. Shu, E. W. Gaylord, and W. F. Hughes, *Trans. ASME*, **86,** 417–422 (1964).

Figures 3.7 (p. 69), 3.8 (p. 69), and 3.9 (p. 70): F. F. Ling and T. E. Simkins, *Trans. ASME*, **85,** 481–487 (1963).

Figures 3.10 (p. 74), 3.11 (p. 76), and 3.12 (p. 81): F. F. Ling, *Trans. ASME*, **91,** 397–405 (1969).

Figure 4.12 (p. 101): M. Hetenyi, *J. Appl. Mech.*, **27,** 289–296 (1960).

Figure 10.9 (p. 216): F. F. Ling and V. C. Mow, *Trans. ASME*, **87,** 729–734 (1965).

Figures 10.52 (p. 252), 10.53 (p. 252), and 10.54 (p. 253): S. C. Hunter, *J. Appl. Mech.*, **28,** 611–617 (1961).

Figure 6.2 (p. 148): G. R. Abrahamson and J. N. Goodier, *J. Appl. Mech.*, **28,** 608–610 (1961).

Figure 7.6 (p. 164): F. F. Ling, *Trans. ASME*, **80,** 1113–1120 (1958).

Figures 8.3 (p. 175) and 8.4 (p. 176): J. Wallach; Figures 8.5 (p. 177), 8.6 (p. 178), and 8.7 (p. 178): J. B. P. Williamson, J. Pullen, and R. T. Hunt, *Surface Mechanics—a symposium volume* (1969).

Figures 10.16 (p. 221), 10.17 (p. 222), 10.18 (p. 223), 10.19 (p. 224), and 10.20 (p. 225): J. J. Kalker, *J. Appl. Mech.*, **37,** 677–688 (1970).

Figures 10.21 (p. 226) and 10.22 (p. 226): Y. C. Pao, T. S. Wu, and Y. P. Chiu, *J. Appl. Mech.*, **38,** 608–614 (1971).

Figures 10.23 (p. 227), 10.24 (p. 228), 10.25 (p. 229), and 10.26 (p. 230): D. C. Gakenheimer, *J. Appl. Mech.*, **38,** 99–110 (1971).

Figures 10.27 (p. 231), 10.28 (p. 232), and 10.29 (p. 232): Y. O. Tu and D. G. Gazis, *J. Appl. Mech.*, **31,** 659–666 (1964).

Figures 10.30 (p. 233) and 10.31 (p. 234): S. L. Pu and M. A. Hussain, *J. Appl. Mech.*, **37**, 859–861 (1970).

Figures 10.32 (p. 235), 10.33 (p. 236), 10.34 (p. 236), and 10.35 (p. 237): G. B. Warburton, J. D. Richardson, and J. J. Webster, *J. Appl. Mech.*, **38**, 148–156 (1971).

Figures 10.47 (p. 248), 10.48 (p. 249), 10.49 (p. 249), and 10.50 (p. 250): H. D. Conway, H. C. Lee, and R. G. Bayer, *J. Appl. Mech.*, **37**, 159–162 (1970).

Figures 10.67 (p. 269), 10.68 (p. 270), 10.69 (p. 271), 10.70 (p. 272), 10.71 (p. 273), 10.72 (p. 274)), and 10.73 (p. 275): C. H. Yew and W. Goldsmith, *J. Appl. Mech.*, **31**, 635–646 (1964).

Figures 10.78 (p. 278), 10.79 (p. 279), and 10.80 (p. 282): M. B. Peterson and F. F. Ling, *Trans. ASME*, **92**, 535–542 (1970).

CONTENTS

CHAPTER 1

CLASSICAL CONTINUUM MECHANICS

Field equations of classical continuum mechanics and various constitutive equations are surveyed in this chapter. No attempts are made to derive these equations except in the cursory review of irreversible thermodynamics of solids which serves as the hub for interacting fields. The degree of generality of the equations shown is governed by that which is required in the body of the text. In the main, consideration of simplicity prevails. For example, Cartesian tensor notation is used for its simplicity and conciseness. More general consideration in this regard, while desirable, is beyond the purpose of this book, which is intended for those who are interested in surface science and who need not be students of mechanics.

A. BALANCE OF MOMENTUM

Cauchy's first law of motion, which is a statement of balance of linear momentum,

$$\rho \ddot{u}_j = \sigma_{ij,i} + \rho f_j , \qquad [1.1]$$

in which $(i, j = 1, 2, 3)$, ρ is the mass density, u_i is the ith component of displacement, the super dot refers to partial differentiation with respect to time, σ_{ij} is the Cauchy stress tensor, the notation $(,i)$ refers to partial differentiation with respect to x_i which is the ith coordinate, repeated indices denote summation over all values of the index, and f_i is the ith component of body forces. There are three equations embodied in [1.1] and f_i may be composed of gravitational, inertial, or electromagnetic forces.

Equation 1.1 in general contains terms involving assigned couples and couple stresses, which become relevant for polar materials. In that case, the stress tensor is asymmetric. For the nonpolar case, Cauchy's second law of motion is a statement of the balance of moment of momentum in a body where linear momentum is already balanced:

$$\sigma_{ij} = \sigma_{ji} \; . \tag{1.2}$$

For detailed treatment of the subject the reader is referred to references elsewhere.[1,2]

B. ENERGY BALANCE

The fundamental law of energy balance or the first law of thermodynamics states

$$(d/dt)(\mathcal{K} + \mathcal{E}) = \mathcal{M} + \mathcal{Q}, \tag{1.3}$$

where $\mathcal{K}$ is the kinetic energy, $\mathcal{E}$ is the internal energy, $\mathcal{M}$ is the mechanical power, and $\mathcal{Q}$ is the nonmechanical power. The $\mathcal{K}$ for a body of volume V is

$$\frac{1}{2} \int_V \rho \, \dot{u}_i \, \dot{u}_i \, dV \; .$$

Defining an internal energy density ϵ, $\mathcal{E}$ may be written as

$$\int_V \rho \, \epsilon \, dV \; .$$

Note that ϵ may contain terms associated with chemical potential; a finite layer of V may be associated with the surface phase. The $\mathcal{M}$, in the absence of assigned couples and couple stresses, consists partly of

$$\int_S \sigma_{ij} \, n_j \, \dot{u}_i \, dS \; ,$$

where S is the boundary of V and n_j is the jth component of the external unit normal. Also from body forces,

$$\int_V \rho\, f_i\, \dot{u}_i\, dV \quad .$$

The Q may be expressed as the sum of two parts. First, there is the afflux of energy

$$\int_S h_i\, n_i\, dS \quad ,$$

where h_i is the ith component of the flux vector. Second, there is the supply of energy

$$\int_V \rho\, q\, dV \quad ,$$

where q is the supply per unit mass due to radiation, thermoelectric, and so on. Inserting the integrals above in [1.3], and invoking the divergence theorem which relates the surface integral to volume integral through

$$\int_S (\ )_j\, n_j\, dS = \int_V (\ )_{j,j}\, dV \quad ,$$

$$\frac{d}{dt}\int_V (\tfrac{1}{2}\, \rho\, \dot{u}_i\, \dot{u}_i + \rho\epsilon)\, dV = \int_V [(\sigma_{ij}\, \dot{u}_i)_{,j} + \rho\, f_i\, \dot{u}_i + \rho q + h_{i,i}]\, dV \quad .$$

Simplifying,

$$\int_V [\sigma_{ij,j} + \rho\, f_i - \rho\, \ddot{u}_i]\dot{u}_i\, dV + \int_V [\sigma_{ij}\, \dot{u}_{i,j} + h_{i,i} + \rho\, q - \rho\, \dot{\epsilon}]\, dV = 0 \quad .$$

$$[1.4]$$

The square-bracketed term in the first integral of [1.4] vanishes in view of [1.1]. The remaining integral, by virtue of the fact that V is arbitrary, leads to

$$\sigma_{ij}\, \dot{u}_{i,j} + h_{i,i} + \rho q = \rho\, \dot{\epsilon} \quad .$$

Using Euler's stretching tensor $\epsilon_{ij} = \tfrac{1}{2}(\dot{u}_{i,j} + \dot{u}_{j,i})$, the spin tensor

$\dot{\omega}_{ij} = \frac{1}{2}(\dot{u}_{i,j} - \dot{u}_{j,i})$, the fact that σ_{ij} is symmetrical, and that ω_{ij} is anti-symmetric, the relationship above becomes

$$\sigma_{ij}\,\dot{\epsilon}_{ij} + h_{i,i} + \rho\,q = \rho\,\dot{\epsilon} \quad . \tag{1.5}$$

This is equivalent to [1.3] and therefore a statement of the balance of energy. It has been assumed in all the considerations above that all functions are sufficiently smooth.

C. ENTROPY

In irreversible thermodynamics, the basic equations of classical thermostatics are applied to elements of volume in a moving material or in a mixture of materials. Thermodynamics is viewed primarily as a study of entropy. According to Gibbs, the thermodynamic substate is regarded as influencing the internal energy density ϵ. The substate is, in general, the parameters $v_\alpha(\alpha = 1, \ldots, n)$. The substate plus a single further dimensionally independent scalar suffices to determine ϵ, independently of time, place, motion, and stress. That is, a function f is assumed such that

$$\epsilon = f(\eta,\, v_1, \ldots, v_n,\, \chi_i) \quad , \tag{1.6}$$

where the parameter η is called the specific entropy for the particle χ_i. Dimensionally, the ratio ϵ to η gives that of temperature T. Equation 1.6 indicates that the thermodynamic state is obtained by adjoining the entropy η to the substate v_α. Also from [1.6], the temperature T and the thermodynamic tensions τ_α are defined by

$$T \equiv \frac{\partial \epsilon}{\partial \eta} \quad , \quad \tau_\alpha \equiv \frac{\partial \epsilon}{\partial v_\alpha} \quad . \tag{1.7}$$

Thus for a given χ_i, [1.6] and [1.7] yield

$$\dot{\epsilon} = T\,\dot{\eta} + \sum^{n} \tau_\alpha\,\dot{v}_\alpha \quad . \tag{1.8}$$

In the simplest case, when v_1 is the specific volume, τ_1 is the thermodynamic pressure. In the case of a homogeneous mixture, when the substate includes both the total volume and the masses of the constituents, then the tension τ_α corresponding to the mass of the constituent is called its chemical potential. When v_α is the deformation gradient, τ_α is the stress vector.

In the absence of chemical potentials and couple stresses, [1.8] takes the form

$$\rho \, \dot{\epsilon} = \rho \, T \, \dot{\eta} + \sigma_{ij} \, \dot{\epsilon}_{ij} \quad , \qquad [1.9]$$

which is the same as [1.5] if $(h_{i,i} + \rho q)$ had been equated to $\rho T \eta$. The equating of the aforementioned quantities comes from the second law of thermodynamics, which is not used here. Instead, [1.9] comes directly from the fundamental postulate 1.6.

It may be worth noting that if f is sufficiently smooth, the first of [1.7] may be inverted so as to give $\eta = \eta(T, v_\alpha)$. This, together with [1.6], gives $\epsilon = \epsilon(T, v_\alpha)$, while it yields with the second of [1.7]

$$\tau_\alpha = \tau_\alpha(T, v_\alpha) \quad . \qquad [1.10]$$

Similarly,

$$v_\alpha = v_\alpha(T, \tau_\alpha) \quad . \qquad [1.11]$$

These are the thermal equations of state.

D. CONSTITUTIVE RELATIONSHIP AND THE ENERGY EQUATION FOR AN ELASTIC SOLID

Defining the free energy function

$$\phi(\epsilon_{ij}, T) = \epsilon(\epsilon_{ij}, T) - T \, \eta(\epsilon_{ij}, T) \quad ,$$

the energy equation 1.9 may be put in the form

$$(\sigma_{ij} - \rho \frac{\partial \phi}{\partial \epsilon_{ij}}) \dot{\epsilon}_{ij} - \rho(\eta + \frac{\partial \phi}{\partial T}) \dot{T} = 0 \quad . \qquad [1.12]$$

Assuming that the bracketed terms are independent of $\dot{\epsilon}_{ij}$ and $\dot{T}$, then

$$\sigma_{ij} = \rho \frac{\partial \phi}{\partial \epsilon_{ij}} \quad , \quad \eta = - \frac{\partial \phi}{\partial T} \quad . \qquad [1.13]$$

Since $\phi(\epsilon_{ij}, T)$ is a scalar valued tensor function of one tensor variable, thus for isotropic elastic material

$$\phi(\epsilon_{ij}, T) = \phi(\vartheta_1, \vartheta_2, \vartheta_3, T) \quad,$$

where ϑ_1, ϑ_2, and ϑ_3 are the invariants of the strain tensor. The first of [1.13] yields

$$\sigma_{ij} = \rho \left(\frac{\partial \phi}{\partial \vartheta_1} \frac{\partial \vartheta_1}{\partial \epsilon_{ij}} + \frac{\partial \phi}{\partial \vartheta_2} \frac{\partial \vartheta_2}{\partial \epsilon_{ij}} + \frac{\partial \phi}{\partial \vartheta_3} \frac{\partial \vartheta_3}{\partial \epsilon_{ij}} \right) .$$

Letting T_0 be the temperature at which the material is stress free, assuming that ϕ may be expanded in a power series in terms of the ϑ's and $T' = (T - T_0)/T_0$, and retaining only the first-order terms,

$$\sigma_{ij} = C_1 \delta_{ij} \epsilon_{kk} + C_2 \epsilon_{ij} + C_3 \delta_{ij} T' \quad, \qquad [1.14]$$

in which C_1, C_2, and C_3 are constants and use has been made of the fact that

$$\frac{\partial \vartheta_1}{\partial \epsilon_{ij}} = \delta_{ij} \quad, \qquad \frac{\partial \vartheta_2}{\partial \epsilon_{ij}} = \delta_{ij} \vartheta_1 - \epsilon_{ji} \quad, \qquad \frac{\partial \vartheta_3}{\partial \epsilon_{ij}} = \epsilon_{ik} \epsilon_{jk} - \epsilon_{ji} \vartheta_1 + \delta_{ij} \vartheta_2$$

and δ_{ij} is the Kronecker delta. The above is of the form

$$\sigma_{ij} = \lambda \delta_{ij} \epsilon_{kk} + 2\mu \epsilon_{ij} - (3\lambda + 2\mu)\delta_{ij} \alpha (T - T_0) , \qquad [1.15]$$

in which $\lambda \equiv Ev/(1 + v)(1 - 2v)$ and $\mu \equiv E/2(1 + v)$ are the Lamé constants, α is the coefficient of linear thermal expansion, E is Young's modulus, and v is Poisson's ratio. In other wrods, [1.15] may be determined empirically once the form is known from [1.14]. This constitutive relationship is known as the Duhamel-Neumman law.

E.   CONSTITUTIVE RELATIONSHIP—HEAT CONDUCTION

Equations 1.5 and 1.9 lead to

$$h_{i,i} + \rho q = \rho T \dot{\eta} \quad .$$

By differentiating the second of [1.13], the above becomes

$$h_{i,i} + \rho q = \rho T \left(\frac{\partial \eta}{\partial \epsilon_{ij}} \dot{\epsilon}_{ij} + \frac{\partial \eta}{\partial T} \dot{T} \right)$$

or

$$-(h_{i,i} + \rho q) = \rho T \left(\frac{\partial^2 \phi}{\partial \epsilon_{ij} \partial T} \dot{\epsilon}_{ij} + \frac{\partial^2 \phi}{\partial T^2} \dot{T} \right) . \quad [1.16]$$

Since the specific heat at constant volume, C_V, is so defined as to give $-(h_{i,i} + \rho q) = \rho C_V \dot{T}$ for $\dot{\epsilon}_{ij} = 0$, therefore $C_V = (\partial^2 \phi / \partial T^2)T$. Also, from the first of [1.13]

$$\frac{\partial^2 \phi}{\partial \epsilon_{ij} \partial T} = \frac{1}{\rho} \frac{\partial \sigma_{ij}}{\partial T} . \quad [1.17]$$

Now the Fourier law of heat conduction is $-h_i = KT_{,i}$. Surface integral over an arbitrary volume leads to

$$-\int h_i n_i \, dA = \int K T_{,i} n_i \, dA \quad ,$$

and the divergence theorem leads to the volume integral

$$\int [h_{i,i} + (K T_{,i})_{,i}] dV = 0 \quad .$$

Since volume is arbitrary,

$$-h_{i,i} = (K T_{,i})_{,i} \quad . \quad [1.18]$$

Thus [1.15], [1.16], [1.17], and [1.18] lead to

$$(K T_{,i})_{,i} = \rho q + \rho C_V \dot{T} - (3\lambda + 2\mu)\alpha T_o \dot{\epsilon}_{kk} \quad , \quad [1.19]$$

in which T_0 has been substituted for T in the first approximation. Equation 1.19 is the generalized heat equation. Note that it is coupled with [1.15] and [1.1] so that they should be solved simultaneously.

Sections B, C, D, and the part of E above have been devoted to a cursory review of irreversible thermodynamics as a focal point for the meeting of the field equation and the constitutive equation for a very simple case of infinitesimal thermoelasticity. All the equations can be made more general, thus yield more general constitutive equations. However, in what follows, only a list of several constitutive relationships and other field equations are given.

F.   CONSTITUTIVE RELATIONSHIP—OTHER DIFFUSION TYPES

Fick's law governing diffusion is similar to Fourier's law, and the diffusion equation, which is similar to the heat equation, is

$$\dot{C} = (D\,C_{,i})_{,i} \quad , \qquad [1.20]$$

where C is the concentration and D is the diffusivity.[3,4]

Starting from Darcy's law governing the capillary flow in porous media, a diffusive-type constitutive equation may be found in the so-called equation of continuity:

$$\Gamma\,e_{,ii} = \dot{e} \quad , \qquad [1.21]$$

where Γ is the coefficient of consolidation and e is the dilation $\epsilon_{kk}/3$. This is coupled with a set of elasticity constitutive equations governing the soil skeleton:

$$\sigma_{ij} - \sigma\,\delta_{ij} = \lambda\,\delta_{ij}\,\epsilon_{kk} + 2\mu\,\epsilon_{ij} \quad , \quad [1.22]$$

where σ, the excess pore pressure, is related to e.[5]

G.   CONSTITUTIVE RELATIONSHIP—LINEARLY VISCOUS FLUID

From Newton's law governing shearing of viscous fluid, the Navier-Stokes law is derived for the isothermal case:

$$\sigma_{ij} = -p\,\delta_{ij} + \lambda\,\delta_{ij}\,\dot{\epsilon}_{kk} + 2\mu\,\dot{\epsilon}_{ij} \quad , \quad [1.23]$$

where $-p$ is the hydrostatic component of stress, μ is the shear viscosity, and $(\lambda + \frac{2}{3}\mu)$ is the bulk viscosity.[1,6]

H. CONSTITUTIVE RELATIONSHIP—PERFECTLY PLASTIC BODIES

For a perfectly plastic body and isothermal case:

$$\sigma_{ij} = \lambda \, \delta_{ij} \, \dot{\epsilon}_{kk} + 2\mu \, \dot{\epsilon}_{ij} \, , \qquad [1.24]$$

where λ and μ are not material constants, but they are functions of $\dot{\epsilon}_{ij}$.
  Defining the deviators such that

$$s_{ij} = \sigma_{ij} - \delta_{ij} \, \sigma$$

$$e_{ij} = \epsilon_{ij} - \delta_{ij} \, e \, , \qquad [1.25]$$

where $\sigma = \sigma_{ii}/3$ and $e = \epsilon_{ii}/3$ are mean stress and strain, respectively, [1.24] and [1.25] lead to

$$\sigma = \alpha \, \dot{e}$$

$$s_{ij} = 2\mu \, \dot{e}_{ij} \, , \qquad [1.26]$$

where α is the compressibility $(3\lambda + 2\mu)/3$. The above gives the flow rule while a yield condition must yet be imposed. One of the simplest of these is Mises' yield condition.[1,7,8]

$$\Sigma_2 = k^2 \, , \qquad [1.27]$$

where k is yield stress in shear and Σ_2 is the second invariant of the stress tensor, $(1/2!)\delta_{lm}{}^{ij}\sigma_{il}\sigma_{jm}$, and $\delta_{lm}{}^{ij}$ is the generalized Kronecker delta. The Tresca yield condition is

$$[(s_2 - s_3)^2 - 4k^2][(s_3 - s_1)^2 - 4k^2][(s_1 - s_2)^2 - 4k^2] = 0 \, , \qquad [1.28]$$

where s_1, s_2, and s_3 are the principal deviatoric stresses.

I. CONSTITUTIVE RELATIONSHIP—VISCOELASTIC BODIES

For simplicity, the whole class of viscoelastic material can be written as an expansion in infinite series of simple materials. When nonlinear terms are neglected from the time flux, the law is given below:

$$\sum_{k=0}^{n} A_k \left(\frac{\partial}{\partial t} \right)^k s_{ij} = \sum_{k=0}^{m} B_k \left(\frac{\partial}{\partial t} \right)^k e_{ij} \, , \qquad [1.28]$$

and a similar one for the mean stress and strain. The A_k's and B_k's are constants.[1] For fluid-like viscoelastic material the constitutive relationships involve the nonlinear terms in the time flux and are more complicated than those intended for this monograph.

J. CONSTITUTIVE RELATIONSHIP—MAXWELLIAN DIELECTRIC

The problem of formulating constitutive equations for moving and deforming bodies are outside of the province of this monograph. Recorded here are the Euclidean invariant constitutive equations of a Maxwellian dielectric:[1,9]

$$D_i = \epsilon_o E_i \quad , \quad H_i = \frac{1}{\mu_o} B_i \quad , \qquad [1.29]$$

where the ith component of the charge potential is shown as D_i, that of the electric field, as E_i, that of the current potential as H_i, that of the magnetic flux density as B_i, and ϵ_0, μ_0 are fundamental electromagnetic constants. The D_i and H_i are for the total charge including that due to polarization and magnetization of the material medium.

K. FIELD EQUATIONS OF CLASSICAL ELECTROMAGNETIC THEORY

The associated field equations for the constitutive equations 1.29 are[1,9]

$$J_{i,i} + \dot{Q} = 0 \quad , \qquad [1.30]$$

$$\epsilon_{ijk} E_{k,j} + \dot{B}_i = 0 \quad , \qquad [1.31]$$

$$B_{i,i} = 0 \quad , \qquad [1.32]$$

$$Q = D_{i,i} \quad , \qquad [1.33]$$

$$J_i = \epsilon_{ijk} H_{k,j} - \dot{D}_i \quad , \qquad [1.34]$$

$$B_i = \epsilon_{ijk} A_{k,j} \quad , \qquad [1.35]$$

$$E_i = -\dot{A}_i - V_{,i} \quad , \qquad [1.36]$$

where J_i is the ith component of the current density, Q is the charge density, A_i is the ith component of the magnetic potential, V is the electric potential, and ϵ_{ijk} is the permutation tensor in Cartesian coordinates.

CHAPTER 2

HEAT CONDUCTION IN SOLIDS

The heat equation, as a representative of the three diffusion types discussed earlier, is chosen for study first for its relative simplicity and for its required use in the treatment of thermoelasticity. Since only a few problems are explored here, the solution of a problem is shown in detail because of its wide estimated range of application. Many times, it is not the most general problem that is exposed.

A. TWO-DIMENSIONAL CASE OF A SEMIINFINITE SOLID

Figure 2.1 shows the model to be examined. Coordinates (x_1', x_2) are fixed with the semiinfinite solid moving in the positive x_1' direction at a constant rate V. Time is denoted by t. Heat is fed to a surface of the solid over a length of $2l$, the distribution $q(x_1)$ being arbitrary. Coordinates (x_1, x_2) are fixed to the center of heat source. Heat is convected away over the remaining portion of the surface. For thermal properties independent of temperature and constant initial temperature, the heat equation in the variable of temperature rise, T, is

$$\frac{\partial^2 T}{\partial x_1'^2} + \frac{\partial^2 T}{\partial x_2^2} = \kappa^{-1} \frac{DT}{Dt} , \qquad [2.1]$$

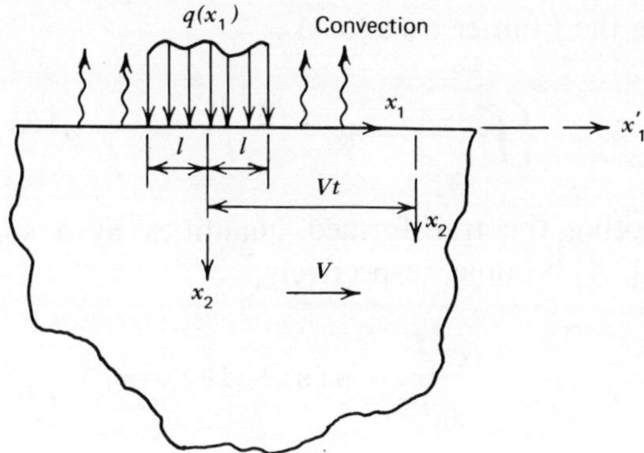

Figure 2.1

where κ is the thermal diffusivity and $D(\ )/Dt$ is the material derivative. Since $x_1 = x_1' + Vt$, $\partial^2 T/\partial x_1'^2 = \partial^2 T/\partial x_1^2$, and $\partial T/\partial t = DT/Dt - (\partial T/\partial x_1)(\partial x_1/\partial t) = DT/Dt - V\partial T/\partial x$. For quasistationary state, $\partial T/\partial t = 0$; therefore, $DT/Dt = V\partial T/\partial x_1$. The boundary conditions are that $-(qx_1) = K\partial T/\partial x_2$ under the heat source and $hT = K\partial T/\partial x_2$ elsewhere on the surface. The K is the thermal conductivity and h is the film coefficient. In addition, the regularity conditions are $\partial T/\partial x_1$, $\partial T/\partial x_2 \to 0$ as $(x_1^2 + x_2^2)^{1/2} \to \infty$. Now [2.1] is changed to the moving coordinates (x_1, x_2) for it is the temperature at the heat input zone that is of interest. For convenience, define the following dimensionless quantities: $\xi \equiv x_1/l$, $\eta \equiv x_2/l$, $P \equiv q/q_0$, $\phi \equiv TK/q_0 l$, $R \equiv Vl/\kappa$, and $H \equiv hl/K$. Equation 2.1 becomes

$$\frac{\partial^2 \phi}{\partial \xi^2} + \frac{\partial^2 \phi}{\partial \eta^2} = R \frac{\partial \phi}{\partial \xi} . \qquad [2.2]$$

The boundary conditions are

$$\frac{\partial \phi}{\partial \eta} = \begin{cases} -P(\xi) & (0 \leq |\xi| \leq 1, \ \eta = 0) \\ \\ H\phi & (1 < |\xi| < \infty, \ \eta = 0) \end{cases}, \qquad [2.3]$$

$$\phi \to 0 \quad \text{as} \quad (\xi^2 + \eta^2)^{1/2} \to \infty \qquad [2.4]$$

Taking the Fourier transform

$$\mathscr{F}\left\{ \ \right\} = \pi^{-1/2} \int_{-\infty}^{\infty} \left\{ \ \right\} e^{is\xi} d\xi \quad , \qquad [2.5]$$

and denoting the transformed quantities by a superposed $\sim$, [2.2] through [2.4] become, respectively,

$$\frac{d^2\tilde{\phi}}{d\eta^2} - s(s + iR)\tilde{\phi} = 0 \quad , \qquad [2.6]$$

$$-\frac{\partial\tilde{\phi}}{\partial\eta} = \tilde{P} \equiv \pi^{-1/2} \int_{-\infty}^{\infty} P^*(\xi)e^{is\xi} d\xi \quad , \qquad [2.7]$$

$$\tilde{\phi} \to 0 \quad \text{as} \quad \eta \to \infty \quad , \qquad [2.8]$$

where

$$P^*(\xi) = \begin{cases} P(\xi) & (0 \le |\xi| \le 1) \\ \\ -H\phi(\xi,0) & (1 < |\xi| < \infty) \end{cases} \qquad [2.9]$$

Equation 2.6 has a solution

$$\tilde{\phi} = Ae^{\sqrt{s^2 + isR}\ \eta} + Be^{-\sqrt{s^2 + isR}\ \eta} \quad , \qquad [2.10]$$

where A and B are constants. Enforcement of [2.8] shows $A = 0$ for $s > 0$; also enforcing [2.7] results for $\eta = 0$,

$$\tilde{\phi}(s) = \frac{\sqrt{1 - iR/s}}{\sqrt{s^2 + R^2}}\ \tilde{P} = f(s)\tilde{P} \quad , \qquad [2.11]$$

with $f(s)$ similarly extended in $s \le 0$. Carrying out the inverse transform

$$\mathscr{F}^{-1}\left\{ \ \right\} = \pi^{-1/2} \int_{-\infty}^{\infty} \left\{ \ \right\} e^{-is\xi} ds \quad , \qquad [2.12]$$

[2.11] becomes

$$\phi(\xi) = \int_{-\infty}^{\infty} f(s) \left[\int_{-\infty}^{\infty} P^*(\xi') e^{is\xi'} d\xi' \right] e^{is\xi} \, ds \quad .$$

Interchanging the order of integration

$$\phi(\xi) = \int_{-\infty}^{\infty} P^*(\xi') K(\xi,\xi') d\xi' \quad , \qquad [2.13]$$

where

$$K(\xi,\xi') = 2\pi^{-1} e^{R(\xi-\xi')} K_0 \left\{ R|\xi-\xi'| \right\} \quad , \qquad [2.14]$$

and $K_0\{\ \}$ is the modified Bessel function of the second kind and zero order. Equation 2.14 is actually the fundamental solution, that is, if a unit of heat P^* is imposed at ξ' the resulting surface temperature at ξ is $K(\xi, \xi')$.

A closer examination of [2.13] shows that [2.14] is the fundamental solution for P^* distributed over $-\infty \leq \xi \leq \infty$. The P^* is a pseudo heat source defined by [2.9]. It will be shown that the fundamental solution for the actual problem is not known and [2.13], with the aid of [2.9], is seen to be an integral equation of the second kind.

$$\phi(\xi) = -H \int_{-\infty}^{-1} \phi(\xi')K(\xi,\xi')d\xi' + \int_{-1}^{1} P(\xi')K(\xi,\xi')d\xi'$$

$$- H \int_{1}^{\infty} \phi(\xi') K(\xi,\xi')d\xi' \quad .$$

$$[2.15]$$

For a kernel as involved as [2.14], iterative procedure may be used.[10]

Two comments about [2.15] are worth mentioning: (1) where $R \gg 1$, [2.14] degenerates into

$$K(\xi,\xi') = \begin{cases} 2[\pi R(\xi-\xi')]^{-1/2} & (\xi-\xi' \geq 0) \\ \\ 0 & (\xi-\xi' < 0) \quad . \end{cases} \qquad [2.16]$$

This means that for $R \gg 1$, not only $K(\xi, \xi')$ is very simple but also the points to the left of the heat source do not get heated at all (i.e., the heat source is moving faster than the speed of propagation). By the same token, the region of heat, which is cooled by the region to the right, is not effected by cooling. Therefore, [2.15] would be well approximated by

$$\phi(\xi) = 2(\pi R)^{-1/2} \int_{-1}^{\xi} P(\xi')(\xi - \xi')^{-1/2} \, d\xi' \qquad (-1 \leq \xi \leq 1) \ .$$

$$[2.17]$$

(2) When $R \to 0$ (i.e., when $v \to 0$). It is well known that for the two-dimensional model shown except $h = 0$, then, when $R \to 0$, $\phi \to \infty$. However, [2.15] shows that this would not happen if $h \neq 0$, for the higher $\phi(\xi)$ is, the more cooling the first and third integrals would induce. Therefore, the consistent two-dimensional model is for $h \neq 0$. Then $K(\xi, \xi')$ for $h = 0$, which may be used for starting iteration, is

$$K(\xi, \xi') = \pi^{-1}[\, N - \ell n \, | \, \xi - \xi' \, | \,] \quad , \qquad [2.18]$$

where N is a large number.

B. THREE-DIMENSIONAL CASE OF A SEMIINFINITE SOLID

Without going into details, the fundamental solution on the surface for the three-dimensional case is

$$K(\xi, \zeta; \xi', \zeta') = 2\pi^{-1} \frac{e^{-R[r - (\xi - \xi')]}}{r} \quad , \qquad [2.19]$$

where the unit heat source P^* is imposed at (ξ', ζ') while the point in question is (ξ, ζ), both on the surface, and $r = [(\xi - \xi')^2 + (\zeta - \zeta')^2]^{1/2}$. For $R \to 0$,

$$K(\xi, \zeta; \xi', \zeta') = \frac{2}{\pi r} \quad . \qquad [2.20]$$

The total surface temperature is

$$\int_A P^*(\xi', \zeta') K(\xi, \zeta; \xi', \zeta') d\xi' \, d\zeta' \quad , \qquad [2.21]$$

where A extends over the whole half plane and $P^*(\xi, \zeta)$ is analogous to the two-dimensional counterpart $P^*(\xi)$.

C. SURFACE TEMPERATURE WITH TEMPERATURE DEPENDENT THERMAL PROPERTIES

The subject is generally discussed for one-dimensional problems.[3,4] In what follows temperature dependent thermal properties are discussed for the model shown in Figure 2. 1 with $h = 0$ and the origin of the (x_1, x_2) system located at the left-hand corner of heat source for a total length l. Also [2.1] now takes the form

$$\frac{\partial}{\partial x_1'} \left(K \frac{\partial T}{\partial x_1'} \right) + \frac{\partial}{\partial x_2} \left(K \frac{\partial T}{\partial x_2} \right) = \rho c \frac{DT}{Dt} \quad , \qquad [2.22]$$

where K as well as ρc are functions of T. For V large compared to $K/\rho c l$, $\partial^2 T/\partial x^2$ may be neglected. This is the very same assumption which led to [2.16]. Let f and g be two functions of the temperature T defined by

$$\rho c (T) = \rho c_0 \, f (T)$$

$$K(T) = K_0 \, g (T) \quad , \qquad [2.23]$$

where c_0 and K_0 are the specific heat and thermal conductivity at a reference temperature T_0, respectively. Introducing the dimensionless quantities: $P_0 \equiv V l \rho c_0 / K_0$, $\xi \equiv x_1 / P_0 l$, $\eta \equiv x_2 / l$ and the dimensional quantities $Q \equiv q l / K_0$ and $\phi \equiv \int_{T_0}^{T} g(T) dT$, [2.22] and its associated boundary conditions become

$$\frac{\partial^2 \phi}{\partial \eta^2} = [1 + \gamma(\phi)] \frac{\partial \phi}{\partial \xi} \quad , \qquad [2.24]$$

$$\frac{\partial \phi}{\partial \eta} = Q^*(\xi) \equiv \begin{cases} 0 & (-\infty < \xi < 0) \\ -Q(\xi) & (0 \leq \xi \leq P_0^{-1}) \\ 0 & (P_0^{-1} \leq g < \infty) \end{cases} \qquad [2.25]$$

and

$$\frac{\partial \phi}{\partial \xi} \, , \, \frac{\partial \phi}{\partial \eta} \to 0 \quad \text{as} \quad (\xi^2 + \eta^2)^{1/2} \to \infty \quad (g \neq 0) \, , \quad [2.26]$$

where $1 + \gamma(\phi) = f[T(\phi)]/g[T(\phi)]$ since by definition $f(T_0) = g(T_0) = 1$.

The nonlinear term in [2.24] plays the role of a distributed heat source or, as the case may be, heat sink. For an arbitraty function $\gamma(\phi)$, an iterative scheme is very feasible as will be developed below.

Fourier cosine transform is

$$\mathcal{F}\left\{ \ \right\} = \int_0^\infty \left\{ \ \right\} \cos s\eta \ d\eta \, , \quad [2.27]$$

with the inverse transform formula

$$\mathcal{F}^{-1}\left\{ \ \right\} = 2\pi^{-1} \int_0^\infty \left\{ \ \right\} \cos s\eta \ ds \, , \quad [2.28]$$

where s is the transform parameter. Recalling conditions 2.25 and 2.26, the application of [2.27] transforms [2.24] into

$$\frac{d\bar{\phi}}{d\xi} + s^2 \bar{\phi} = - \gamma \overline{\frac{d\phi}{d\xi}} \, , \quad [2.29]$$

where the super bar denotes transformed quantities.

Equation 2.29 possesses a solution

$$\bar{\phi} = e^{-s^2 \xi} \int_0^\xi [Q^*(\xi') - \gamma \overline{\frac{d\phi}{d\xi}}] \, e^{s^2 \xi'} \, d\xi' \, , \quad [2.30]$$

or defining

$$\bar{\phi} = \bar{\phi}_1 + \bar{\phi}_2 \, , \quad [2.31]$$

where

$$\bar{\phi}_1 = \int_0^\xi Q^*(\xi') e^{-s^2 (\xi - \xi')} d\xi' \, , \quad [2.32]$$

and

$$\overline{\phi}_2 = -\int_0^{\xi} \gamma \frac{\overline{d\phi}}{d\xi} e^{-s^2(\xi-\xi')} d\xi' \quad . \qquad [2.33]$$

Equations 2.32 and 2.33 are the Fourier transforms of ϕ_1 and ϕ_2, respectively. Of course,

$$\phi = \phi_1 + \phi_2 \quad . \qquad [2.34]$$

Equations 2.32 and 2.28 lead to

$$\phi_1 = 2\pi^{-1} \int_0^{\infty} \int_0^{\xi} Q^*(\xi') e^{-s^2(\xi-\xi')} d\xi' \cos s\eta \ ds \quad .$$

Interchanging the order of integration,

$$\phi_1 = 2\pi^{-1} \int_0^{\xi} Q^*(\xi') d\xi' \int_0^{\infty} e^{-s^2(\xi-\xi')} \cos s\eta \ ds \quad ,$$

and carrying out the infinite integral,

$$\phi_1 = \pi^{-1/2} \int_0^{\xi} Q^*(\xi') \frac{e^{-\eta^2/4(\xi-\xi')}}{\sqrt{\xi-\xi'}} d\xi' \qquad (\xi-\xi' \geq 0) \quad . \qquad [2.35]$$

Now [2.33] and [2.28] lead to

$$\phi_2 = -2\pi^{-1} \int_0^{\infty} \int_0^{\xi} \int_0^{\infty} \gamma \frac{\partial\phi}{\partial\xi} \cos s\eta' \ d\eta' \ e^{-s^2(\xi-\xi')} d\xi' \cos s\eta \ ds \quad .$$

Recalling $\cos s\eta \cos s\eta' = \frac{1}{2}[\cos s(\eta + \eta') + \cos s(\eta - \eta')]$, the last infinite integral can be integrated as before. Thus

$$\phi_2 = -2\pi^{-1/2} \int_0^{\xi} \frac{d\xi'}{\sqrt{\xi-\xi'}} \int_0^{\infty} \gamma \frac{\partial\phi}{\partial\xi} [e^{-(\eta+\eta')^2/4(\xi-\xi')} + e^{-(\eta-\eta')^2/4(\xi-\xi')}] d\eta' \ ,$$

$$[2.36]$$

Part of the boundary conditions 2.25 and 2.26, which have not been imposed, can easily be shown to be true, that is, $\partial\phi/\partial\xi \rightarrow 0$ as $(\xi^2 + \eta^2)^{1/2} \rightarrow \infty$.

Equation 2.34, in conjunction with [2.35] and [2.36], constitutes a nonlinear integral equation for ϕ. For the evaluation, the following iterative integrals are defined:

$$\phi^{(0)} = \pi^{-1/2} \int_0^{\xi} \frac{Q*(\xi')}{\sqrt{\xi-\xi'}} \, e^{-\eta^2/4(\xi-\xi')} d\xi' \quad ,$$

$$\phi^{(1)} = \phi^{(0)} - \int_0^{\xi} \int_0^{\infty} \gamma[\phi^{(0)}] \, \frac{\partial \phi^{(0)}}{\partial \xi} \, H(\xi,\eta;\xi',\eta') d\xi' \, d\eta' \quad ,$$

$$\vdots$$

$$\text{and} \quad \phi^{(n)} = \phi^{(0)} - \int_0^{\xi} \int_0^{\infty} \gamma[\phi^{(n-1)}] \, \frac{\partial \phi^{(n-1)}}{\partial \xi} \, H(\xi,\eta;\xi',\eta') d\xi' \, d\eta'$$

$$(0 \leq \xi \leq P_0^{-1}) \quad , \qquad [2.37]$$

where

$$H(\xi,\eta;\xi',\eta') = \frac{1}{2\sqrt{\pi}\,(\xi-\xi')} \, [e^{-(\eta-\eta')^2/4(\xi-\xi')} + e^{-(\eta-\eta')^2/4(\xi-\xi')}]$$

$$(\xi-\xi' \geq 0) \quad .$$

In [2.37], $\phi^{(n)}$ represents the nth iterative value of $\phi(\xi, \eta)$. Once $\phi(\xi, \eta)$ is found to the degree of accuracy desired, $T(\xi, \eta)$ may be found by returning to the original definition

$$\phi(T) = \int_{T_0}^{T} g(T')dT' \quad , \qquad [2.38]$$

and the given data on $g(T)$.[11]

An examination of thermal physical properties shows that $f \geq 1$ and g may be either greater or less than 1. A detail study for this problem shows the following:

1. $f = g = 1 = $ Constant

This is the classical case; that is, $\rho c = \rho c_0 = $ constant and $K = K_0 = $ constant. For a special case of heat input, $q_0 = $ constant in $0 \leq x \leq l$, the surface temperature is

$$T = 2(\pi V)^{-1/2} q_0 (\rho c_0 K_0)^{-1/2} x^{1/2} \qquad (0 \leq x \leq \ell) ,$$

[2.39]

where T_0, the reference temperature, is set at zero.

2. $f = 1$, g Generally Increases with Temperature

Local decreases are not excluded in this definition of generally increasing function of temperature. In seeking the temperature, [2.38] is used to transform the variable T to that of ϕ, and the iterative method is not needed in general. Once the solution is found, [2.38] should be used to transform the solution in ϕ back to T.

For this case it may be shown quantitatively that the temperature is generally lower than that given by [2.39], that is, that solution neglecting any temperature dependence of thermal properties so to speak. Depending on the nature of g, [2.39] may give a maximum surface temperature which is up to 20% too high.

3. $f = 1$, g Generally Decreases with Temperature

The discussion in Section 2.C.2 applies except that [2.39] may give a maximum surface temperature which is up to 30% too low.

4. f Generally Increases with Temperature, $g = 1$

For this category, transformation of [2.38] is not necessary. In general, the iterative method may be used effectively, except that [2.37] applies to T instead of ϕ. Also, generally, it may be shown that surface temperatures calculated are lower than those provided by [2.39]. Again, depending on the nature of f, there may be a substantial discrepancy between the correct values and those given by the simple formula [2.39].

5. Both f and g Generally Increase with Temperature

In general, the iterative method may be used effectively. It may be shown that the temperature is generally higher than that of [2.39].

6.   Generally *f* Increases While *g* Decreases with Temperature

In general, the iterative method may be used effectively for this case. The temperature may be either higher or lower than that provided by [2.39].

The importance of the function $(fg)^{-1/2}$ or $(\rho cK)^{-1/2}$ in characterizing the degree of importance of temperature dependence of thermal properties is identified.[11] As a drastic demonstration, although the thermal diffusivity of $1/[1 + \gamma(\phi)]$ in dimensionless form, may vary as much as two-thirds, this change may be ignored for computing the quasi-stationary surface temperature for high Peclet number, P_0, provided $(fg)^{-1/2}$ is approximately constant.

It should be noted that the Peclet number is large for most engineering problems in which temperature dependence of thermal properties are at issue. If the high Peclet number assumption were not adopted, generally the $(fg)^{-1/2}$ function will not be the pertinent characterizing function for determining whether temperature dependence may be ignored in surface temperature calculations. The point to note, however, is that these characterizing functions can be worked out for many cases.

D.   TWO-DIMENSIONAL CASE OF A CIRCULAR DISC

Now that a measure of the consequences of temperature dependent thermal properties has been examined, additional models using temperature independent properties will be studied. One of these is shown in Figure 2.2. This model[12] is akin to that shown in Figure 2.1. Here (r, Θ) are the coordinates fixed in the material and (r, ϕ) are fixed with respect to the arbitrary heat source. The heat equation in polar coordinates (r, Θ) is

$$\frac{\partial^2 T}{\partial r^2} + r^{-1} \frac{\partial T}{\partial r} + r^{-2} \frac{\partial^2 T}{\partial \theta^2} = \kappa^{-1} \frac{DT}{Dt} \ . \qquad [2.40]$$

The two coordinate systems are related by $\phi = \Theta - \omega t$. In an analogous manner as before, for quasistationary state, $DT/Dt = \omega \partial T/\partial \phi$. The boundary conditions are that $q(\phi) = K\partial T/\partial r$ under the heat source while $K\partial T/\partial r = -hT$ over the remaining portion of the boundary. Moreover, T should remain bounded at $r = 0$. Define the following dimensionless

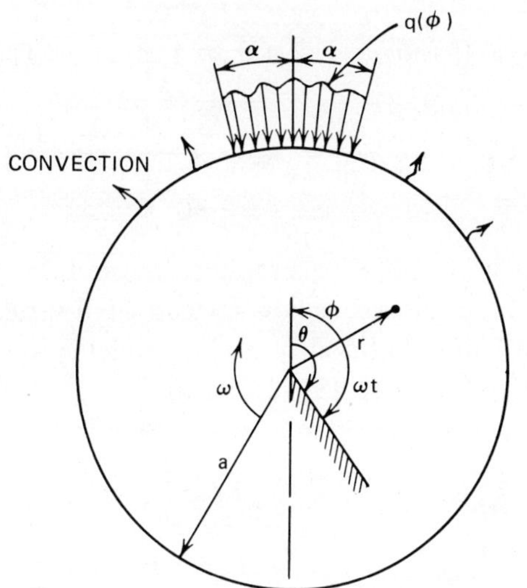

Figure 2.2

quantities $\rho \equiv r/a$, $N \equiv \omega a^2/2\kappa$, $B \equiv ah/K$, $f(\phi) \equiv q/q_0$, and $u \equiv 2\pi hT/q_0$. Equation 2.40, changed to the (r, ϕ) coordinates, for quasistationary state, and in the dimensionless variables, is

$$\frac{\partial^2 u}{\partial \rho^2} + \rho^{-1} \frac{\partial u}{\partial \rho} + \rho^{-2} \frac{\partial^2 u}{\partial \phi^2} = 2N \frac{\partial u}{\partial \phi} \quad , \qquad [2.41]$$

subject to the boundary conditions

$$\frac{\partial u}{\partial \rho} = \begin{cases} -Bu & (-\pi < \phi < -\alpha, \ \rho=1) \\[1em] 2\pi Bf(\phi) & (-\alpha \leq \phi \leq \alpha, \ \rho=1) \\[1em] -Bu & (\alpha < \phi < \pi, \ \rho=1) \end{cases} \qquad [2.42]$$

$$u \quad \text{bounded} \qquad (\rho = 0) \quad . \qquad [2.43]$$

For this geometry, use is made of the finite Fourier transform

$$\mathscr{F}_f\left\{ \ \right\} = \epsilon_p (2\pi)^{-1} \int_{-\pi}^{\pi} \left\{ \ \right\} e^{-ip\phi} \, d\phi \quad , \qquad [2.44]$$

where $\epsilon_p = 1$ if $n = 0$ and $\epsilon_p = 2$ if $n = 1, 2, \ldots$. The inverse is

$$\mathcal{F}_f^{-1}\left\{\ \right\} = \text{Re} \sum_{p=0}^{\infty} \left\{\ \right\} e^{ip\phi} \ . \qquad [2.45]$$

Note that [2.45] is applicable for simply connected regions. The formal use of transform for a closed region instead of Fourier series, gives a priori assurance of differentiability and completeness. Application of [2.44] to [2.41], [2.42] and [2.43] results in

$$\frac{\partial^2 \tilde{u}}{\partial \phi^2} - \rho^{-1}\frac{\partial \tilde{u}}{\partial \rho} - [2Nip + (\frac{p^2}{\rho^2})]\ \tilde{u} = 0 \ , \qquad [2.46]$$

$$\frac{\partial \tilde{u}}{\partial \rho} = -B\tilde{u} + \epsilon_p(2\pi)^{-1}\int_{-\infty}^{\infty} B[u + 2\pi\ f(\phi)]e^{-ip\phi}d\phi \quad (\rho = 1) \ , \qquad [2.47]$$

$$\tilde{u} \quad \text{bounded} \qquad (\rho = 0) \ , \qquad [2.48]$$

where the super $\sim$ denotes the transformed quantities. Letting $\omega_p \equiv (2pN)^{1/2}$, [2.46] subject to [2.47] gives

$$\tilde{u} = C\ I_p(i^{1/2}\omega_p)\rho \ , \qquad [2.49]$$

where $I_p\{\ \}$ is the modified Bessel function of the first kind of order p. Now enforcement of [2.47] leads to

$$\tilde{u} = \frac{\tilde{Q}(i^{1/2}\omega_p)I_p(i^{1/2}\omega_p\rho)}{[i^{1/2}\omega_p B^{-1}I_p'(i^{1/2}\omega_p) + I_p(i^{1/2}\omega_p)} \ ,$$

where

$$\tilde{Q}(i^{1/2}\omega_p) = \epsilon_p(2\pi)^{-1}\int_{-\infty}^{\infty} [u(i^{1/2}\omega_p,\phi) + 2\pi\ f(\phi)]e^{-ip\phi}d\phi \ .$$

Applying the inverse formula 2.45 and interchanging the summation sign with that of the integral, for $\rho = 1$,

$$u(1,\phi) = \int_{-\infty}^{\infty} [f(\phi') + (2\pi)^{-1}u(1,\phi')]K(\phi,\phi')d\phi' \quad ,$$

$$[2.50]$$

where

$$K(\phi,\phi') = 1 + \sum_{p=1}^{\infty} \left\{ \frac{I_p(i^{1/2}\omega_p)e^{ip(\phi-\phi')}}{[i^{1/2}\omega_p B_p^{-1}I_p'(i^{1/2}\omega_p) + I_p(i^{1/2}\omega_p)]} + \text{conjugate} \right\} \quad .$$

$$[2.51]$$

Note that for $\rho \neq 1$, $K(\rho, \phi, \phi')$ is the same as above except in the numerator $I_p(i^{1/2}\omega_p)$ should be replaced by $I_p(i^{1/2}\omega_p\rho)$. Again the surface value for u, that is, $u(1, \phi)$, can be found by inverting the integral equation [2.50]. For N, the Peclet number, large, [2.51] for any ρ degenerates into

$$K(\rho,\phi;\phi') = 1 + [(2/N\rho)^{1/2}B] \sum_{p=1}^{\infty} (pT_p)^{-1/2} e^{-(1-\rho)\omega_p}$$

$$\cdot \cos[p(\phi'-\phi) - (1-\phi')\omega_p - \psi_p] \quad ,$$

$$[2.52]$$

where

$$T_p = 1 + (B/\omega_p)^2 + (2B/\omega_p)\cos \pi/4 \quad ,$$

$$\psi_p = \cos^{-1}[(B/\omega_p + \cos \pi/4)/T_p] \quad ,$$

and provided $h \neq 0$.

E. TWO-DIMENSIONAL CASE OF A RING SECTOR

For this model,[12] [2.41] applies and the boundary conditions, see Figure 2.3, are

$$\frac{\partial u}{\partial \rho} = \begin{cases} -2\pi\, Bf(\phi) & (\rho = 1) \\ -Bu & (\rho = \lambda^{-1}) \end{cases}, \qquad [2.53]$$

$$\frac{\partial u}{\partial \phi} = 0 \qquad (|\phi| = \alpha), \qquad [2.54]$$

where $\lambda = a/b$.

For $2n\alpha = \pi$, where n may be any integer, the problem above is equivalent to a ring subject to [2.53], with $f(\phi)$ having n-fold symmetry in $0 \le \theta \le 2\pi$. Thus in a similar fashion as in D,

$$u(\rho) = \int_{-\alpha}^{\alpha} f(\phi')K(\rho,\phi;\phi')d\phi', \qquad [2.55]$$

where

$$K(\rho,\phi;\phi') = 2S_0 + 4m \sum_{j=1}^{\infty} \cos\, jm\,\psi \cos\, jm\,\psi'\, S_{jm},$$

$$S_0 = -B\, \ell n(\lambda\rho) + \lambda,$$

$$S_n = \frac{(B/n)[-(\lambda\rho)^n(-n\lambda + B) + (\lambda\rho)^{-n}(n\lambda + B)]}{\lambda^4(-n\lambda + B) + \lambda^{-n}(n\lambda + B)},$$

$$\psi \equiv \phi + \alpha, \qquad \psi' \equiv \phi' + \alpha \qquad (n=1,2,\ldots).$$

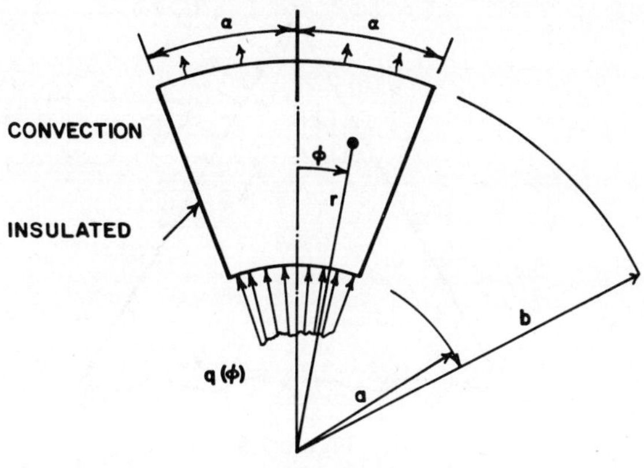

Figure 2.3

F.   THE TRUNCATED CONE

An arbitrarily distributed but rotationally symmetric heat input on the flat face of a truncated right circular cone is shown in Figure 2.4. The sides are insulated. The natural coordinates to use are spherical coordinates (r, Θ, ϕ). As shown in Figure 2.5, for this case, the azimuthal angle would not be relevant. The semiopening angle is $\Theta_0 (0 < \Theta_0 < \pi/2)$

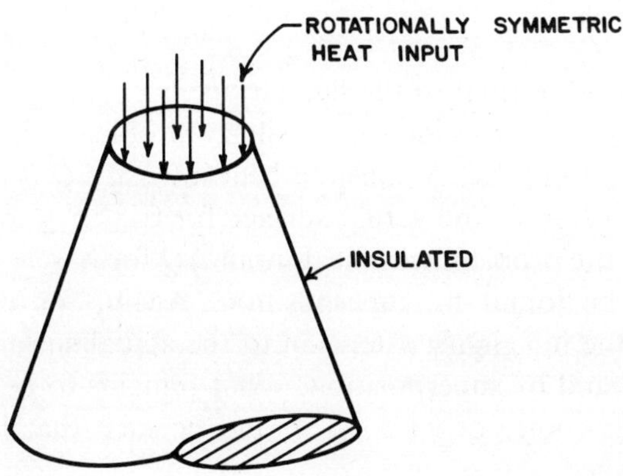

Figure 2.4

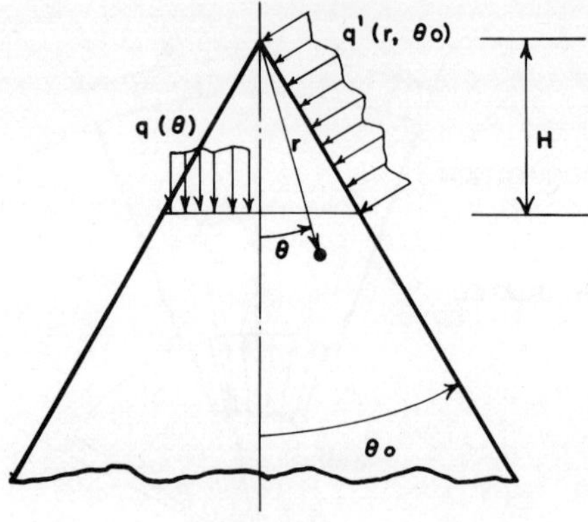

Figure 2.5

and the height of the truncated portion is H. For steady state, the heat equation is the Laplace equation. The boundary conditions are

$$-K \frac{\partial T}{\partial n} = q(\theta) \qquad (r = \frac{H}{\cos \theta} , \; 0 \leq \theta < \theta_0) \quad , \qquad [2.56]$$

$$Kr^{-1} \frac{\partial T}{\partial \theta} = 0 \qquad (r_0 < r < \infty , \; \theta = \theta_0) \quad , \qquad [2.57]$$

$$\frac{\partial T}{\partial r} \to 0 \quad \text{as} \quad r \to \infty , \qquad [2.58]$$

where n is inward normal to the flat surface.

It is convenient to consider an auxiliary problem which is the heat conduction in a full cone. A constant heat intensity Q is applied on the imaginary extension to the actual surface for $0 \leq r \leq r_0$ and $\theta = \theta_0$. It follows that the problem with heat input ΔQ for $r_1 \leq r \leq r + \Delta r$ and $\theta = \theta_0$ can be found by superposition. Again, for the heat input $q'(r, \theta_0)$ over the imaginary extension to the actual surface, the answer may also be found by superposition. What remains to be carried out is that the proper choice of $q'(r, \theta_0)$ should be such that [2.56] through [2.58] are satisfied.

It may be seen that the aforementioned procedure outlined would be quite lengthy[13]; therefore details are not given here. It should be noted that the auxiliary problem is best attacked by resorting to Mellin transform

$$\mathfrak{M}\left\{\;\right\} = \int_0^\infty \left\{\;\right\} r^{s-1} dr \;, \qquad [2.59]$$

where s is the transform parameter while the inverse formula is

$$\mathfrak{M}^{-1}\left\{\;\right\} = (2\pi i)^{-1} \int_{c-i\infty}^{c+i\infty} \left\{\;\right\} r^{-s} ds \;, \qquad [2.60]$$

where c is a suitably chosen constant so that the position of the line of integration is varied within one and the same strip of regularity of the function undergoing inverse transform.

Letting $p = \cos\Theta$, $p_0 = \cos\Theta_0$, $\bar{p}_0 = \sin\Theta_0$, $P_s(p)$ be the Legendre function of the first kind and degree s and $s_i(p_0)$ be the zeros of $P_s'(p_0)$, the solution for the full cone with heat input $q'(r, \Theta_0)$ is

$$T(r,p) = \frac{1}{K\,\bar{p}_0} \sum_{i=1}^{\infty} \frac{P_{s_i}(p)}{\dfrac{d^2 P_{s_i}(p_0)}{ds\;dp}} \left[\int_0^{r-\epsilon} q'(r',\Theta_0)\left(\frac{r'}{r}\right)^{s_i+1} dr' \right.$$

$$\left. + \int_{r+\epsilon}^{r_0} q'(r',\Theta_0)\left(\frac{r}{r'}\right)^{s_i} dr' \right] \qquad (0 \le r \le r_0-\epsilon,\ \epsilon > 0,\ p_0 < p \le 1) \;,$$

$$T(r,p) = \frac{1}{K\,\bar{p}_0} \sum_{i=1}^{\infty} \frac{P_{s_i}(p)}{\dfrac{d^2 P_{s_i}(p_0)}{ds\;dp}} \int_0^{r_0} q'(r',\Theta_0)\left(\frac{r'}{r}\right)^{s_i+1} dr'$$

$$(r_0 + \epsilon \le r < \infty,\ \epsilon > 0,\ p_0 < p \le 1) \;.$$

$$[2.61]$$

Given the above, it is a simple matter to express $q(p)$, the given heat input at $r = H/\cos\Theta$, in terms of [2.61]. This results in an integral equation of the first kind. Its inversion, in general, is not possible. However, the inversion may be carried out approximately by resorting to a numerical procedure. Consider $q'(r, \Theta_0)$ as a set of piecewise continuous functions h_j: $q'(r, \Theta) = h_j$ for $(j - 1)(r_0/N) < r < j(r_0/N)$, $j = 1, 2, \ldots . N$. Letting $m = rN/r_0$ and M be the integer part of real number m ($m = M + \epsilon$, $a < \epsilon < 1$),

$$T(r,p) = \frac{r}{K\,\bar{p}_0}\left[-ph_{M+1} + \sum_{i=1}^{\infty} \frac{P_{s_i}(p)}{\dfrac{d^2 P_{s_i}(p)}{ds\,dp}} \right.$$

$$\cdot \left\{ \frac{1}{s_i + 2} \sum_{j=1}^{M} h_j \left[\left(\frac{j}{m}\right)^{s_i + 2} - \left(\frac{j-1}{m}\right)^{s_i + 2} \right] \right.$$

$$+ \frac{1}{s_i - 1} \sum_{j=M+2}^{N} h_j \left[\left(\frac{m}{j-1}\right)^{s_i - 1} - \left(\frac{m}{j}\right)^{s_i - 1} \right]$$

$$\left. \left. - h_{M+1} \left[\frac{1}{s_i - 1}\left(\frac{m}{M+1}\right)^{s_i - 1} + \frac{1}{s_i + 2}\left(\frac{M}{m}\right)^{s_i + 2} \right] \right\} \right]$$

$$(0 \leq r < r_0 \, , \; p_0 < p \leq 1) \, , \qquad [2.62]$$

$$T(r,p) = \frac{r}{K\,\bar{p}_0} \sum_{i=1}^{\infty} \frac{1}{s_i + 2} \frac{P_{s_i}(p)}{\dfrac{d^2 P_{s_i}(p_0)}{ds\,dp}} \sum_{j=1}^{N} h_j \left[\left(\frac{j}{m}\right)^{s_i + 2} - \left(\frac{j-1}{m}\right)^{s_i + 2} \right]$$

$$(r_0 < r < \infty \, , \; p_0 < p \leq 1) \, , \qquad [2.63]$$

Letting $p_k = \cos \Theta_k$, $r_k = (H/p_k) = m(r_0/N)$, $m = b + \lambda$ for $p_k = 1$ ($b = $ integer, $\lambda \geq 0$),

$$q(p_k) = \frac{1}{\bar{p}_0} \left[h_{M+1} + \sum_{i=1}^{\infty} \left[\frac{d^2 P_{s_i}(p_0)}{ds\, dp} \right]^{-1} \right.$$

$$\cdot \left\{ \frac{F_{s_i}(p_k)}{s_i + 2} \sum_{j=1}^{M} h_j \left[(\tfrac{j}{m})^{s_i + 2} - (\tfrac{j-1}{m})^{s_i + 2} \right] \right.$$

$$+ \frac{G_{s_i}(p_k)}{s_i - 1} \sum_{j=M+2}^{N} h_j \left[(\tfrac{m}{j})^{s_i - 1} - (\tfrac{m}{j-1})^{s_i - 1} \right]$$

$$\left. \left. + h_{M+1} \left[\frac{G_{s_i}(p_k)}{s_i - 1} (\tfrac{m}{M+1})^{s_i - 1} - \frac{F_{s_i}(p_k)}{s_i + 2} (\tfrac{M}{m})^{s_i + 2} \right] \right\} \right]$$

$$(b + \lambda + \epsilon \leq m \leq b + 1 - \epsilon \,,\; \epsilon > 0) \,, \qquad [2.64]$$

where

$$F_{s_i}(p_k) = (s_i + 1)p_k\, P_{s_i}(p_k) - (1 - p_k^2)P'_{s_i}(p_k) \,,$$

$$G_{s_i}(p_k) = s_i\, p_k\, P_{s_i}(p_k) + (1 - p_k^2)P'_{s_i}(p_k) \,.$$

Some numerical results are given in the appropriate sections later. It should be mentioned that as $\theta_0 \to \pi/2$, $T(r,z)$ in polar coordinates (r, z) for rationally symmetric case becomes

$$T(r,z) = \int_0^{\infty} r'\, q(r')K(r,r')dr' \,, \qquad [2.65]$$

where $K(r, r') = 2E(k)/[\pi K\sqrt{(r + r')^2 + r^2}]$, $K(k)$ is the complete elliptical integral of the first kind; $q(r)$ is the heat distribution; and $k = 2\sqrt{rr'}/[(r + r')^2 + r^2]$. This result is most directly found by resorting to the Hankel transform method with

$$\mathcal{H}\left\{\ \right\} = \int_0^\infty r\left\{\ \right\} J_0(\lambda r)\,dr \quad , \qquad [2.66]$$

where λ is the transform parameter and, $J_0(\ )$ is the Bessel function of the first kind and order zero. The inverse formula is

$$\mathcal{H}^{-1}\left\{\ \right\} = \int_0^\infty \left\{\ \right\} \lambda\, J_0(\lambda r)\,d\lambda \quad . \qquad [2.67]$$

G.   THE SPHERE

Figure 2.6 shows the cross-section of a sphere through its center. Over the polar cap an arbitrarily distributed, but rotationally symmetric heat source is located. The remaining surface convects heat away. An expression of the surface temperature is sought. Using the dimensionless quantities defined in Section 2.D and $u \equiv hT/q$, the heat equation for

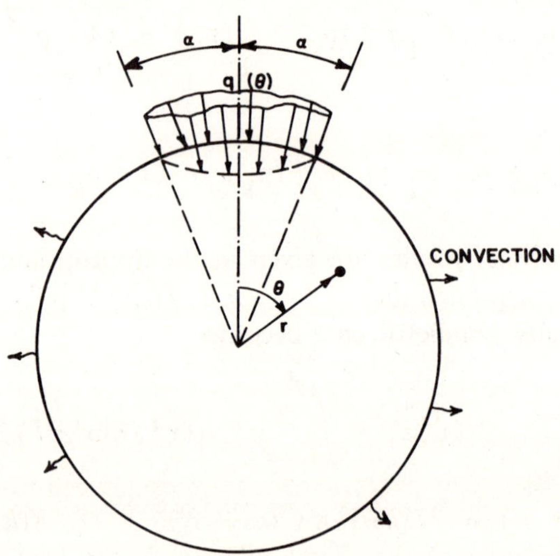

Figure 2.6

steady state and in spherical coordinates with rotational symmetry is

$$\frac{\partial^2 u}{\partial \rho^2} + 2\rho^{-1} \frac{\partial u}{\partial \rho} + \rho^{-2} \frac{\partial^2 u}{\partial \theta^2} + \rho^{-2} \cot \theta \frac{\partial u}{\partial \theta} = 0 \quad . \quad [2.68]$$

The boundary conditions are

$$\frac{\partial u}{\partial \rho} = \begin{cases} -Bu & (-\pi < \theta < -\alpha \ , \ \rho=1) \\ Bf & (-\alpha \le \theta \le \alpha, \ \rho=1) \\ -Bu & (\alpha < \theta < \pi \ , \ \rho=1) \end{cases} \quad [2.69]$$

$$u \quad \text{bounded} \quad (\rho=0) \quad . \quad [2.70]$$

Letting $\mu = \cos \theta$, [2.68] can be put in the standard Legendre form

$$\frac{\partial^2 u}{\partial \rho^2} + 2\rho^{-1} \frac{\partial u}{\partial \rho} + \rho^{-2} \left[(1 - \mu^2) \frac{\partial^2 u}{\partial \mu^2} - 2\mu \frac{\partial u}{\partial \mu} \right] = 0, \quad [2.71]$$

and the limits in [2.69] would be $\mu = -1$ corresponding to $\theta = \pm\pi$, $\mu = 1$ corresponding to $\theta = 0$, and $\mu = \mu_\alpha$ corresponding to $\theta = \pm\alpha$.
Define the Legendre transform

$$L \left\{ \quad \right\} = (n + \tfrac{1}{2})^{1/2} \int_{-1}^{1} \left\{ \quad \right\} P_n(\mu) d\mu \quad , \quad [2.72]$$

where $P_n(\ )$ is the Legendre function of the first kind. The inverse is

$$L^{-1} \left\{ \quad \right\} = \sum_{n=0}^{\infty} (n + \tfrac{1}{2})^{1/2} \left\{ \quad \right\} P_n(\mu) \quad . \quad [2.73]$$

Denoting the transformed quantities by a super $\sim$, the transform of [2.71] becomes

$$\frac{d^2 \tilde{u}}{d\rho^2} + 2\rho^{-1} \frac{d\tilde{u}}{d\rho} - \rho^{-2} n(n+1)\tilde{u} = 0 \quad , \quad [2.74]$$

and that of [2.69] and [2.70] become, respectively,

$$\frac{d\tilde{u}}{d\rho} = -B\tilde{u} + \left(n + \frac{1}{2}\right)^{1/2} \int_{\mu\alpha}^{1} B[u + f(\mu')]P_n(\mu')d\mu' \quad (\rho=1) \quad , \quad [2.75]$$

$$\tilde{u} \quad \text{bounded} \quad (\rho=0) \quad . \qquad [2.76]$$

Equation 2.74, considering [2.76], has the solution

$$\tilde{u} = A \, \rho^{\,-\frac{1}{2} + \sqrt{n(n+1) + \frac{1}{4}}} \quad ,$$

where A is a constant of integration.

   Enforcement of [2.75] on [2.77] leads to the integral equation which, upon the application of [2.73], becomes

$$u(1,\mu) = \int_{\mu\alpha}^{1} [u(1,\mu') + f(\mu')]K(\mu,\mu')d\mu' \quad , \quad [2.77]$$

where

$$K(\mu,\mu') = \sum_{n=0}^{\infty} \frac{\left(n + \frac{1}{2}\right)^{1/2} P_n(\mu)P_n(\mu')}{B - \frac{1}{2} + \sqrt{n(n+1) + \frac{1}{4}}}$$

H.   TRANSIENT CASE OF A MOVING SEMIINFINITE SOLID

Figure 2.7 is the generalization of that shown in Figure 2.1 with the exception that the flat surface outside of the heating area is nonconvecting. The $q(x_1, x_3)$ is instantaneous. In fact, the reason for not showing any details in arriving at [2.19], and consequently [2.20], is that (2.19] may be obtained from the present case by integration. Using the same variables as in Section 2.A with the addition that $\tau = \kappa t/l^2$ and

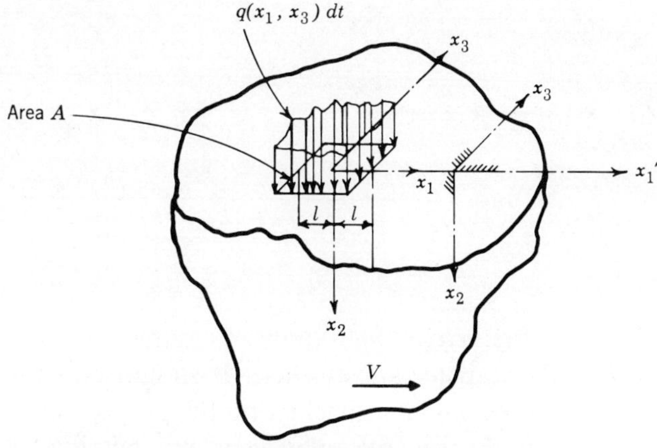

Figure 2.7

$\zeta \equiv x_3/l$, where x_3 is the third coordinate, and t is the time, [2.2] generalizes to

$$\frac{\partial^2 \phi}{\partial \xi^2} + \frac{\partial^2 \phi}{\partial \eta^2} + \frac{\partial^2 \phi}{\partial \zeta^2} = \frac{\partial \phi}{\partial \tau} + R \frac{\partial \phi}{\partial \xi} \quad . \qquad [2.78]$$

The boundary conditions are

$$\frac{\partial \phi}{\partial \eta} = -P(\xi, \zeta) \qquad (\text{in } A, \ \eta = 0) \quad , \qquad [2.79]$$

$$\frac{\partial \phi}{\partial \eta} \rightarrow 0 \quad \text{as} \quad (\xi^2 + \eta^2 + \zeta^2)^{1/2} \rightarrow \infty \quad . \quad [2.80]$$

Equation 2.78 with [2.79] and [2.80] is most readily solved in the following manner: by the successive use of Fourier transform [2.5] in the ξ, η, ζ variables, [2.78] is reduced to a total differential equation in τ. This can be solved immediately through inverse transform of the kind [2.12] with due consideration to the transformation $e^{-i(\xi-R\tau)s}$ instead of e^{-is} in the inverse transform.

The result is

$$\phi(\xi, \eta, \zeta, \tau) = \int_A P'(\xi', \zeta')\, d\tau \ K(\xi, \eta, \zeta, \tau; \xi', \zeta')\, dA \ , \qquad [2.81]$$

where

$$K(\xi,\eta,\zeta,\tau;\xi',\zeta') = \frac{1}{4}\left[\pi(\tau-\tau')\right]^{-3/2}$$

$$\cdot \exp\left\{-\frac{[(\xi-\xi') - R(\tau-\tau')]^2 + (\eta-\eta')^2 + (\zeta-\zeta')^2}{4(\tau-\tau')}\right\} .$$

$$[2.82]$$

Equation 2.82 is, of course, a well-known fundamental solution sometimes known as the Laplace substitution. It should be noted that the present method is direct and consistent with the purpose of finding only surface quantities if the general solution is not possible. Integration with respect to time and the limit $\tau \to \infty$ gives [2.19].

I. TRANSIENT CASE OF A CYLINDRICAL ROD

Figure 2.8 shows the case of an infinitely long cylindrical rod with a band of heat input, which possesses cylindrical symmetry.[14] Using the dimensionless quantities: $\rho \equiv r/a$, $\zeta \equiv z/a$, $\tau \equiv \kappa t/a^2$, $\beta \equiv b/a$, $\phi \equiv KT/lq_0$, and $P \equiv q/q_0$, the heat in cylindrical coordinates, for the case of rotational symmetry is

$$\frac{\partial^2 \phi}{\partial \rho^2} + \rho^{-1}\frac{\partial \phi}{\partial \rho} + \frac{\partial^2 \phi}{\partial \zeta^2} = \frac{\partial \phi}{\partial \tau} , \qquad [2.83]$$

subject to the condition

$$\frac{\partial \phi}{\partial \rho} = \begin{cases} P(\zeta) & (|\zeta| \leq \beta, \quad \rho=1) \\ 0 & (|\zeta| > \beta, \quad \rho=1) \end{cases} , \qquad [2.84]$$

$$\frac{\partial \phi}{\partial \rho} \to 0 \quad \text{as} \quad (\rho^2 + \zeta^2)^{1/2} \to \infty . \qquad [2.85]$$

Initially,

$$\phi = 0 . \qquad [2.86]$$

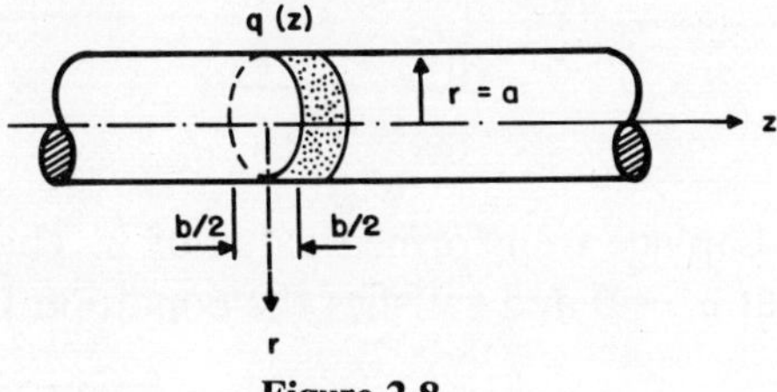

Figure 2.8

Let P be symmetric with respect to the ζ axis and let $\bar{\phi}$ denote the Fourier transform with parameter γ, then [2.83] becomes

$$\frac{\partial^2 \bar{\phi}}{\partial \rho^2} - \rho^{-1} \frac{\partial \bar{\phi}}{\partial \rho} - \gamma^2 \bar{\phi} = \frac{\partial \bar{\phi}}{\partial t} \quad , \qquad [2.87]$$

while [2.85] becomes

$$\frac{\partial \bar{\phi}}{\partial \rho} = \bar{P} \equiv (2/\pi)^{1/2} \int_0^{\beta/2} P(\zeta) \cos \gamma \, \zeta \, d\zeta \quad , \quad [2.88]$$

and [2.86] becomes

$$\bar{\phi} = 0 \quad . \qquad [2.89]$$

Define the Laplace transform

$$\mathscr{L}\{\ \} = \int_0^\infty \{\ \} e^{-s\tau} \, d\tau \quad , \qquad [2.90]$$

which has an inverse

$$\mathscr{L}^{-1}\{\ \} = (2\pi i)^{-1} \int_{p-i\infty}^{p+i\infty} \{\ \} e^{s\tau} \, ds \quad , \quad [2.91]$$

in which the line of integration $\text{Re}[s] = p$ is to be chosen to the right of the singularities of the integrand. Applying [2.90] to [2.87] and [2.88] and using [2.89],

$$\frac{d^2 \tilde{\phi}}{d\rho^2} + \rho^{-1} \frac{d \tilde{\phi}}{d\rho} - (\gamma^2 + s)\tilde{\phi} = 0 \quad , \qquad [2.92]$$

$$\frac{d\tilde{\phi}}{d\rho} = \frac{\bar{P}}{s} \, ,$$

[2.93]

where $\tilde{\phi}$ is the Laplace transformed value of $\bar{\phi}$. The solution of [2.92] which is finite at $\rho = 0$ and satisfies the condition [2.93] is

$$\tilde{\phi} = \frac{\bar{P}}{s\sqrt{\gamma^2 + s}} \cdot \frac{I_0(\sqrt{\gamma^2 + s}\,\rho)}{I_1(\sqrt{\gamma^2 + s}\,)} \, ,$$

[2.94]

where I_1 is the modified Bessel function of the first kind and order one. For inversion, apply [2.91] to [2.94]. It is apparent that the poles of $\tilde{\phi}(\rho, \gamma, s)$ are located at $s = 0$ and at the zeros of $I_1(\sqrt{\gamma^2 + s})$ which occurs at

$$s = -(\omega_j^2 + \gamma^2) \qquad (j = 1, 2, \ldots) \, .$$

[2.95]

Here ω_j are the positive ordered roots of the equation $J_1(x) = 0$, where J_1 is the Bessel function of the first kind of order one. Inversion of [2.94], that is, [2.94] with [2.91] and the accompanying integration,

$$\bar{\phi} = \bar{P}\left\{ \frac{I_1(\gamma\rho)}{\gamma I_1(\gamma)} - \sum_{j=1}^{\infty} \frac{4}{\omega_j^2 + \gamma^2} \cdot \frac{J_1(\omega_j\rho)}{J_0(\omega_j) + J_2(\omega_j)} \exp[-(\omega_j^2 + \gamma^2)\tau] \right\} \, .$$

[2.96]

Inverse Fourier cosine transform leads to

$$\phi = \int_0^{\beta/2} P(\zeta')K(\zeta,\rho;\zeta')d\zeta' \, ,$$

[2.97]

where

$$K(\zeta,\rho;\zeta') = \pi^{-1}\int_0^{\infty} \left\{ \frac{I_0(\gamma\rho)}{\gamma I_1(\gamma)} - \frac{4e^{-\gamma^2\tau}}{\gamma^2} - 4\sum_{j=1}^{\infty} \frac{J_0(\omega_j\rho)e^{-(\omega_j^2 + \gamma^2)\tau}}{(\omega_j^2 + \gamma^2)[J_0(\omega_j) + I_2(\omega_j)]} \right\}$$

$$\cdot [\cos\gamma(\zeta - \zeta') + \cos\gamma(\zeta + \zeta')]d\gamma \, .$$

[2.98]

The $K(\varsigma, \rho; \varsigma')$ is the fundamental solution in integral form for the ring source of infinitesimal width.

J. GENERAL THREE-DIMENSIONAL TRANSIENT CASE OF A CYLINDRICAL ROD

Of interest here is the problem shown in Figure 2.2, generalized so that the disc is an infinitely long rod and that q is a function of ϕ and z, which is the coordinate in the axial direction. Of course, the model is more general than that shown in Figure 2.8. Equation 2.41 is generalized to

$$\frac{\partial^2 T}{\partial r^2} + r^{-1} \frac{\partial T}{\partial r} + r^{-2} \frac{\partial^2 T}{\partial \theta^2} + \frac{\partial^2 T}{\partial z^2} = \kappa^{-1} \frac{DT}{Dt} \quad . \quad [2.99]$$

The initial condition is that

$$T = 0 \quad , \qquad\qquad [2.100]$$

and that the boundary conditions are

$$K \frac{\partial T}{\partial r} = \begin{cases} -hT & (-\pi < \theta - \omega t < -\alpha, \; \rho=1 \; , \; z=z) \\[2mm] q(\rho,z) & (-\alpha \leq \theta - \omega t \leq \alpha \; , \; \rho=1 \; , \; |z| \leq \ell) \; [2.101] \\[2mm] -hT & (-\alpha \leq \theta - \omega t < \pi \; , \; \rho=1 \; , \; z=z) \end{cases}$$

$$T \quad \text{bounded at} \quad \gamma = 0 \quad , \qquad\qquad [2.102]$$

and

$$\frac{\partial T}{\partial z} = 0 \quad \text{as} \quad (r^2 + z^2)^{1/2} \rightarrow \infty \quad . \quad [2.103]$$

In the above the heat source covers a length $2l$.

Taking the Fourier cosine transform with transform parameter γ and variable $\varsigma \equiv z/a$ and using the dimensionless variables used in Section 2.C, [2.99], [2.101], and [2.102] become

$$\frac{\partial^2 \bar{u}}{\partial \rho^2} + \rho^{-1} \frac{\partial \bar{u}}{\partial \rho} + \rho^{-2} \frac{\partial^2 \bar{u}}{\partial \theta^2} - \gamma^2 \bar{u} = \frac{D\bar{u}}{D\tau} \quad , \qquad [2.104]$$

where u is the transformed quantity, $\tau \equiv \kappa t/a^2$,

$$\frac{\partial \bar{u}}{\partial \rho} = \begin{cases} -B\bar{u} & (\pi < \theta - \omega t < -\alpha, \quad \rho=1) \\[2mm] 2\pi \, B\bar{f}(\phi,\gamma) & (-\alpha \leq \phi \leq \alpha, \quad \rho=1) \\[2mm] -B\bar{u} & (\alpha < \phi < \pi, \quad \rho=1) \end{cases} \qquad [2.105]$$

$$\bar{u} \quad \text{bounded} \quad (\rho = 0)\, \phi\,, \qquad\qquad [2.106]$$

and $\bar{f} = (2/\pi)^{-1/2} \int_{-\delta}^{\delta} q(\phi, \zeta) \cos \gamma\zeta d\zeta$, $\delta \equiv l/a$.

Examining first the subproblem of a stationary case, taking the Laplace transform of [2.104] with transform parameters and considering [2.100],

$$\frac{\partial^2 \tilde{u}}{\partial \rho^2} + \rho^{-1} \frac{\partial \tilde{u}}{\partial \rho} + \rho^{-2} \frac{\partial^2 \tilde{u}}{\partial \theta^2} - q^2 \tilde{u} = 0 \,, \qquad [2.107]$$

where $q = \sqrt{\gamma^2 + s}$. Let the source be located at $(\rho' < 1, \theta')$ and the whole boundary along $\rho = 1$ be convecting. Also let $\bar{G}(\rho, \theta; \rho', \theta')$ be the fundamental solution to this subproblem for a source of strength unity; of course, $\tilde{G}$, the Laplace transform of $\bar{G}$, satisfies [2.107]. The boundary condition should read

$$\frac{\partial \tilde{G}}{\partial \rho} + B\tilde{G} = 0 \qquad (\rho = 1) \quad . \qquad [2.108]$$

Now let $\tilde{G} = \tilde{G}_1 + \tilde{G}_2$, where $\tilde{G}_1$ is the fundamental solution for the infinite plane. This is known to be $Bq^{-2}K_0(qR)$, where $R = [\rho^2 + \rho'^2 - 2\rho\rho' \cos(\theta - \theta')]^{1/2}$. The $\tilde{G}_2$ is the residual part of $\tilde{G}$ for the ultimate enforcement of [2.108]. By the addition theorem,

$$\tilde{G}_1 = Bq^{-2} \sum_{n=0}^{\infty} \epsilon_n \, I_n(q\rho)K_n(q\rho') \cos n(\theta-\theta') \qquad (\rho < \rho')\,,$$
$$[2.109]$$

where $\epsilon_n = 1$ if $n = 0$, and $\epsilon_n = 2$ if $n = 1, 2, \ldots$. If $\rho > \rho'$, [2.109] is true when ρ and ρ' interchange. A general solution of [2.107], and therefore of $\tilde{G}_2$, is

$$\tilde{G}_2 = Bq^{-2} \sum_{n=0}^{\infty} [a_n I_n(q\rho) + b_n K_n(q\rho)] \cos n(\theta-\theta') \cdot \quad [2.110]$$

In view of [2.106], $b_n = 0$ in [2.110]. In the region $0 \leq \rho < 1$, [2.109], [2.110], and [2.108] lead to a trivial solution of a_n. However, when ρ and ρ' are interchanged in [2.109], [2.110] and [2.108] lead to a nontrivial solution of a_n. Thus.

$$\tilde{G} = Bq^{-2} \sum_{n=0}^{\infty} \epsilon_n [K_n(q\rho') + D I_n(q\rho')] I_n(q\rho) \cos n(\theta-\theta'),$$

$$[2.111]$$

where $D = -[qK_n'(q) + BK_n(q)]/[qI_n'(q) + I_n(q)]$. The actual problem, which is one with heat sources distributed on a portion of the boundary, can be replaced by one in which the boundary condition is homegeneous but the sources distributed just inside the boundary.[15] The sources in the neighborhood of the boundary may be found from $\tilde{G}$. Let the strength of the source (surface Green's function) be denoted $\bar{G}_s$. The Laplace transform of $\bar{G}_s$, $\tilde{G}_s$, is obtained from $\tilde{G}$ by a limiting process $\rho' \to 1$ where the boundary condition is of the radiation and the Neumann type. Letting $\rho' \to 1$ in the appropriate fashion in [2.111] and using the identity $K_n(z)I_n'(z) - I_n(z)K_n'(z) = z^{-1}$, $\tilde{G}_s$ is[15]

$$\tilde{G}_s = q^{-2} \sum_{n=0}^{\infty} \epsilon_n \tilde{G}_{sn} \cos n(\theta-\theta') \quad , \quad [2.112]$$

where

$$\tilde{G}_{sn} = \frac{B I_n(q\rho)}{q I_n'(q) + B I_n(q)} \quad . \quad [2.113]$$

It is seen that [2.112] satisfies

$$\frac{\partial \tilde{G}_s}{\partial \rho} + B \tilde{G}_s = \delta(\theta-\theta') \quad (\rho = 1) \quad , \quad [2.114]$$

where $\delta(\ )$ is the Dirac-delta function. Note that [2.114] embodies all the requirements of the problem.

The inverse of $\bar{G}_s$ is expected to be of the form

$$\bar{G}_s = B \sum_{n=0}^{\infty} \epsilon_n \, G_{sn} \, \cos n(\theta - \theta') \quad , \qquad [2.115]$$

although G_{sn} are yet to be evaluated.

Now turning to the actual problem, which has moving sources, define $\bar{G}_s^*$ as the counterpart of $\bar{G}_s$ for the actual problem. The $\bar{G}_s^*$ may be found by integrating $\bar{G}_s$. In this regard, let the initial position of the source be at $(1, \beta)$; after time t, then, the position will be $(1, \omega t + \beta)$. Application of the convolution theorem,

$$\tilde{G}_s^* = \frac{B \, I_0(q\rho)}{q^2 [q \, I_0'(q) + B \, I_0(q)]}$$

$$+ 2 \sum_{n=1}^{\infty} \frac{[q^2 \cos n(\theta-\beta) + n \, \omega \, \sin n(\theta-\beta)] I_n(q\rho)}{(q^4 + n^2 \omega^2)[q \, I_n'(q) + B \, I_n'(q)]} \quad .$$

$$[2.116]$$

Inverse Laplace of [2.116] leads to

$$\bar{G}_s^* = \frac{B \, I_0(\gamma\rho)}{\gamma \, I_0'(\gamma) + B \, I_0(\gamma)}$$

$$+ \sum_{n=1}^{\infty} \left[\frac{I_n(i^{1/2} \rho \omega_n) e^{i n(\phi - \theta)} e^{-\gamma^2 t}}{i^{1/2} \bar{\omega}_n I_n'(i^{1/2} \bar{\omega}_n) + B \, I_n(i^{1/2} \bar{\omega}_n)} + \text{conjugate} \right]$$

$$+ B \sum_{j=1}^{\infty} \frac{I_0(\omega_j \rho) e^{(\omega_j^2 - \gamma^2)t}}{\omega_j^2 [\omega_j \, I_0'(\omega_j) + B \, I_0(\omega_j)]}$$

$$+ 2 \sum_{n=1}^{\infty} \sum_{k=1}^{\infty} \frac{[\omega_{nk}^2 \cos(\theta-\beta) + n \, \omega \, \sin n(\theta-\beta)] I_n(\omega_{nk}\rho) e^{(\omega_{nk}^2 - \gamma^2)t}}{(\omega_{nk}^4 + n^2 \omega^2)[\omega_{nk} \, I_n'(\omega_{nk}) + B \, I_n(\omega_{nk})]} \quad ,$$

$$[2.117]$$

where $\omega_n = (n\omega)^{1/2}$, $\phi = \beta + \omega t$, $\omega_j^2 = \gamma^2 + s$ with ω_j as the zeros of $[x I_0'(x) + I_0(x)] = 0$, and $\omega_{nk}^2 = \gamma^2 + s$ with ω_{nk} as the zeros of $[x I_n'(x) + B I_n(x)] = 0$.

Note that the nontransient terms in [2.117], for $\gamma = 0$, check with [2.51]. The position of B is related to the ways heat inputs are handled in the two cases. Upon multiplication to [2.117] of the heat input f, the application of inverse Fourier transform and integrating the effect of $f(\phi, \zeta)$ in $-\alpha < \phi < \alpha$, $u(\rho, \Theta, \zeta, t)$ may be expressed in terms of infinite integrals.

K.   LAYERED HALF-SPACE

Of interest is the case shown in Figure 2.1 except that there is a layer of thickness d on the surface. Also the body is not convecting heat. The governing equation for isotropic heat conduction wherein the properties are temperature independent in the film for the fast moving and quasistationary cases is

$$\frac{\partial^2 T_1}{\partial x_2^2} = (V/\kappa_1)\,\frac{\partial T_1}{\partial x_1} \quad , \qquad [2.118]$$

where T_1 is the temperature in the layer, κ_1 is the thermal diffusivity, (x_1, x_2) are the Lagrangian, Cartesian coordinates with x_1 pointing in the direction of motion and parallel to the surface, and x_2 into the surface. The speed V is a constant. For the substrate, to which the film adheres, a similar equation applies except that T_2 and κ_2 replace T_1 and κ_1, respectively.

The boundary conditions are those of heat balance on the surface, heat balance and continuity at the layer-substrate interface and regularity conditions. These are, with K's denoting thermal conductivity and d the film thickness,

$$-K_1\frac{\partial T_1}{\partial x_2} = q^*(x_1) = \begin{cases} q(x_1) & (0 \le x_1 \le \ell,\; x_2 = 0) \\ 0 & (x_1 < 0,\; x_1 > \ell,\; x_2 = 0) \end{cases} \quad ,$$

$$K_1\frac{\partial T_1}{\partial x_2} = K_2\frac{\partial T_2}{\partial x_2} \qquad (x_1 = x_1,\; x_2 = d) \quad , \qquad [2.119]$$

$$T_1 = T_2 \qquad (x_1 = x_1,\; x_2 = d) \quad ,$$

$$T_1,\, T_2 \to 0 \qquad [(x_1^2 + x_2^2)^{1/2} \to \infty] \quad .$$

Define the dimensionless quantities: $\xi \equiv x_1/l$, $\eta \equiv x_2/l$, $Q \equiv q/q_0$, $\phi_1 \equiv T_1 K_1 q_0 l$, $\phi_2 \equiv T_2 K_1/q_0 l$, $R_1 \equiv Vl/\kappa_1$, $R_2 \equiv Vl/\kappa_2$, $D \equiv d/l$, $\alpha^2 \equiv \kappa_2/\kappa_1 = R_1/R_2$, and $\beta \equiv K_2/K_1$. The q_0 is the maximum value of $q(x_1)$. With the above, [2.118], the companion one for the substrate and conditions [2.119] may be written:

$$\frac{\partial^2 \phi_1}{\partial \eta^2} = R_1 \frac{\partial \phi_1}{\partial \xi} \quad , \qquad [2.120]$$

$$\frac{\partial^2 \phi_2}{\partial \eta^2} = R_2 \frac{\partial \phi_2}{\partial \xi} \quad , \qquad [2.121]$$

$$-\frac{\partial \phi_1}{\partial \eta} = Q^* \equiv \begin{cases} Q(\xi) & (0 \leq \xi \leq 1, \ \eta = 0) \\ 0 & (\xi < 0, \ \xi > 1, \ \eta = 0) \end{cases} , \qquad [2.122]$$

$$\frac{\partial \phi_1}{\partial \eta} = \beta \frac{\partial \phi_2}{\partial \eta} \qquad (\xi = \xi, \ \eta = D) \quad , \qquad [2.123]$$

$$\phi_1 = \phi_2 \qquad (\xi = \xi, \ \eta = D) \quad , \qquad [2.124]$$

$$\phi_1, \phi_2 \to 0 \qquad [(\xi^2 + \eta^2)^{1/2} \to \infty] \quad . \qquad [2.125]$$

Applying the Fourier transform, [2.5], with s representing the transform parameter and super $\sim$ denoting the transformed quantities, [2.120] and [2.121] with the aid of [2.125], yield

$$\frac{d^2 \tilde{\phi}_1}{d\eta^2} = -isR_1 \tilde{\phi}_1 \quad ,$$

$$\frac{d^2 \tilde{\phi}_2}{d\eta^2} = -isR_2 \tilde{\phi}_2 \quad .$$

The above possess the general solution,

$$\tilde{\phi}_1 = A_1 e^{-\sqrt{R_1 s/2}\,(1-i)\eta} + B_1 e^{\sqrt{R_1 s/2}\,(1-i)\eta}$$

$$\tilde{\phi}_2 = A_2 e^{-\sqrt{R_2 s/2}\,(1-i)\eta} + B_2 e^{\sqrt{R_2 s/2}\,(1-i)\eta} \quad . \tag{2.126}$$

In view of condition [2.125], $B_2 = 0$. Conditions [2.122], [2,123], and [2.124] lead to

$$\sqrt{R_1 s/2}\,(1-i)(A_1 - B_1) = \tilde{Q}^*$$

$$A_1 e^{-\sqrt{R_1 s/2}\,(1-i)D} + B_1 e^{\sqrt{R_1 s/2}\,(1-i)D} = A_2 e^{-\sqrt{R_2 s/2}\,(1-i)D} \tag{2.127}$$

$$A_1 e^{-\sqrt{R_1 s/2}\,(1-i)D} - B_1 e^{\sqrt{R_1 s/2}\,(1-i)D} = \frac{\beta}{\alpha} A_2 e^{-\sqrt{R_2 s/2}\,(1-i)D} \quad .$$

For the surface temperature, only A_1 and B_1 are important. Solving [2.127] for A_1 and B_1, [2.126] gives

$$\tilde{\phi}_1 = \frac{\tilde{Q}^*(1+i)}{\sqrt{2R_1 s}\,(1+\gamma)} \left\{ e^{\sqrt{R_1 s/2}\,(1-i)\eta} - \gamma e^{-\sqrt{R_1 s/2}\,(1-i)\eta} \right\} \quad ,$$

which becomes, for the surface $\eta = 0$,

$$\tilde{\phi}_1 = \frac{\tilde{Q}^*(1+i)}{\sqrt{2R_1 s}} \frac{(1-\gamma)}{(1+\gamma)} \quad , \tag{2.128}$$

where

$$\gamma = A e^{-\sqrt{2R_1 s}\,(1-i)D} \quad ,$$

$$A = \frac{\beta/\alpha - 1}{\beta/\alpha + 1} \quad .$$

Inverse transform, by convolution theorem, gives

$$\phi_1 = \frac{1}{2\pi} \int_{-\infty}^{\infty} \frac{Q^*(\xi')(1+i)}{\sqrt{2R_1}} \int_{-\infty}^{\infty} \frac{e^{-is(\xi-\xi')}}{\sqrt{s}} \frac{1-\gamma}{1+\gamma} \, ds \, d\xi' \,. \qquad [2.129]$$

It can be shown that [2.129] is expressible, after integrating once, as

$$\phi_1 = \begin{cases} 0 & (\xi < 0) \\[2ex] \displaystyle\int_0^{\xi} Q(\xi')H(\xi,\xi')d\xi' & (0 \leq \xi' \leq \xi) \\[2ex] \displaystyle\int_0^{1} Q(\xi')H(\xi,\xi')d\xi' & (\xi > 1) \end{cases} \,, \qquad [2.130]$$

where

$$H(\xi,\xi') = [\pi R_1(\xi-\xi')]^{-1/2} \left\{ 1 + 2 \sum_{n=1}^{\infty} (-1)^n A^n e^{-\frac{n^2 D^2 R_1}{(\xi-\xi')}} \right\} d\xi' \,,$$

and $\sum_{n=1}^{\infty} (-1)^n A^n e^{-(n^2 D^2 R_1)/(\xi-\xi')}$ is an absolutely convergent series for $(\xi - \xi') \geq 0$.

Figure 2.9 shows results of a layer material consisting of graphite on steel as a substrate for a uniform heat input of intensity q_0. The top curve shows the case of a thick layer as expressed by $D \to \infty$, and the lowest curve shows the case of no layer as expressed by $D \to 0$. The intermediate cases represent the ϕ's for various dimensionless D's. Figure 2.10 is a similar one for a copper layer on steel as a substrate. Again there is the case of a thick layer of copper as expressed by $D \to \infty$, the case of no layer as expressed by $D \to 0$, and intermediate cases.

Although much needs to be done to see all the effects of a layer on surface temperatures, including examples on the layered cylinder and layered sphere, it is already clear from the examples above that a thin layer does a great deal in changing the surface temperatures. Referring to Figure 2.9, for a sliding contact of 1/10 in., the case $D = 0.0001$ corresponds to film thickness of 10 μin.

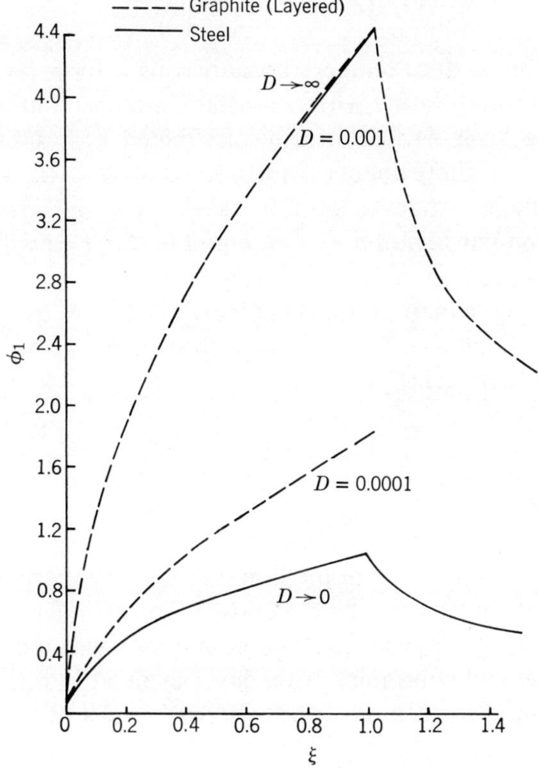

Figure 2.9

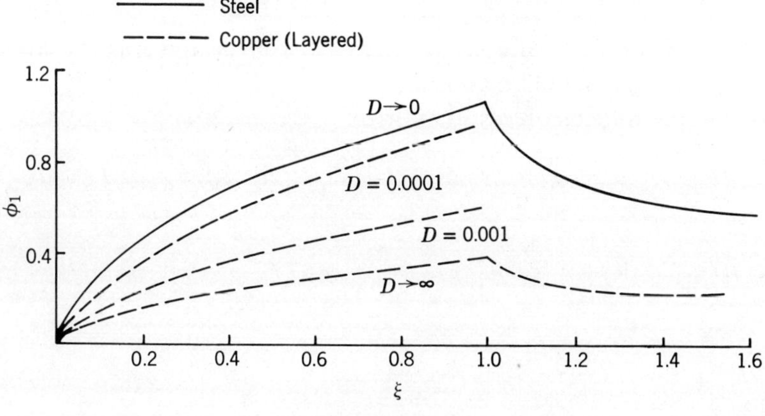

Figure 2.10

L. LAYERED CYLINDER

In this section, surface temperatures are sought for a layered circular cylinder composite rotating with a constant speed ω, that is, Figure 2.2 with a surface layer. An arbitrarily distributed heat source over the periphery, $q(\theta)$, of finite angular width, say $0 \leq \theta \leq \theta_0$, applies to the composite cylinder while convection takes place outside of the heat source. The annular region $a \leq r \leq b$ and $0 \leq r \leq a$ will be referred to as region 1 and region 2, respectively.

The governing equation for heat conduction in the Lagrangian, cylindrical coordinates (r, θ) for the high speed and quasistationary case, and comparable to [2.118] is

$$\frac{\partial^2 T_1}{\partial r^2} + r^{-1} \frac{\partial T_1}{\partial r} + (\omega/\kappa_1) \frac{\partial T_1}{\partial \theta} = 0 \quad , \qquad [2.131]$$

where T_1 is the temperature in the film and κ_1 is the thermal diffusivity. A similar equation governs the heat flow in the cylinder with T_2 and κ_2 replacing T_1 and κ_1, respectively. The boundary conditions, in addition to heat balance and continuity at the layer-cylinder interface and regularity condition at the center of the cylinder, consist of

$$- K_1 \frac{\partial T_1}{\partial r} = q^* = \begin{cases} q(\theta) & (r = b, \ 0 \leq \theta \leq \theta_0) \\ -h T_1 & (r = b, \ \theta_0 < \theta < 2\pi) \end{cases} \quad ,$$

where K_1 is the thermal conductivity of the layer and h is the film coefficient, being assumed a constant.

Define the dimensionless quantities:

$$\rho \equiv r/b \ , \quad A \equiv a/b \ , \quad \phi_1 \equiv T_1 K_1/q_0 b \ , \quad \phi_2 \equiv T_2 K_1/q_0 b \ ,$$

$$Q \equiv q/q_0 \ , \quad R_1 \equiv \omega b^2/\kappa_1 \ , \quad R_2 \equiv \omega b^2/\kappa_2 \ , \quad H \equiv hb/K_1 \ ,$$

$$\alpha^2 \equiv R_1/R_2 \ , \quad \text{and} \quad \beta \equiv K_2/K_1 \ .$$

With the above, the governing equations become

$$\frac{\partial^2 \phi_1}{\partial \rho^2} + \rho^{-1} \frac{\partial \phi_1}{\partial \rho} + R_1 \frac{\partial \phi_1}{\partial \theta} = 0 \quad (A \leq \rho \leq 1, \ \phi = \phi) \ , \quad [2.132]$$

$$\frac{\partial^2 \phi_2}{\partial \rho^2} + \rho^{-1} \frac{\partial \phi_2}{\partial \rho} + R_2 \frac{\partial \phi_2}{\partial \theta} = 0 \quad (0 < \rho < A, \ \phi = \phi) \ , \quad [2.133]$$

and the boundary conditions are

$$-\frac{\partial \phi_1}{\partial \rho} = Q^* = \begin{cases} Q(\theta) & (\rho = 1, \ 0 \leq \theta \leq \theta_0) \\ \\ -H\phi_1 & (\rho = 1, \ \theta_0 < \theta < 2\pi) \end{cases} \quad [2.134]$$

$$\frac{\partial \phi_1}{\partial \rho} = \beta \frac{\partial \phi_2}{\partial \rho} \quad (\rho = A) \ , \quad [2.135]$$

$$\phi_1 = \phi_2 \quad (\rho = A) \ , \quad [2.136]$$

$$\phi_2 \ \text{is finite} \quad (\rho = 0) \ . \quad [2.137]$$

Applying the finite Fourier transform, [2.27], with n representing the transform parameter and super $\sim$ denoting the transformed quantities, [2.132] and [2.133] become

$$\frac{\partial^2 \tilde{\phi}_1}{\partial \rho^2} + \rho^{-1} \frac{\partial \tilde{\phi}_1}{\partial \rho} - in R_1 \tilde{\phi}_1 = 0 \ , \quad [2.138]$$

$$\frac{\partial^2 \tilde{\phi}_2}{\partial \rho^2} + \rho^{-1} \frac{\partial \tilde{\phi}_2}{\partial \rho} - in R_2 \tilde{\phi}_2 = 0 \ . \quad [2.139]$$

Equations 2.138 and 2.139 possess the solution,

$$\tilde{\phi}_1 = c_1 \, I_0(\sqrt{inR_1}\,\rho) + d_1 \, K_0(\sqrt{inR_1}\,\rho)$$

$$\tilde{\phi}_2 = c_2 \, I_0(\sqrt{inR_2}\,\rho) + d_2 \, K_0(\sqrt{inR_2}\,\rho) \quad , \quad [2.140]$$

where positive square roots are chosen; $I_0(\ )$ and $K_0(\ )$ are the modified Bessel functions of the first and second kind of order zero, respectively. And c_1, c_2, d_1, and d_2 are constants of integration, determinable through conditions [2.134] through [2.137].

For the transformed temperature, set $\rho = 1$ in [2.140]. With the appropriate constants and inverse transformation [2.122],

$$\phi_1(\theta) = \int_0^{\theta_0} Q(\theta')M(\theta,\theta')d\theta' - H \int_{\theta_0}^{2\pi} \phi_1(\theta')M(\theta,\theta')d\theta' \,, \quad [2.141]$$

where

$$M(\theta,\theta') = \mathcal{R}e \sum_{n=-\infty}^{\infty} \frac{fe^{-in(\theta-\theta')}}{2\pi\sqrt{inR_1}} \,,$$

$$f = \frac{I_0(\Omega_1)U + K_0(\Omega_1)V}{-I_1(\Omega_1)U + K_1(\Omega_1)V} \,,$$

$$U = K_1(\Omega_1 A) + \beta \, I_1(\Omega_2 A)K_0(\Omega_1 A)/\alpha \, I_0(\Omega_2 A) \,,$$

$$V = I_1(\Omega_1 A) - \beta \, I_1(\Omega_2 A)I_0(\Omega_1 A)/\alpha \, I_0(\Omega_2 A) \,,$$

$$\Omega_1 = (inR_1)^{1/2} \quad \text{and} \quad \Omega_2 = (inR_2)^{1/2} \,.$$

Equation 2.141 is a singular integral equation of the first kind, whose inversion has yet to be executed. To obtain the fundamental solution, we express the solution in the form

$$\phi_1 = \int_0^{\theta_0} Q(\theta')L(\theta,\theta')d\theta' \quad , \qquad [2.142]$$

where the kernel $L(\theta, \theta')$, of course, is the fundamental solution.

It should be noted here that, although [2.141] is formally correct, $M(\theta, \theta')$ is unbounded. However, because of conservation of energy in the quasistationary state, that is,

$$\int_0^{\theta_0} Q(\theta')d\theta' = H \int_{\theta_0}^{2\pi} \phi_1(\theta')d\theta' \quad ,$$

[2.141] may be expressed as

$$\phi_1(\theta) = \int_0^{\theta_0} Q(\theta')N(\theta,\theta')d\theta' - H \int_{\theta_0}^{2\pi} \phi_1(\theta')N(\theta,\theta')d\theta' \quad , \qquad [2.143]$$

where

$$N(\theta,\theta') = \mathcal{R}e \sum_{\substack{n=-\infty \\ n\neq 0}}^{\infty} \frac{fe^{-in(\theta-\theta')}}{2\pi\sqrt{inR_1}} \quad . \qquad [2.144]$$

Equation 2.144 is the correct integral equation and can be used in practice.

Returning to the fundamental solution it may be seen with the help of [2.143] that $L(\theta, \theta')$ is expressible as an infinite series of iterated integrals:

$$L(\theta,\theta') = N(\theta,\theta') - H \int_{\theta_0}^{2\pi} N(\theta,\theta'')N(\theta',\theta'')d\theta''$$

$$+ \; H^2 \int_{\Theta_0}^{2\pi} N(\Theta,\Theta''') \int_{\Theta_0}^{2\pi} N(\Theta',\Theta'')N(\Theta'',\Theta''')d\Theta'' \; d\Theta'''$$

$$+ \; \ldots \qquad \qquad [2.145]$$

It should be mentioned that the high speed approximation is used for its accuracy and its wide range of applicability. In this problem, the results for all speeds may be obtained much as in those embodied in [2.145]. The high speed approximation would make computation with $N(\Theta, \Theta')$ much easier. Of course, for surface temperature computation, in practice, integral equation 2.143 may be inverted and the result expressed as a Neumann iterative series.

M.   LAYERED SPHERE

In this section, surface temperatures are sought for a layered sphere composite rotating with a constant speed ω, that is, Figure 2.6 with a surface of thickness $b - a$. An arbitrarily distributed heat source over the periphery, $q(\Theta, \phi)$, of finite surface region, say $\Theta_1 \leq \Theta \leq \Theta_2$ and $0 \leq \phi \leq \phi_0$, is applied to the sphere while convection takes place outside the region above in the hemisphere where the heat source is located. The remaining hemisphere is insulated. The regions $a \leq r \leq b$ and $0 \leq r < a$ will be referred to as region 1 and region 2, respectively.

The governing equation for heat conduction in the Lagrangian spherical coordinates (r, Θ, ϕ) for the high speed and quasistationary case and comparable to [2.118] is

$$\frac{\partial^2 T_1}{\partial r^2} + \frac{2}{r}\frac{\partial T_1}{\partial r} + \frac{1}{r^2}\frac{\partial}{\partial \mu}[1 - \mu^2)\frac{\partial T_1}{\partial \mu}] = -(\omega/\kappa_1)\frac{\partial T_1}{\partial \phi} \; , \quad [2.146]$$

where ϕ is the azimuthal angle, T_1 the temperature in the film, κ_1 the thermal diffusivity, and $\mu = \cos \Theta$. A similar equation governs the heat conduction in the sphere with T_2 and κ_2, replacing T_1 and κ_1, respectively.

The boundary conditions, in addition to heat balance and continuity at the layer-sphere interface and regularity condition at the center, consist of

$$-K_1 \frac{\partial T_1}{\partial r} = q^* = \begin{cases} 0 & (r = b, \ 0 < \theta < \frac{\pi}{2}, \ \phi = \phi) \\[2ex] q(\theta,\phi) & (r = b, \ \theta_1 \leq \theta \leq \theta_2, \ 0 \leq \phi \leq \phi_0) \\[2ex] -h\,T_1 & (r = b, \ \theta_1 \leq \theta \leq \theta_2, \ \phi_0 < \phi < 2\pi, \\[1ex] & \quad \frac{\pi}{2} \leq \theta \leq \theta_1, \ \phi = \phi, \\[1ex] & \quad \theta_2 \leq \theta \leq \pi, \ \phi = \phi) \end{cases} ,$$

where K_1 is the thermal conductivity of the layer and h is the film coefficient, assumed a constant.

Define the dimensionless quantities: $\rho \equiv r/b$, $A \equiv a/b$, $v_1 \equiv T_1 K_1/q_0 b$, $v_2 \equiv T_2 K_1/q_0 b$, $Q \equiv q/q_0$, $R_1 \equiv \omega b^2/\kappa_1$, $R_2 \equiv \omega b^2/\kappa_2$, $H \equiv hb/K_1$, $\alpha^2 \equiv R_1/R_2$, and $\beta \equiv K_2/K_1$.

With the above, the governing equations become

$$\frac{\partial^2 v_1}{\partial \rho^2} + 2\rho^{-1} \frac{\partial v_1}{\partial \rho} + \rho^{-2} \frac{\partial}{\partial \mu} \left[(1 - \mu^2) \frac{\partial v_1}{\partial \mu} \right] = -R_1 \frac{\partial v_1}{\partial \phi} , \qquad [2.147]$$

$$\frac{\partial^2 v_2}{\partial \rho^2} + 2\rho^{-1} \frac{\partial v_1}{\partial \rho} + \rho^{-2} \frac{\partial}{\partial \mu} \left[(1 - \mu^2) \frac{\partial v_2}{\partial \mu} \right] = -R_2 \frac{\partial v_2}{\partial \phi} , \qquad [2.148]$$

and the boundary conditions are

$$-\frac{\partial v_1}{\partial \rho} = Q^* = \begin{cases} 0 & (\rho = 1, \ 0 < \mu < 1, \ \phi = \phi) \\[2ex] Q(\mu,\phi) & (\rho = 1, \ \cos\theta_2 \leq \mu \leq \cos\theta_1, \ 0 \leq \phi \leq \phi_0) \\[2ex] -H\,v_1 & (\rho = 1, \ \cos\theta_2 \leq \mu \leq \cos\theta_1, \ \phi_0 < \phi < 2\pi, \\[1ex] & \quad \cos\theta_1 < \mu < 0, \ \phi = \phi \\[1ex] & \quad -1 < \mu < \cos\theta_2, \ \phi = \phi) \end{cases} ,$$

$$[2.149]$$

$$\frac{\partial v_1}{\partial \rho} = \beta \frac{\partial v_2}{\partial \rho} \qquad (\rho = A) \quad , \qquad [2.150]$$

$$v_1 = v_2 \qquad (\rho = A) \quad , \qquad [2.151]$$

$$v_2 \text{ is finite} \qquad (\rho = 0) \quad . \qquad [2.152]$$

Applying the finite Fourier transform as before, [2.147] and [2.148] become

$$\frac{\partial^2 \tilde{v}_1}{\partial \rho^2} + 2\rho^{-1} \frac{\partial \tilde{v}_1}{\partial \rho} - \text{in} R_1 \tilde{v}_1 + \rho^{-2} \frac{\partial}{\partial \mu} [(1-\mu^2)\frac{\partial \tilde{v}_1}{\partial \mu}] = 0 , \quad [2.153]$$

$$\frac{\partial^2 \tilde{v}_2}{\partial \rho^2} + 2\rho^{-1} \frac{\partial \tilde{v}_2}{\partial \rho} - \text{in} R_2 \tilde{v}_2 + \rho^{-2} \frac{\partial}{\partial \mu} [(1-\mu^2)\frac{\partial \tilde{v}_2}{\partial \mu}] = 0 . \quad [2.154]$$

Furthermore, taking the finite Legendre transform [2.72], where m is the transform parameter, P_m is the Legendre function, and super $\sim$'s denote the transformed quantities, [2.153] and [2.154] become

$$\frac{\partial^2 \tilde{\tilde{v}}_1}{\partial \rho^2} + \rho^{-1} \frac{\partial \tilde{\tilde{v}}_1}{\partial \rho} - [\text{in} R_1 + (m + \tfrac{1}{2})^2 \rho^{-2}] \tilde{\tilde{v}}_1 = 0 , \quad [2.155]$$

$$\frac{\partial^2 \tilde{\tilde{v}}_1}{\partial \rho^2} + \rho^{-1} \frac{\partial \tilde{\tilde{v}}_2}{\partial \rho} - [\text{in} R_2 + (m + \tfrac{1}{2})^2 \rho^{-2}] \tilde{\tilde{v}}_2 = 0 , \quad [2.156]$$

where $\tilde{v}_1 = \tilde{\tilde{v}}_1 \rho^{-1/2}$ and $\tilde{v}_2 = \tilde{\tilde{v}}_2 \rho^{-1/2}$.

Upon inverse transform,

$$v_1(\theta,\phi) = \int_\Gamma Q(\theta',\phi')M(\theta,\phi;\theta',\phi')d\theta'\,d\phi'$$

$$- H\int_{\Gamma_r} v_1(\theta',\phi')M(\theta,\phi;\theta',\phi')d\theta'\,d\phi' \quad , \qquad [2.157]$$

where

$$M(\theta,\phi;\theta',\phi') = \mathcal{R}e \sum_{\substack{n=-\infty \\ n\neq 0}}^{\infty} \sum_{m=0}^{\infty} \frac{2m+1}{4\pi} \cdot$$

$$\left\{ \frac{[I_{m+\frac{1}{2}}(\Psi_1)f_1 - K_{m+\frac{1}{2}}(\Psi_1)f_2]P_m(\cos\theta)P_m(\cos\theta')e^{in(\phi-\phi')}}{[\Psi_1 I'_{m+\frac{1}{2}}(\Psi_1) - \frac{1}{2}I_{m+\frac{1}{2}}(\Psi_1)]f_1 - [\Psi_1 K'_{m+\frac{1}{2}}(\Psi_1) - \frac{1}{2}K_{m+\frac{1}{2}}(\Psi_1)]f_2} \right\} ,$$

and

$$f_1 = K'_{m+\frac{1}{2}}(\Psi_1 A) - K_{m+\frac{1}{2}}(\Psi_1 A) \quad ,$$

$$f_2 = I'_{m+\frac{1}{2}}(\Psi_1 A) - I_{m+\frac{1}{2}}(\Psi_1 A) \quad ,$$

$$= \frac{\beta}{\alpha} \frac{I'_{m+\frac{1}{2}}(\Psi_2 A)}{I_{m+\frac{1}{2}}(\Psi_2 A)} - \frac{1-\beta}{2\Psi_1 A} \quad ,$$

$I_{m+\frac{1}{2}}(\ )$ and $K_{m+\frac{1}{2}}(\ )$ are the modified Bessel functions of the first and second kind of order $m + \frac{1}{2}$, respectively. Prime refers to differentiation with respect to the argument. The Ψ_1 and Ψ_2 are the same as defined in the last section. Also, in the above, use has been made of the fact that

$$\int_{\Gamma} Q(\theta',\phi')d\theta' \ d\phi' = H \int_{\Gamma_r} v_1(\theta',\phi')d\theta' \ d\phi' \quad .$$

The area covered by the heat source is Γ, and Γ_r is the remaining of the hemisphere containing Γ.

Again, [2.157] is a singular integral equation whose inversion may be accomplished by iterated integrals. This form is suitable for computation. For the fundamental solution, that is,

$$v_1(\theta,\phi) = \int_{\Gamma} Q(\theta',\phi')L(\theta,\phi;\theta',\phi')d\theta' \ d\phi' \quad , \qquad [2.158]$$

L may be shown to be expressible as an infinite series of iterated integrals:

$$L(\theta,\phi;\theta',\phi') = M(\theta,\phi;\theta',\phi') - H \int_{\Gamma_r} M(\theta,\phi;\theta'',\phi'') \cdot$$

$$M(\theta',\phi';\theta'',\phi'')d\theta'' \ d\phi''$$

$$+ H^2 \int_{\Gamma_r} M(\theta,\phi;\theta''',\phi''') \int_{\Gamma_r} M(\theta',\phi';\theta''',\phi''') \cdot$$

$$M(\theta'',\phi'';\theta''',\phi''')d\theta''' \ d\phi'''$$

$$+ \ . \ . \ . \ . \qquad\qquad\qquad [2.159]$$

It is again noted that while the high speed approximation is convenient and has a wide range of applicability, the general case may also be derived as [2.159].

N.   RANDOM HEAT CONDUCTION

In this section a simple analysis is made in which the linear system involves Gaussian processes as forcing functions. Figure 2.11 shows a

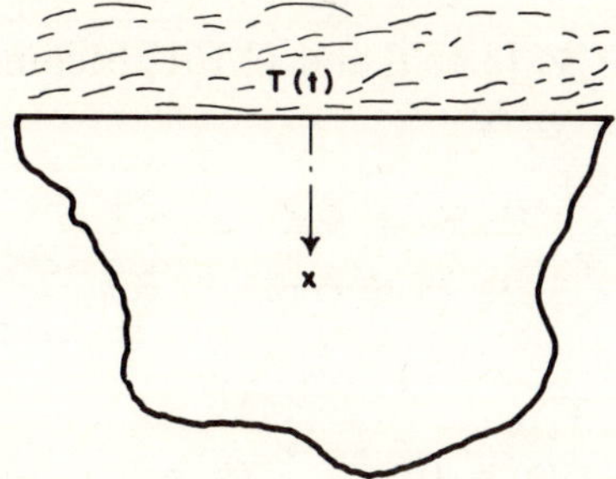

Figure 2.11

semiinfinite solid whose flat surface is exposed to a heat source $T(t)$ [first order radiation or convection] which is random. The central aim in the analysis of Gaussian random processes is to obtain the auto-correlation function of the output, which is the resulting temperature in this case. For zero mean, this quantity determines the entire output process. Also for stationary Gaussian processes for the exciting functions, the autocorrelation function of the temperature gives the mean square temperature.[16] Limitation of space does not permit the input area of heat conduction in random media.[17]

The one-dimensional heat equation is

$$\frac{\partial^2 v}{\partial \xi^2} = \frac{\partial v}{\partial \tau} , \qquad\qquad [2.160]$$

where $v \equiv \theta/T_m$, $\xi \equiv x/l$, $\tau \equiv \kappa t/l^2$, and θ is the temperature within the body, T_m is the mean value of the given $T(t)$, l is some reference distance, κ is the thermal diffusivity, and t is the time.

The boundary conditions are

$$\frac{\partial v}{\partial \xi} - H(v - \psi) = 0 \qquad (\xi = 0, \quad \tau = \tau) ,$$

$$[2.161]$$

$$\frac{\partial v}{\partial \xi} = 0 \qquad (\xi \to \infty) ,$$

where $\Psi \equiv T/T_m$ and H is the dimensionless heat transfer coefficient which is hl/κ in the case of convection while h is the film coefficient.

Letting $u = v - \Psi$, [2.160] and [2.161] become, respectively,

$$\frac{\partial^2 u}{\partial \xi^2} = \frac{\partial u}{\partial \tau} + \frac{\partial \psi}{\partial \tau} \quad , \qquad [2.162]$$

$$\frac{\partial u}{\partial \xi} - Hu = 0 \qquad (\xi = 0, \quad \tau = \tau) \quad ,$$

$$[2.163]$$

$$\frac{\partial u}{\partial \xi} = 0 \qquad (\xi \to \infty, \quad \tau = \tau) \quad .$$

Using the integral transform

$$u_k(\eta, \tau) = \int_0^\infty u(\xi, \tau)\, K(\eta, \xi)\, d\xi$$

$$u(\xi, \tau) = \int_0^\infty u_k(\eta, \tau)\, K(\eta, \xi)\, d\eta \quad ,$$

where

$$K(\eta, \xi) = \sqrt{\frac{2}{\pi}} \left[\frac{H \sin(\eta \xi) + \eta \cos(\eta \xi)}{(\eta^2 + H^2)^{1/2}} \right] \quad ,$$

[2.162] becomes

$$\frac{du_k}{d\tau} + \eta^2 u_k = -\frac{d\psi}{d\tau} \int_0^\infty K(\eta, \xi)\, d\xi \quad . \qquad [2.164]$$

Now taking the Fourier transform [2.5] or [2.164] and denoting the transformed quantity by a super bar,

$$\bar{u}_k = (2\pi)^{-1/2} \int_{-\infty}^\infty e^{is\tau} u_k(\eta, \tau)\, d\tau = \frac{-is\bar{\psi}(\zeta s) \int_0^\infty K(\eta, \xi)\, d\xi}{is + \eta^2} \quad ,$$

where s is the transform parameter. Applying the inversion theorems 2.12,

$$u(\xi,\tau) = -(2\pi)^{-1} \int_0^\infty \int_{-\infty}^\infty e^{is\tau} \frac{is\bar{\psi}(s) \int_0^\infty K(\eta,\xi)d\xi \; ds \; K(\eta,\xi)d\eta}{is + \eta^2}$$

or

$$u(\xi,\tau) = -\int_0^\infty \left\{ \int_{-\infty}^\infty \psi(\tau-\beta)w(\beta,\eta)d\beta \right\} \lambda(\eta) \; K(\eta,\xi)d\eta \quad , \qquad [2.165]$$

$$w(\beta,\eta) = (2\pi)^{-1} \int_{-\infty}^\infty e^{isp} \frac{is \; ds}{is + \eta^2} = \left\{ \begin{array}{ll} -\eta^2 e^{-\eta^2\beta} + \delta(\beta^+) \; , & (\beta > 0) \\ 0 & (\beta < 0) \end{array} \right. \quad ,$$

and

$$\int_0^\infty K(\eta,\xi)d\xi = \lim_{\alpha \to \infty} \int_0^\infty e^{-\alpha\xi} K(\eta,\xi)d\xi$$

$$= \frac{H\sqrt{2/\pi}}{\eta(\eta^2 + H^2)^{1/2}} \equiv \lambda(\eta) \quad .$$

The autocorrelation function of u is

$$R_u(\xi,\xi',\tau,\tau+d) = \int_0^\infty \int_0^\infty \left\{ \int_{-\infty}^\infty \int_{-\infty}^\infty R_\psi(\tau-\beta, \; \tau+d-\beta')w(\beta,\eta)w(\beta',\eta')d\beta \; d\beta' \right\}$$

$$\cdot \lambda(\eta)\lambda(\eta')K(\eta,\xi)K(\eta',\xi')d\eta \; d\eta' \quad , \qquad [2.166]$$

where

$$R_\psi(\tau,\tau+d) = E\left\{ \psi(\tau) \; \psi(\tau+d) \right\}$$

is the autocorrelation function of the external temperature. If $T(\tau)$ is a stationary process in time, then

$$R_u(\xi,\xi',d) = \int_0^\infty \int_0^\infty \left\{ \int_{-\infty}^\infty \int_{-\infty}^\infty R_\psi(d+\beta-\beta')w(\beta,\eta)w(\beta',\eta')d\beta\, d\beta' \right\}$$

$$\cdot\, \lambda(\eta)\lambda(\eta')K(\eta,\xi)K(\eta',\xi')d\eta\, d\eta'$$

and

$$R_v(\xi,\xi',d) = \int_0^\infty \int_0^\infty \left\{ \int_{-\infty}^\infty \int_{-\infty}^\infty R_\psi(d+\beta-\beta')\tilde{w}(\beta,\eta)\tilde{w}(\beta',\eta')d\beta\, d\beta' \right\}$$

$$\cdot\, \lambda(\eta)\lambda(\eta')K(\eta,\xi)K(\eta',\xi')d\eta\, d\eta' \quad , \qquad [2.167]$$

where

$$\tilde{w}(\beta) = \begin{cases} -\eta^2\, e^{-\eta^2\beta} & (\beta > 0) \\[2mm] 0 & (\beta < 0) \end{cases} .$$

For purely random process for $\Psi(\tau)$,

$$R_\psi(d) = S_0\, \delta(d) \quad ,$$

where S_0 is the constant spectral density. The mean square of $v(\xi, \tau)$ is then given by

$$\langle v^2(\xi) \rangle = S_0 \int_0^\infty \int_0^\infty \left.\frac{e^{-\eta'^2 d}}{\eta'^2 + \eta^2}\right|_{d=0^+} \lambda(\eta)\lambda(\eta')K(\eta,\xi)K(\eta',\xi')d\eta\, d\eta'$$

or

$$\langle v^2(\xi) \rangle = \frac{4S_0 H^2}{\pi^2} \int_0^\infty \int_0^\infty \frac{\eta\, \eta'}{\eta^2 + \eta'^2}$$

$$\cdot\, \frac{[h\, \sin\, \eta\xi + \eta\, \cos\, \eta\xi][H\, \sin\, \eta'\xi + \eta'\, \cos\, \eta'\xi]d\eta\, d\eta'}{(\eta^2 + H^2)(\eta'^2 + H^2)} .$$

$$[2.168]$$

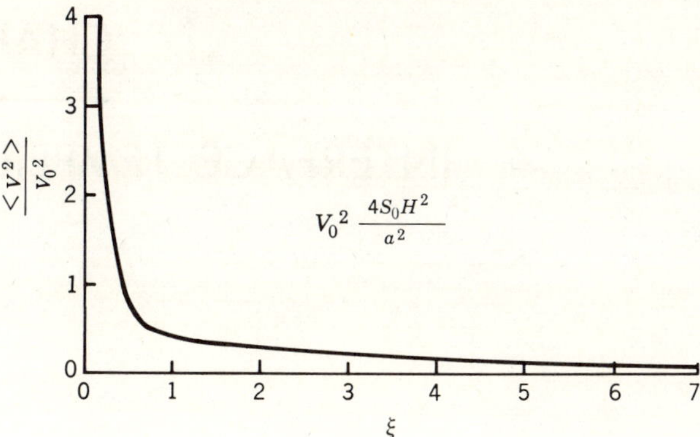

Figure 2.12

The mean square value of v at the surface $\xi = 0$ is given by

$$< v^2(0) > = \frac{4S_0 H^2}{\pi^2} \int_0^\infty \int_0^\infty \frac{\eta^2 \, \eta'^2 \, d\eta \, d\eta'}{(\eta^2 + H^2)(\eta'^2 + H^2)(\eta^2 + \eta'^2)} \, .$$

$$[2.169]$$

This expression can be evaluated by use of a digital computer. A plot of [2.168] is shown in Figure 2.12, where v_0 denotes $v(0)$ from [2.169].

CHAPTER 3

INTERFACE TEMPERATURE

A.   DYNAMIC THERMOCOUPLE

When a wire of metal A and one of metal B are fused together at two ends so as to form a closed loop, it is known that current will flow if the junctions are kept at different temperatures, T_1 and T_2. This thermo-electric effect may be described by the equation:

$$E = \pi_{AB}(T_2 - T_1) + \int_{T_1}^{T_2} (\sigma_A - \sigma_B)dT \quad , \qquad [3.1]$$

where the voltage E is known as the Seebeck emf, $\pi_{AB}(T_2 - T_1)$ is the Peltier emf, π_{AB} is the Peltier coefficient, $\int_{T_1}^{T_2} (\sigma_A - \sigma_B)dT$ is the Thomson emf, and σ's are the Thomson coefficients. Figure 3.1 shows that metal block A and metal block B are in sliding contact. Since electrons move much faster than sliding speeds generally encountered, the bodies may be considered stationary insofar as electric fields are concerned. The diagram is self-explanatory.

[62]

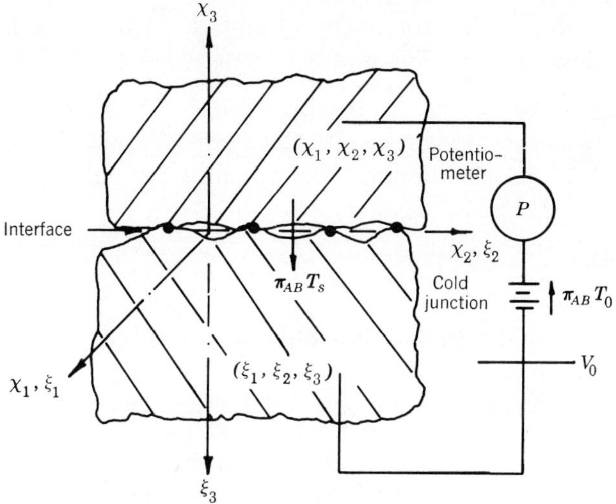

Figure 3.1

Within a body with resistivity γ, a temperature field $T(x_1, x_2, x_3)$ and an electric field $V(x_1, x_2, x_3)$,

$$\gamma\, J_i = \sigma\, T_{,i} - V_{,i} \quad , \qquad [3.2]$$

where J_i is the ith component of the current vector. The left-hand term is, of course, the Joule effect. For steady state

$$J_{i,i} = 0 \quad . \qquad [3.3]$$

Equations 3.2 and 3.3 lead to

$$V_{,ii} = (\sigma\, T_{,i})_{,i} \quad , \qquad [3.4]$$

which applies equally well for bodies A and B for the appropriate σ's. Referring to Figure 3.1,

$$V = P + \int_{T_0}^{T} \sigma\, dT + V_0 - \pi_{AB}\, T_0 \quad . \qquad [3.5]$$

In view of [3.4],

$$P_{,ii} = 0 \qquad [3.6]$$

which, again, applies for bodies A and B. At the boundaries with no current flow

$$J_i \, n_i = 0 \quad , \qquad\qquad [3.7]$$

where n_i is the ith component of the unit normal to the boundaries. Equations 3.2 and 3.7 lead to

$$\sigma \, T_{,n} - V_{,n} = 0 \quad , \qquad\qquad [3.8]$$

where $(,n)$ denotes partial derivative with respect to the normal of the boundaries. Equations 3.8 and 3.5 yield

$$P_{,n} = 0 \quad . \qquad\qquad [3.9]$$

For those portions of the boundary where there is contact, Kirchhoff's law applies.

$$P = \int_{T_0}^{T_s} (\sigma_B - \sigma_A) dT + \pi_{AB}(T_s - T_0)$$

$$- \int_{(x_1,x_2,0)}^{(x_1,x_2,x_3)} r^{-1} \, J \, ds - \int_{(\xi_1,\xi_2,0)}^{(\xi_1,\xi_2,\xi_3)} r^{-1} \, J \, ds \quad .$$

$$[3.10]$$

But the first two terms of [3.10] are the Seebeck emf corresponding to the temperature on the boundary. In general, P is a function of x_i and ξ_i. For the present purpose, let x_i and ξ_i be symmetrical with respect to the interface. Moreover, let both blocks be semiinfinite solids, and x_i, ξ_i be placed an infinite distance away from the interface. Thus the Laplacian with respect to the x_i system is the same as that with respect to the ξ_i system. Also, if $x_i = (x_1, x_2, 0)$ and $\xi_i = (\xi_1, \xi_2, 0)$, [3.10] gives

$$P = E(T_s) \quad . \qquad\qquad [3.11]$$

Thus mathematically the thermoelectric potential problem has been reduced to a potential problem in which the boundary value is equal to the Seebeck emf when contact is made, and $\partial P/\partial n = 0$ elsewhere along the boundary.

If there are n circular contact areas, each of area A_n with associated Seebeck emf E_n, then for sparsely located areas of the kind described,[18]

$$P(\infty) = \sum_n E_n \sqrt{A_n} \Big/ \sum_n \sqrt{A_n} \, . \qquad [3.12]$$

That is, the potentiometer reading for the leads placed at a large distance away from the interface shows the average Seebeck emf weighted by the square root of the area and not the area.

For a continuous contact area, in particular a circular area, it has been found that[19] the $P(\infty)$ reads the average Seebeck emf weighted by the area divided by its distance from the perimeter of the circle. This leads to the conclusion that the temperature near the perimeter of a contact area has a greater influence on the thermocouple emf than that in the interior of the contact area. For a rectangular contact area in sliding contact, the interface temperature based on continuity at the interface has been computed numerically.[20] A typical result is shown in Figure 3.2. Also the weights are shown in Figure 3.3.

B.   JUNCTURE CONDITION AT MOVING INTERFACES

Figure 3.3 refers to the calculated temperature using continuum theory. In the calculation, it has been tacitly assumed that the temperatures on

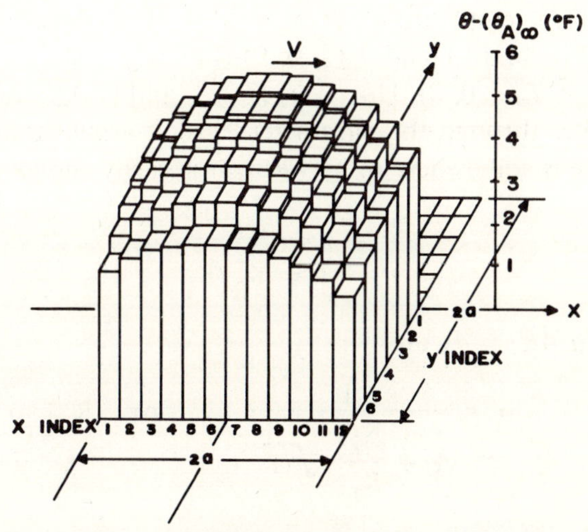

Figure 3.2

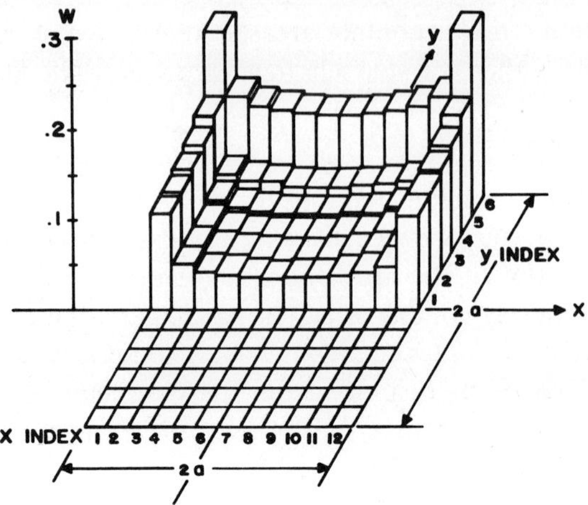

Figure 3.3

both surfaces in contact are the same. This condition, which is certainly valid for smooth surface without contaminating films, was first used in 1937.[21] Essentially,[22] using the heat equation solutions already found, the surface temperature on both bodies shown in Figure 3.4 within the squares can be found. Uniform and constant heat flux is used: σq for the stationary body 1 and $(1 - \sigma)q$ for the moving body 2. Examination of the surface temperatures at $y = \eta = 0$ as shown in Figure 3.5, $T_1(x, 0, 0)$ for body 1 and $T_2(\xi, 0, 0)$ for body 2 shows that generally $T_1(x, 0, 0) \neq T_2(\xi, 0, 0)$. However, $(T_1)_{\max}$ and $(T_2)_{\max}$ can be set equal to each other through the parameter σ. Blok argues that the probable temperature is somewhere in between. It is easily shown that

$$\sigma = \frac{1}{1 + K_2/K_1} \, , \qquad [3.13]$$

for $R \equiv Va/4K_2 = 0$, and

$$\sigma = \frac{1}{1 + \dfrac{K_2}{K_1}\sqrt{2R}} \qquad \text{for} \quad R > 5 \, , \qquad [3.14]$$

while, in between, the expression would be more complicated, in fact,

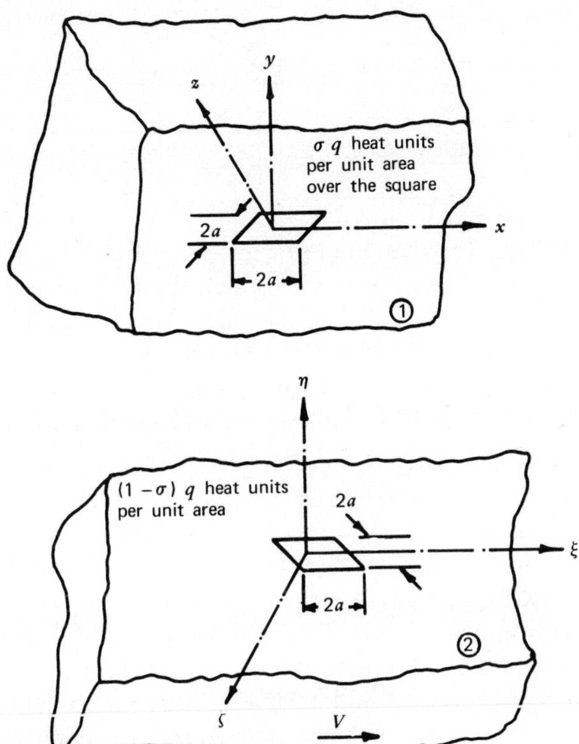

Figure 3.4

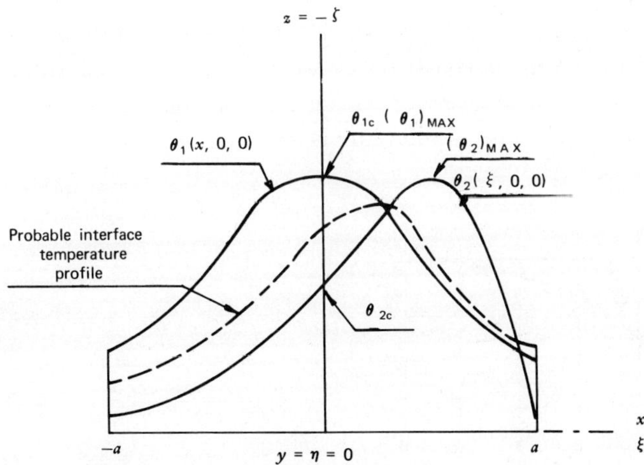

Figure 3.5

for the whole range of R,[22]

$$\sigma \doteq \cfrac{1}{1 + \cfrac{K_2}{K_1} \cfrac{2\pi}{[1 + .414(1-e^{-1.3R})I(R)]}} \qquad [3.15]$$

where $I(R)$ is shown in Figure 3.6.

Strictly speaking, the matching of $T_1(x, y, 0)$ and $T_2(\xi, \eta, 0)$ leads to an integral equation, the inversion of which should give σ as a function of (x, y). This has been done in one instance [10] and it was found that the picture shown in Figure 3.5 is not too far off.

So much for the condition of perfect matching of temperature. What of the actual situation? For static contacts, experiments in 1948 and 1963 have shown that there is a temperature break at the interface,[23,24] that is, temperature, considering surfaces smooth. In other words, when surfaces are considered smooth, then the temperatures extrapolated by means of continuum theory to the surfaces are not necessarily the same. The basis of extrapolation is measured temperature away from the surface in one instance.[25] The solid curves in Figure 3.7 show the data of h_c, thermal conductance versus average pressure for various values of surface roughness between 3 and 3320 μin. The h_c is so defined that the heat flux is equal to the product of h_c and the temperature jump between the surfaces in contact.

For moving contacts, the measurement of h_c was first carried out[25] as shown in schematic diagram, Figure 3.8. The essential components of the apparatus are the friction wheel, A, and the friction arm assembly, B. Affixed to the drive shaft, C, between layers of insulation, is a thin ring plate, A, whose exterior surface is in sliding contact with the stationary rider, D. The round plate is the slider and is insulated from the

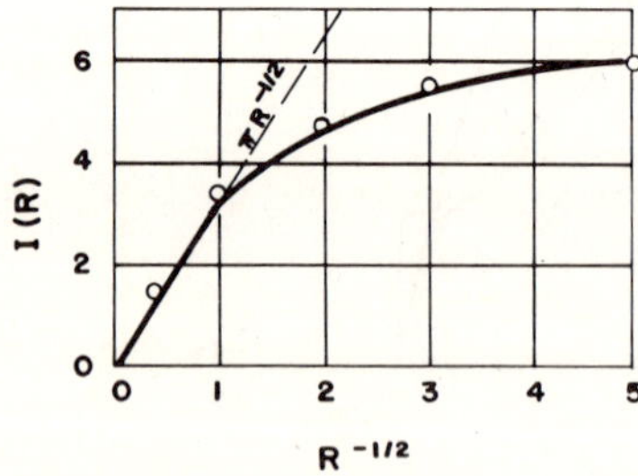

Figure 3.6

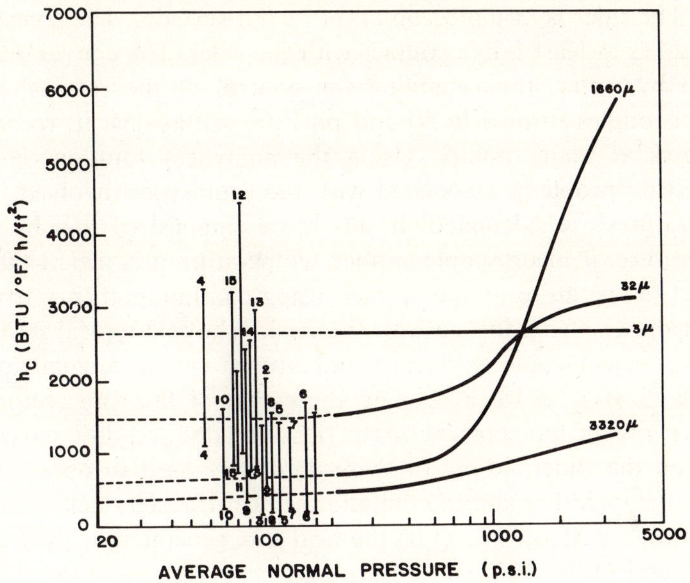

Figure 3.7

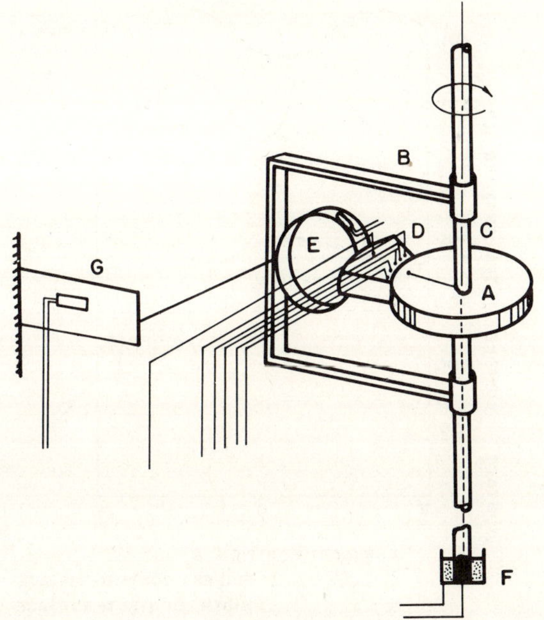

Figure 3.8

shaft. The rider is insulated on all of its flat surfaces. The concave surface makes at least a line contact with the rider. The convex surface is cooled by water, thus enabling the control of the rate of cooling. Thermocouples formed by D and pins pressed against it, record temperatures at many points. Using the analytical solutions[12] of heat conduction problems associated with the geometries involved, surface temperatures are calculated. It should be emphasized that by surface temperature or macroscopic surface temperature it is meant the extrapolated temperature to the surface using continuum theory from data obtained away from the surface. Figure 3.9 shows a typical set of data, that is, macroscopic surface temperatures versus angular position $-\pi/18 \leq \theta \leq \pi/18$, $\pi/9$ being the extent of the ring sector. Solid curves show the temperature on the rider and dot-and-dash curves show those on the slider. Many of these data were used to obtain average surface temperature and to compute the heat transfer coefficient across the sliding interface. Let Q be the heat flux generated at the interface. Let Q_1 and Q_2 be the heat flux through the periphery of the slider and

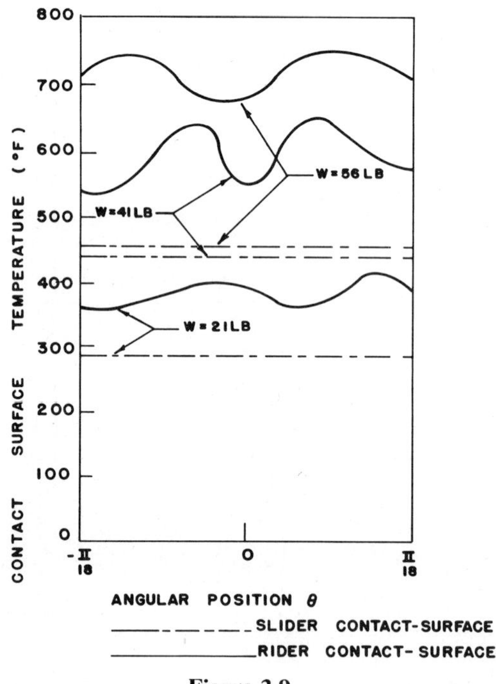

Figure 3.9

the back of the rider, respectively. Of course, $Q = Q_1 + Q_2$. If ΔT is the area-average macroscopic temperature jump at the interface and h_c is the average heat-transfer coefficient, then the heat transmitted across the interface, Q_h, by conduction, convection, and first order radiation is $h_c\Delta T$. Since Q_h is not exactly known, bounds on h_c can be calculated from experimental data: $\bar{h}_c$ and $\mathbf{h}_c$ for the upper and lower bounds, respectively. Then $\bar{h}_c = \bar{Q}_h/\Delta T$ and $\mathbf{h}_c = \mathbf{Q}_h/\Delta T$, where $\bar{Q}_h$ and $\mathbf{Q}_h$ are the upper and lower bound, respectively. If Q_1 were entirely generated on the surface or within the rider, then $\bar{Q}_h = Q_1$ is certainly true. For the same materials in sliding contact, the rider wears more than the slider based on the amount of exposure to sliding contact. Now if using an equal probability argument, let half of the Q be generated on each body, a lower bound is then $\mathbf{Q}_h = Q_1 - Q/2$. In this way $\bar{h}_c$ and $\mathbf{h}_c$ for 15 sets of data are shown in Figure 3.7 together with data for static situation. The horizontal bars of the elongated I's represent the upper and lower bounds of h_c.

Returning to the question of heat generated on the surface or within the body, let λq denote that associated with the stationary body. Then $(1 - \lambda)q$ is that associated with the moving body, and q is as defined earlier in connection with the heat removal fraction σ. The following observations have been made:[26-28]

1. In general, frictional heat is not generated in the space between two bodies in sliding contact. Most of the heat is generated on the surface and in the surface layer immediately below the surface of both bodies. This is because friction derives from either the breaking of adhered junctions, or from the thermodynamically irreversible process of plastic deformation of asperities and the bulk body. That is $\lambda \neq \sigma$ in general.

2. Whenever the capacity of one of the bodies to remove heat away from the interfacial zone is less than the amount of heat generated on that surface, there will be a temperature jump across the interface. The surface will then have a higher macroscopic temperature. There is still another mechanism for temperature jumps which is discussed later.

3. Observations 1 and 2 may be used as a quantitative tool for selecting sliding material pairs.[28]

Note that the above is distinct from the previously mentioned reason for temperature jumps at the interface.

C. SOME METHODS OF MEASURING INTERFACE TEMPERATURES

1. Indirect method for macroscopic temperature with the aid of analytical solution of the heat equation—as above.

2. Dynamic thermocouple method—for very high loads it has found success in metal cutting investigations.[29–31] In these applications theoretical calculation checks with experiment. It has also been used for gear temperature studies.[32] For friction studies the check between theoretical calculations and experiments has not always been good, for example, the case of ball on cylinder.[33] Choice of material pair is not altogether free.

3. Change of color method.[34] This requires transparent mating surface.

4. Infrared radiation method.[35] This requires transparent mating surface.

5. Imbedding thermocouple beneath the surface. This is widely used although not very accurate in general.

6. Kripton gas detection method. Here surfaces have to be separated for detection.

D. SURFACE TEMPERATURE TRANSIENTS

It is clear from the discussion above that, experimentally, macroscopic surface temperature may be measured indirectly. The usefulness of this measure of surface temperature may be limited or not depending on the purposes at hand. From the fundamental as well as utility points of view, however, it would be desirable to have some measure of the actual surface temperature at points of contact as time of sliding goes on. Qualitatively, several of the methods listed may be useful in this regard; but, for metal systems, there does not seem to be any direct method for measuring these transients. All the available methods are indirect. In all cases, the manner of indirect measurement depends on the model of surfaces in contact.

Models become important as soon as surface details are being discussed, for example, the dynamic thermocouple in Section 3.A. Recently, much attention is put in the profilometric description of surfaces. The main strength of profilometric representation lies in the reproduction, to the degree of accuracy of the instrumentation, of the actual surfaces before or after sliding engagement. Its drawback is at least twofold. First, it does not give the situation while sliding is taking place. Second, a reproduction of actual events, even with true integrity, is not exactly useful if the events are complicated. If statistical treatment of these events becomes necessary, then a typical basic event is all the more important. This leads to a more basic modeling of surfaces insofar as temperatures are concerned.

Modeling of surfaces for deformation studies among others is treated in detail in a later chapter. A model for interface surface temperatures, however, is given here. For this purpose, it is tacitly assumed that, insofar as heat transfer is concerned, surfaces of the material bodies are smooth but contacts are specified only where and when physical contacts are supposed to be taking place. At this point it may be appropriate to digress for a moment and see what sort of error this assumption may lead to. Assume that surface asperities are truncated right circular cones of various semiopening angles θ_0. Then according to the method for the cone stated in Section 2.F, the following conclusion[13] is reached:

$$\overline{T} = \overline{T}_0(1 + \beta/2) \quad , \qquad\qquad [3.16]$$

where $\overline{T}$ is the average surface temperature on the truncated face for a uniform heat input, $\beta \equiv \pi/2 - \theta_0$ is measured in radians and $\overline{T}_0$ is $\overline{T}$ for $\beta = 0$, that is, a flat surface. This formula is found empirically from calculated solution; it is good for β not too large, a condition satisfied by most surfaces. For $\theta_0 = 80°$, for example, [3.16] indicated that $\overline{T}_0$ is 8% higher than $\overline{T}$.

Based on the premise that whereas the population of numerous contact points may move in position and change in temperature magnitude as time marches on and consequently is very cumbersome to monitor, its statistics may be very useful. To this end, a simple stochastic model has been constructed:[36]

1. Given a geometric contact area, Figure 3.10 shows one of a square ($l \times l$) although it need not be.

2. Given a load W, based on existing friction theory, the fraction of the geometric area in actual contact does not change with time.

3. Subdivide the geometric area into equal unit areas, each of which represents the smallest possible areas.

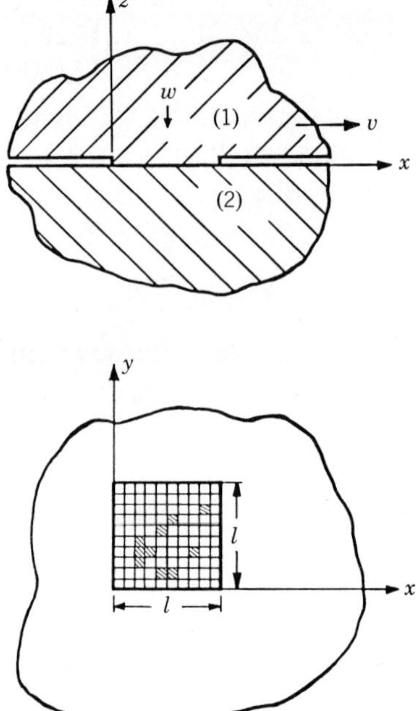

Figure 3.10

4. For a given time interval, suitable for the speeds of sliding and computational requirements, the number of units in contact, n, is fixed as discussed above.

5. By random processes, n units are populated.

6. Coalescence of units makes larger areas of contact than the basic unit.

7. Same process is carried out for each succeeding time interval.

Given such a model, heat conduction analyses using basic solutions outlined in Chapter 2 may be applied.

Define the following dimensionless quantities: $\xi \equiv x/l$, $\eta \equiv y/l$, $\zeta \equiv z/l$, $\tau \equiv 4\alpha t/l^2$ and $f(\xi', \eta', 0, \tau')d\xi'd\eta' \equiv Q/q_0$. The q_0 is the heat flux per unit area, $\mu w V/A$, with μ as the coefficient of friction, w as the total load over the apparent area of contact, V as the sliding velocity, and A as the actual area of contact. If the simple relationship $w = Ap_m$, where p_m is the average yield pressure, is assumed, then $q_0 \equiv \mu p_m V$. With the above, a dimensionless temperature may now be defined:

$\phi \equiv 2\pi^{3/2}KT/\mu V p_m l$, where K is the thermal conductivity. Equation 2.81 leads to

$$d\phi = f(\xi',\eta',0,\tau')d\xi'd\eta'd\tau'(\tau-\tau')^{-3/2}\exp[r^2/(\tau-\tau')],$$

where [3.17]

$$r^2 = (\xi-\xi')^2 + (\eta-\eta')^2 + \zeta^2,$$

and the total temperature change up to time τ over the apparent area is

$$\phi = \int_0^\tau \int_0^1 \int_0^1 d\phi . \qquad [3.18]$$

Let the apparent area be divided into $m \times m$ squares, each of size $\Delta\xi \times \Delta\eta$, and a discrete measure of time be $k\Delta\tau$. Then the discrete f may be defined as

$$f_k^{ij} \equiv f(i\Delta\xi, j\Delta\eta, k\Delta\tau) \qquad 0 \leq i \leq m-1, \quad 0 \leq j \leq m-1.$$

Figure 3.11 shows one of the programs of discrete heat input function f_k^{ij} as stochastic processes discussed earlier.

With reference to a moving frame, [3.17] holds except $r^2 = [\xi - \xi' +u(\tau - \tau')]^2 + (\eta - \eta')^2 + \zeta^2$ and $u = Vl/4\alpha$.

For a rectangular heat source over $\Delta\xi \times \Delta\eta$ with its lower and left corner located at (ξ_i, η_j) generating heat for a period τ_k to $\tau_k + \Delta\tau$, the temperature change at (ξ, η) and τ is

$$\Delta\phi_{\xi\eta\zeta\tau} = \int_{\tau_k}^{\tau_k+\Delta\tau} f_k^{ij}(\tau-\tau')^{3/2}\exp[-\zeta^2/(\tau-\tau')]d\tau'$$

$$\cdot \int_{\xi_i}^{\xi_i+\Delta\xi} \exp\left\{-[(\xi-\xi') + u(\tau-\tau')]^2/(\tau-\tau')\right\} d\xi'$$

$$\cdot \int_{\eta_j}^{\eta_j+\Delta\eta} \exp[-(\eta-\eta')^2/(\tau-\tau')]d\eta' . \qquad [3.19]$$

$0 \leq \tau \leq \Delta\tau$
$\Delta\tau \leq \tau \leq 2\Delta\tau$
$2\Delta\tau \leq \tau \leq 3\Delta\tau$
$3\Delta\tau \leq \tau \leq 4\Delta\tau$
$4\Delta\tau \leq \tau \leq 5\Delta\tau$
$5\Delta\tau \leq \tau \leq 6\Delta\tau$
$6\Delta\tau \leq \tau \leq 7\Delta\tau$
$7\Delta\tau \leq \tau \leq 8\Delta\tau$
$8\Delta\tau \leq \tau \leq 9\Delta\tau$
$9\Delta\tau \leq \tau \leq 10\Delta\tau$

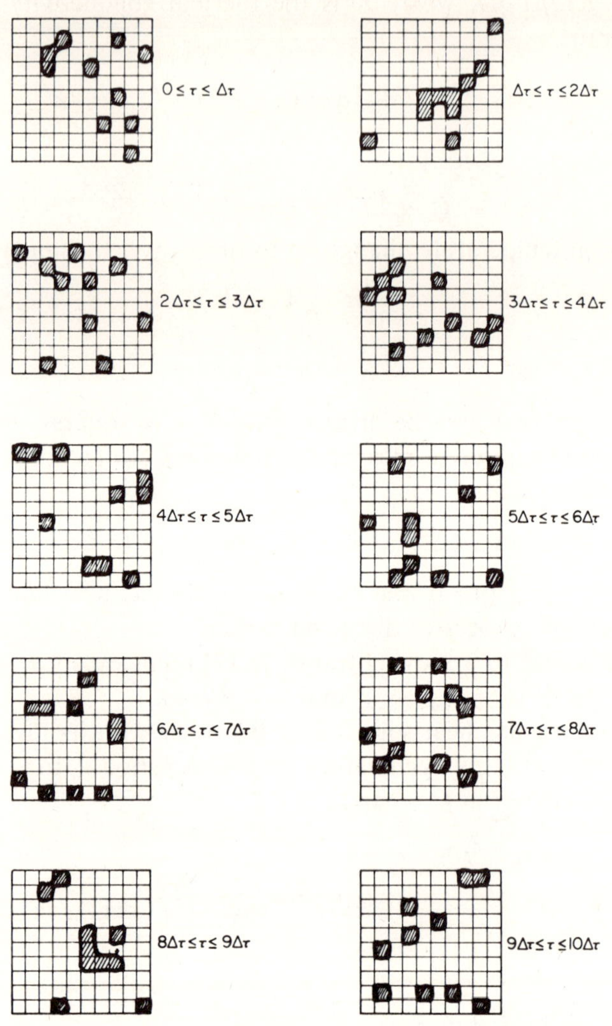

Figure 3.11

Upon area integration, [3.19] becomes

$$\phi_{\xi\eta\zeta\tau} = \frac{\pi}{4} \int_{\tau_k}^{\tau_k + \Delta\tau} f_k^{ij} \, H(\xi,\eta,\xi_i,\eta_j,\tau,\tau') \exp[-\zeta^2/(\tau-\tau')] d\tau' \quad,$$

[3.20]

in which

$$H(\xi,\eta,\xi_i,\eta_j,\tau,\tau') = \frac{1}{(\tau-\tau')^{1/2}} \left[\operatorname{erf} \frac{\eta-\eta_j}{(\tau-\tau')^{1/2}} - \frac{\eta-\eta_j - \Delta\eta}{(\tau-\tau')^{1/2}} \right]$$

$$\times \left[\operatorname{erf} \frac{\xi-\xi_i + u(\tau-\tau')}{(\tau-\tau')^{1/2}} - \operatorname{erf} \frac{\xi-\xi_i - \Delta\xi + u(\tau-\tau')}{(\tau-\tau')^{1/2}} \right]$$

[3.21]

The temperature change on the surface, that is, $\zeta = 0$, from [3.20] is

$$\Delta\phi_{\xi\eta 0\tau} = \frac{\pi}{4} \int_{\tau_k}^{\tau_k + \Delta\tau} f_k^{ij} \, H(\xi,\eta,\xi_i,\eta_j,\tau,\tau') d\tau' \quad.$$

[3.22]

The average value of ϕ over a small rectangular area $\Delta\xi \times \Delta\eta$ with its left and lower corner at $(\xi_\alpha, \eta_\alpha)$ may be obtained from

$$\Delta\bar{\phi}_{\xi_\alpha\eta_\alpha 0\tau} = \int_{\xi_\alpha}^{\xi_\alpha + \Delta\xi} \int_{\eta_\alpha}^{\eta_\alpha + \Delta\eta} (\Delta\xi \cdot \Delta\eta)^{-1} \phi_{\xi\eta 0\tau} \, d\xi \, d\eta \quad.$$

[3.23]

Area integration of [3.23] yields

$$\Delta\bar{\phi}_{\xi_\alpha\eta_\alpha 0\tau} = \pi(4\Delta\xi \cdot \Delta\eta)^{-1} \int_{\tau_k}^{\tau_k + \Delta\tau} f_k^{ij}(\tau-\tau')^{1/2} \, \Psi(\xi_d,\tau') \, \Omega(\eta_d,\tau') d\tau' \quad,$$

[3.24]

where

$$\xi_d = \xi_i - \xi_\alpha \quad , \quad \eta_d = \eta_j - \eta_\alpha \quad , \qquad [3.25]$$

$$\Psi(\xi_d,\tau') = i \text{ erf } c \left[- \frac{\xi_d + \Delta\xi_1 + u(\tau-\tau')}{(\tau-\tau')^{1/2}} \right] + i \text{ erf } c \left[- \frac{\xi_d - \Delta\xi + u(\tau-\tau')}{(\tau-\tau')^{1/2}} \right]$$

$$- i \text{ erf } c \left[- \frac{\xi_d + \Delta\xi_1 - \Delta\xi + u(\tau-\tau')}{(\tau-\tau')^{1/2}} \right] - i \text{ erf } c \left[- \frac{\xi_d + u(\tau-\tau')}{(\tau-\tau')^{1/2}} \right] \quad ,$$

$$[3.26]$$

$$\Omega(\eta_d,\tau') = i \text{ erf } c \left[- \frac{\eta_d + \Delta\eta_1}{(\tau-\tau')^{1/2}} \right] + i \text{ erf } c \left[- \frac{\eta_d - \Delta\eta}{(\tau-\tau')^{1/2}} \right]$$

$$- i \text{ erf } c \left[- \frac{\eta_d + \Delta\eta_1 - \Delta\eta}{(\tau-\tau')^{1/2}} \right] - i \text{ erf } c \left[- \frac{\eta_d}{(\tau-\tau')^{1/2}} \right] \quad ,$$

$$[3.27]$$

and in which

$$i \text{ erf } c( ) = \int_{( )}^{\infty} \text{ erf } c \, x \quad , \qquad [3.28]$$

where erf c is the complimentary error function.

For a program of heat sources, [3.24] leads to the cumulative average

$$\overline{\Phi}_{\xi_\alpha \eta_\alpha 0 \tau} = \pi (4 \Delta\xi \cdot \Delta\eta)^{-1} \sum_{i=0}^{m-1} \sum_{j=0}^{m-1} \sum_{k=0}^{n-1} f_k^{ij} \int_{k\Delta\tau}^{(k+1)\Delta\tau} (\tau-\tau')^{1/2}$$

$$\times \Psi(\xi_d,\tau')\Omega(\eta_d,\tau')d\tau' \quad . \qquad [3.29]$$

The body in the sliding pair will be referred to as being stationary if it is fixed with respect to the apparent contact area and that body will be designated as moving if it has relative motion with respect to the apparent contact area. Define the heat partition factor, σ, as the fraction of heat generated which enters the stationary body over actual contact area. Obviously, σ is a function of ξ, η, and τ in general and $\sigma = 0$ when the point $(\xi, \eta, 0, \tau)$ is not in actual contact. Furthermore, over a generic contact area, σ is a continuous function of ξ, η, and τ for the time period when the area remains in contact. If both the contact area and the time period are small enough, for example, $\Delta\xi \times \Delta\eta$ and $\Delta\tau$, then an average value of heat partition factor may be used without introducing serious error. Denoting the mean value of the function σ by $\sigma_k{}^{ij}$ as averaged over the rectangular contact area $\Delta\xi \times \Delta\eta$ with left and lower corner at (ξ_i, η_j) for the time interval between $\tau_k = k\Delta\tau$ and $\tau_k + \Delta\tau$, and defining

$$\delta_k^{ij} = 1 \quad \text{when} \quad (\xi_i, \eta_j, 0, \tau_k) \quad \text{in actual contact}$$

$$= 0 \quad \text{when} \quad (\xi_i, \eta_j, 0, \tau_k) \quad \text{not in actual contact}$$

then the average surface temperature change for the stationary and the moving body, respectively, can be found from [3.29]. Thus

$$\text{st.} \; \bar{\phi}_{\xi_\alpha \eta_\alpha 0 \tau} = \pi(4\,\Delta\xi \cdot \Delta\eta)^{-1} \sum_{i=0}^{m-1} \sum_{j=0}^{m-1} \sum_{k=0}^{n-1} \sigma_k^{ij}\, f_k^{ij} \int_{k\Delta\tau}^{(k+1)\Delta\tau} (\tau-\tau')^{1/2}$$

$$\times \Psi(\xi_d, \tau')\,\Omega(\eta_d, \tau')\,d\tau' \;, \quad [3.30]$$

$$\text{mo.} \; \bar{\phi}_{\xi_\alpha \eta_\alpha 0 \tau} = \pi(4\,\Delta\xi \cdot \Delta\eta)^{-1} \sum_{i=0}^{m-1} \sum_{j=0}^{m-1} \sum_{k=0}^{n-1} (\delta_k^{ij} - \sigma_k^{ij})\, f_k^{ij} \int_{k\Delta\tau}^{(k+1)\Delta\tau}$$

$$\times (\tau-\tau')^{1/2}\, \Psi\Omega \; d\tau' \;, \quad [3.31]$$

Consider the case that there are always s rectangular areas, each of the size $\Delta\xi \times \Delta\eta$, in actual contact and that the life of contact is $\Delta\tau = \Delta\xi/u$. Let the left and lower corners of actual contact rectangles for the period $(n-1)\Delta\tau$ to $n\Delta\tau$ be denoted by $(\xi_\alpha{}^n, \eta_\alpha{}^n)$, $\alpha = 1, 2, 3, \ldots, s$. Values of $\sigma_1{}^{ij}$ are solvable from the s simultaneous linear algebraic equations

$$\text{st.} \; \bar{\phi}_{\xi_\alpha \eta_\alpha 0 \tau} = \text{mo.} \; \bar{\phi}_{\xi_\alpha \eta_\alpha 0 \tau} \qquad \alpha = 1, 2, \ldots, s \;, \quad [3.32]$$

for some value of τ, $0 < \tau \leq \Delta\tau$. Equation 3.32 is a statement that temperatures of surfaces in contact must be continuous. Inserting values of $\sigma_1{}^{ij}$ found from [3.32] into [3.30] and [3.31], and using [3.32] again for some value of τ, $\Delta\tau < \tau \leq 2\Delta\tau$, the set of values of $\sigma_2{}^{ij}$ can be solved. The set $\sigma_n{}^{ij}$, $n = 1, 2, 3, \ldots$, are obtained by the aforementioned iterative method. From the property of diffusion, it is found that the choice of the value τ, $(n - 1)\Delta\tau < \tau \leq n\Delta\tau$, does not affect the values of $\sigma_n{}^{ij}$ to any great extent. Therefore, in numerical computation $\tau = n\Delta\tau$, $n = 1, 2, 3, \ldots$, have been used in obtaining $\sigma_n{}^{ij}$, $n = 1, 2, 3, \ldots$. With the values $\sigma_n{}^{ij}$ thus found, [3.30] and [3.31] are used to compute the average temperature of each basic rectangular area of the whole apparent area of contact and for consecutive time intervals. The temperature variations over the apparent area will be referred to as temperature distribution.

For numerical evaluations, it has been assumed that $\Delta\xi = \Delta\eta = 0.1$, $\Delta\tau = \Delta\xi/u = 0.1/u$, where u, the dimensionless sliding velocity or Peclet number, equals 1, 10, and 100 has been studied separately. Five different discrete stochastic heat input programs have been studied.

From the data computed[36] the following observations can be made:

1. The average surface temperature of the stationary body differs in general from that of the moving one.

2. The average surface temperatures, those of the stationary and the moving bodies, are low relative to the maximum as well as the actual contact temperatures.

3. The ratios of standard deviations of surface temperatures, those of the stationary and the moving bodies, to the respective average surface temperatures, are between 0.3 and 3.7 with the added observation that (a) the smaller the speeds, the smaller the ratios; and (b) the larger the contact areas, the smaller the range level from the ratios and the lower the levels at which the ratios occur.

4. The ratios of standard deviations of surface temperatures over contact areas to the average of contact surface temperatures are between 0 and 0.37.

5. The spectral information on temperatures is insensitive to the particular program of $f_k{}^{ij}$ for a given percentage of contact, a fact which gives credence to the simplified stochastic process.

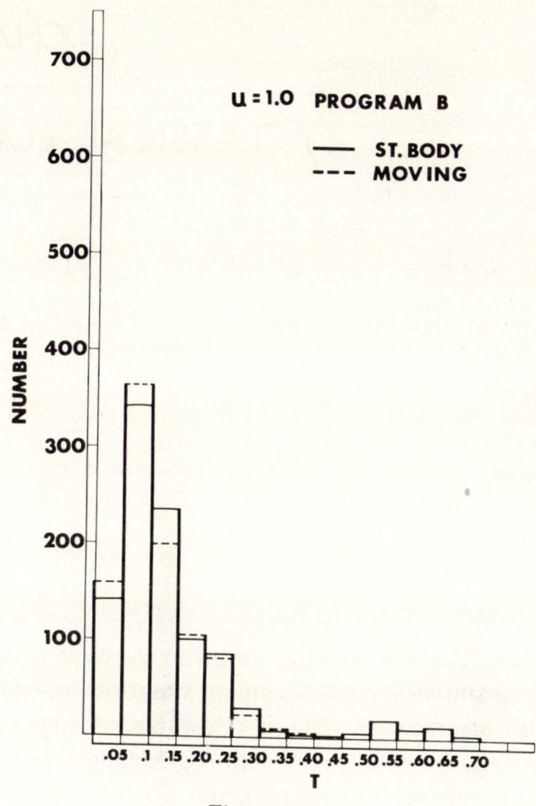

Figure 3.12

Figure 3.12 shows a typical histogram of the surface temperatures population. These general characteristics have been found to be similar to one constructed from an experiment in which:

1. A nonconductor is slid on a metal surface.
2. The microthermocouple probe protrudes from the nonconductor, thus picking up a qualitative measure of the heated metal surface.
3. A time trace of the pickup is made.
4. The time lag of the microthermocouple is corrected.
5. The histogram of the above similar to that of Figure 3.12 is prepared.

CHAPTER 4

CLASSICAL ELASTICITY

Unlike Chapter 2, where moving coordinates were used for moving situations, elastostatics is good for many situations in which the bodies under question are moving. This is allowable so long as the speed is much less than the speed of propagation of elastic waves. Applications of this kind are referred to as quasistatic.

A.   THE FLAMANT PROBLEM

Figure 4.1 shows a semiinfinite solid in plane strain (i.e., the displacement in the x_3 direction, u_3, is zero) and under an arbitrary loading $p(x_1)$ for $-l \leq x_1 \leq l$. On the flat surface, there are to be no shear stresses and normal stresses for $|x_1| > l$. In what follows, the solution to the classical problem is given in some detail by the Fourier transform method. This is done so as to facilitate the understanding of other more complex problems to be discussed. The equation of equilibrium in the absence of body forces, [1.1] without inertia and body forces, and the constitutive relation for homogeneous isotropic elastic solids, [1.15] without the temperature term, constitute the governing equations of the

[82]

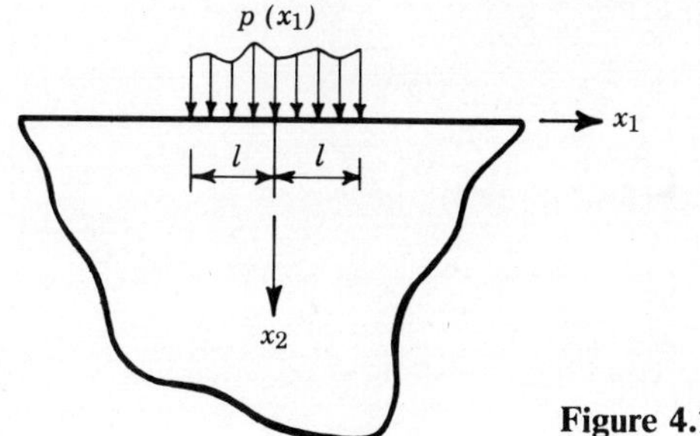

Figure 4.1

problem at hand if it is to be solved through displacements. In this case, compatibility is not a question. Also in the classical theory, the strain tensor ϵ_{ij} is related to the displacement vector through

$$\epsilon_{ij} = \frac{1}{2} (u_{i,j} + u_{j,i}) \ . \tag{4.1}$$

Using the dimensionless quantities $\xi \equiv x_1/l$, $\eta \equiv x_2/l$, $r \equiv \lambda/\mu$, $q \equiv p/\mu$, $u \equiv u_1/l$, $v \equiv u_2/l$, $\sigma_\xi \equiv \sigma_{11}/\mu$, $\sigma_{\xi\eta} \equiv \sigma_{12}/\mu$, and $\sigma_\eta \equiv \sigma_{22}/\mu$, [1.1], [1.15], and [4.1] lead to

$$(r+2) \frac{\partial^2 u}{\partial \xi^2} + \frac{\partial^2 u}{\partial \eta^2} + (r+1) \frac{\partial^2 v}{\partial \xi \partial \eta} = 0$$

$$\tag{4.2}$$

$$\frac{\partial^2 v}{\partial \xi^2} + (r+2) \frac{\partial^2 v}{\partial \eta^2} + (r+1) \frac{\partial^2 u}{\partial \xi \partial \eta} = 0 \ ,$$

$$\sigma_\xi = (r+2) \frac{\partial u}{\partial \xi} + r \frac{\partial v}{\partial \eta}$$

$$\sigma_\eta = r \frac{\partial u}{\partial \xi} + (r+2) \frac{\partial v}{\partial \eta} \tag{4.3}$$

$$\sigma_{\xi\eta} = \frac{\partial u}{\partial \eta} + \frac{\partial v}{\partial \xi} \ ,$$

subject to the boundary conditions

$$\sigma_{\xi\eta} = 0 \qquad (\xi = \xi, \ \eta = 0)$$

$$\sigma_{\eta} = \begin{cases} q(\xi) & (|\xi| \le 1, \ \eta = 0) \\ 0 & (1 < |\xi| < \infty, \ \eta = 0) \end{cases} , \qquad [4.4]$$

and the regularity conditions

$$\sigma_{\xi} , \ \sigma_{\eta} , \ \sigma_{\xi\eta} \to 0 \quad \text{as} \quad (\xi^2 + \eta^2)^{1/2} \to \infty . \qquad [4.5]$$

Taking the Fourier transform [2.5], [4.2] through [4.5] becomes, respectively,

$$\frac{\partial^2 \tilde{u}}{\partial \eta^2} - s^2 (r+2)\tilde{u} - i s (r+1) \frac{\partial \tilde{v}}{\partial \eta} = 0$$

$$[4.6]$$

$$(r+2) \frac{\partial^2 \tilde{v}}{\partial \eta^2} - s^2 \tilde{v} - i s (r+1) \frac{\partial \tilde{u}}{\partial \eta} = 0 ,$$

$$\tilde{\sigma}_{\xi} = -i s (r+2)\tilde{u} + r \frac{\partial \tilde{v}}{\partial \eta}$$

$$\tilde{\sigma}_{\eta} = -i s r \tilde{u} + (r+2)\frac{\partial \tilde{v}}{\partial \eta} \qquad [4.7]$$

$$\sigma_{\xi\eta} = \frac{\partial \tilde{u}}{\partial \eta} - i s \tilde{v} ,$$

$$\tilde{\sigma}_{\xi\eta} = 0 \qquad\qquad\qquad (\eta = 0)$$

$$[4.8]$$

$$\tilde{\sigma}_{\eta} = \tilde{q} \equiv \pi^{-1/2} \int_{-1}^{1} q(\xi')e^{i\xi\xi'} d\xi' \qquad (\eta = 0) ,$$

$$\tilde{\sigma}_{\xi} , \ \tilde{\sigma}_{\eta} , \ \tilde{\sigma}_{\xi\eta} \to 0 \qquad\qquad (\eta \to \infty) . \qquad [4.9]$$

Equation 4.6 yields

$$i\,s\,\tilde{u} = s^{-2}(r+1)^{-1}\frac{d^3\tilde{v}}{d\eta^3} + r(r+1)\frac{d\tilde{v}}{d\eta} \quad . \qquad [4.10]$$

and

$$\frac{d^4\tilde{v}}{d\eta^4} - 2s^2\frac{d^2\tilde{v}}{d\eta^2} + s^4\tilde{v} = 0 \quad . \qquad [4.11]$$

Equation 4.11 with due consideration to [4.9] has a solution

$$\tilde{v} = A\,e^{-s\eta} + B\,\eta\,e^{-s\eta} \quad , \qquad [4.12]$$

where A and B are complex constants of integration.

Enforcement of [4.7] leads to

$$\tilde{v} = (r+2)(r+1)^{-1}(2s)^{-1}\tilde{q} \quad ,$$

for $\eta = 0$. And, carrying out the inverse Fourier transform [2.12], the above yields

$$v(\xi,0) = \int_{-1}^{1} q(\xi')\,K(\xi,\xi')d\xi' \quad , \qquad [4.13]$$

where

$$K(\xi,\xi') = \pi^{-1}(r+2)(r+1)^{-1}(C - \ell n|\xi-\xi'|) \quad . \qquad [4.14]$$

Equation 4.14 is the Flamant solution which is the fundamental solution for the displacement at the surface.[37] Note that it violates the regularity condition [4.5] unless $C \to \infty$, in which case the solution is meaningless. This phenomenon represents a degeneracy of the classical three-dimension theory. This point is discussed later. In practice, however, [4.14] with an indefinite C may be used for finding the relative displacements near the contact zone. For plane stress $(r + 2)(r + 1)^{-1}$ in [4.14] is to be replaced by $4(r + 1)(3r + 2)^{-1}$.

B. THE BOUSSINESQ PROBLEM

The generalization of the Flamant problem is shown in Figure 4.2, where an arbitrary load distribution $p(x_1, x_2)$ is applied to a semiinfinite solid. Using successive Fourier transform, as above in the x_1 and x_2 directions, to the appropriate equations leads to

$$w(\xi,\eta,0) = \int_A q(\xi',\eta')K(\xi,\eta;\xi',\eta')d\xi'\,d\eta' \quad , \quad [4.15]$$

where

$$K(\xi,\eta;\xi',\eta') = (4\pi)^{-1}(r+2)(r+1)^{-1}[(\xi-\xi')^2 + (\eta-\eta')^2]^{-1/2} . \quad [4.16]$$

The A is the load region which may be arbitrary although formally a rectangular one was easier to work out and $w \equiv u_3/l$ is in the u_3 direction. Equation 4.16 is the fundamental solution of Boussinesq.[38] Limitation of space does not permit going into anisotropy of elastic bodies. The counterpart of [4.16] for transversely isotropic material[39] shows the functional form of [4.16] is preserved but the coefficient is different.

C. THE CERRUTI PROBLEM

This is the companion problem[40] to the one shown in Figure 4.2 except the normal and shear stresses are everywhere zero but arbitrarily distributed shear stress, $s(x_1, x_2)$, is given. Letting $\Sigma(\xi, \eta) \equiv s/\mu$,

$$w(\xi,\eta,0) = \int_A \Sigma(\xi',\eta')K(\xi,\eta;\xi',\eta')d\xi'\,d\eta' \quad , \quad [4.17]$$

where

$$K(\xi,\eta;\xi',\eta') = [4\pi(r+1)]^{-1}\,\frac{\xi-\xi'}{(\xi-\xi')^2 + (\eta-\eta')^2} . \quad [4.18]$$

Again, since K is the fundamental solution to the problem, the integral [4.17] may be more general than that shown in Figure 4.2.

D. SEMIINFINITE SOLID WITH AXI-SYMMETRIC NORMAL LOAD

This case, shown in Figure 4.3, can be solved by integrating the result in [4.15] or by working with the equations for the axi-symmetric case

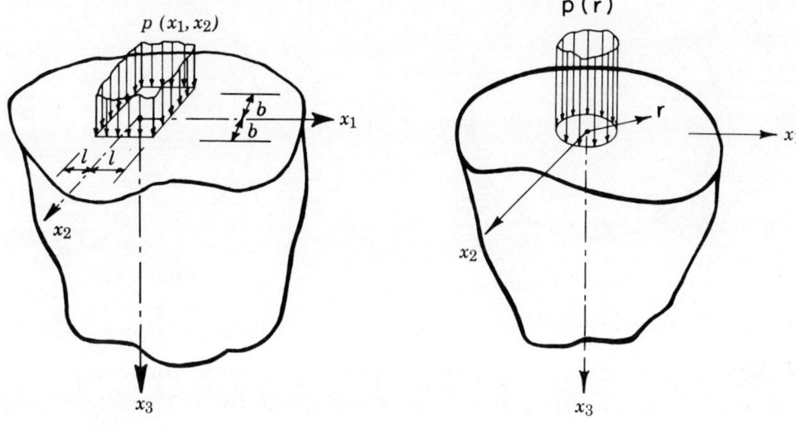

Figure 4.2   **Figure 4.3**

and using Hankel transform [2.66]. Letting $\kappa \equiv \lambda/\mu$, $w \equiv u_3/a$, $\rho \equiv r/a$, $q(\rho) \equiv p(r)/\mu$,

$$w(\rho,0) = \int_0^1 q(\rho')K(\rho,\rho')d\rho' \quad , \qquad [4.19]$$

where

$$K(\rho,\rho') = \pi^{-1}(\kappa+2)(\kappa+1)^{-1}K(k)\left(1 + \frac{\rho}{\rho'}\right)^{-1} \quad , [4.20]$$

K is the complete elliptical integral of the first kind and argument k, $k = 2\sqrt{\rho/\rho'}\,[1 + (\rho/\rho')]^{-1}$.

E.   DISC WITH NORMAL EDGE LOAD

Consider the disc in Figure 4.4. A distributed edge load $p(\theta)$ is applied and it is in equilibrium with a concentrated central load. The periphery is free of shear traction. The problem is the finite analog of the one discussed in Section 4.A. Letting $\kappa \equiv \lambda/\mu$, $u \equiv u_r/a$, $\rho \equiv r/a$, and $q \equiv p/\mu$, where u_r is the radial displacement, then

$$u(1,\theta) = \int_C q(\theta')K(\theta,\theta')d\theta' \quad , \qquad [4.21]$$

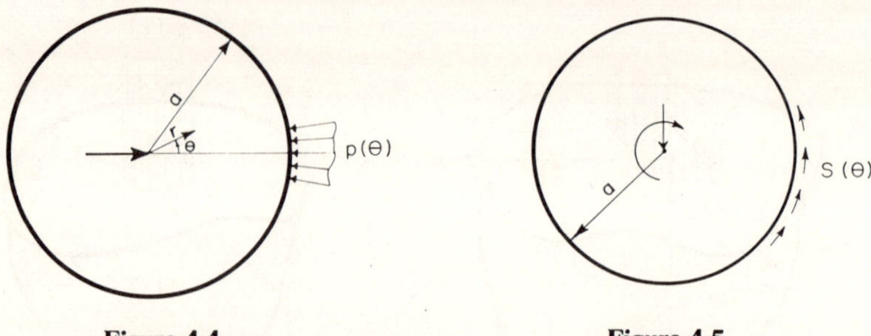

Figure 4.4   Figure 4.5

where C represents the arc over which $q(\theta)$ is prescribed, and the fundamental solution $K(\theta, \theta')$ is best obtained by the complex variable method[41] which yields

$$K(\theta,\theta') = (4\pi)^{-1} \left\{ \cos\phi + 2X \left[\frac{\cos\phi}{2} \ell n\left|2(1 - \cos\phi)\right| \right. \right.$$

$$\left. \left. - \sin\phi \tan^{-1}\frac{\sin\phi}{1 - \cos\phi} \right] + X \right\},$$

$$[4.22]$$

where $\phi = \theta - \theta'$, $\chi = 4(\kappa + 1)(3\kappa + 2)^{-1}$ for plane stress, $\chi = (\kappa + 2)(\kappa + 1)^{-1}$ for plane strain.

F.   DISC WITH TANGENTIAL EDGE LOAD

Figure 4.5 shows the same disc as in Section 4.E which has a distributed tangential edge load. This edge load, $s(\theta)$, is in equilibrium with a central concentrated reaction and a concentrated moment. Letting $\Sigma \equiv s/\mu$, the solution of surface displacement by complex variable method[41] is

$$K(\theta,\theta') = (4\pi)^{-1} \left\{ 2X \frac{\sin\phi}{2} \ell n\left|2(1 - \cos\phi)\right| \right.$$

$$\left. + 2(1 - X)\cos\phi \tan^{-1}\frac{\sin\phi}{1 - \cos\phi} + \sin\phi \right\}.$$

$$[4.24]$$

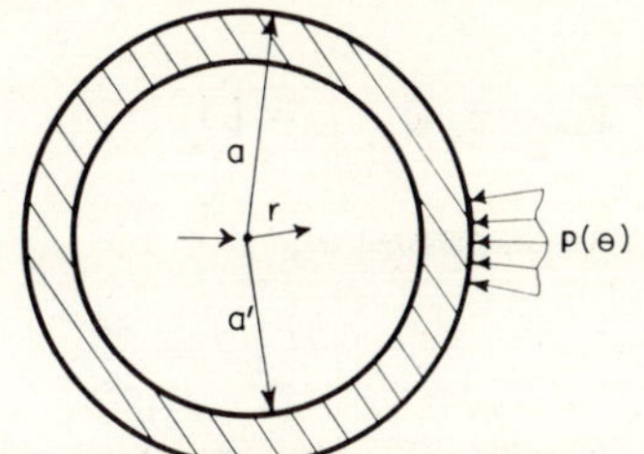

Figure 4.6

G.   LAYERED DISC WITH NORMAL EDGE LOAD

Akin to the problem in Section 4.E is the layered disc with normal edge
load shown in Figure 4.6. The solution to the problem will be carried
out in plane strain.[42] For plane stress solution, replace $k = (\lambda + \mu)/\mu$
by $(3\lambda + 2\mu)/(2\mu + \lambda)$. In addition to the parameters used before, let
$\delta \equiv a'/a$, $v \equiv u_\theta/a$, $\sigma_\rho \equiv \sigma_r/\mu$, and $\sigma_{\rho\theta} \equiv \sigma_{r\theta}/\mu$, where u_θ is the displacement component in the θ direction. Super stars will denote quantities
associated with the inner core $0 < \rho < \delta$.

The applicable plane strain equations 4.2 and 4.3, in polar coordinates, are

$$(k+1)\left(\frac{\partial^2 u}{\partial \rho^2} + \rho^{-1}\frac{\partial u}{\partial \rho} - \rho^{-2}u\right) + \rho^{-2}\frac{\partial^2 u}{\partial \theta^2} + k\rho^{-1}\frac{\partial^2 v}{\partial \rho \partial \theta} - (k+2)\rho^{-2}\frac{\partial v}{\partial \theta} = 0$$

$$\frac{\partial^2 v}{\partial \rho^2} + \rho^{-1}\frac{\partial v}{\partial \rho} - \rho^{-2}v + (k+1)\rho^{-2}\frac{\partial^2 v}{\partial \theta^2} + k\rho^{-1}\frac{\partial^2 u}{\partial \rho \partial \theta} + (k+2)\rho^{-2}\frac{\partial u}{\partial \theta} = 0 \quad ,$$

$$[4.25]$$

with the stress-displacement relationships

$$\sigma_\rho = (k+1)\frac{\partial u}{\partial \rho} + (k-1)\rho^{-1}\left(u + \frac{\partial v}{\partial \theta}\right)$$

$$\sigma_{\rho\theta} = \rho^{-1}\left(\frac{\partial u}{\partial \theta} - v\right) + \frac{\partial v}{\partial \rho} \quad . \qquad [4.26]$$

Equations 4.25 and 4.26, with the appropriate designation, apply in
$0 < \rho < \delta$.

The boundary conditions are

$$\sigma_\rho = \begin{cases} -q(\theta) & \text{(over C, } \rho = 1) \\ 0 & \text{(elsewhere, } \rho = 1) \end{cases},$$

$$\sigma_\rho^* = \begin{cases} q(\theta)/\epsilon & \text{(over C}, \ \rho = \epsilon) \\ 0 & \text{(elsewhere, } \rho = \epsilon) \end{cases},$$

$$\sigma_{\rho\theta} = 0 \qquad\qquad (\rho = 1),$$

$$\sigma_{\rho\theta}^* = 0 \qquad\qquad (\rho = \epsilon),$$

$$u^* = u, \ v^* = v, \ \sigma_\rho^* = \sigma_\rho, \ \sigma_{\rho\theta}^* = \sigma_{\rho\theta} \qquad (\rho = \delta) \ .$$

$$[4.27]$$

Taking the finite Fourier transform 2.44, equations 4.25, 4.26, and 4.27 become, respectively,

$$(k+1)\frac{\partial^2 \tilde{u}}{\partial\rho^2} + (k+1)\rho^{-1}\frac{\partial\tilde{u}}{\partial\rho} - (p^2+k+1)\rho^{-2}\tilde{u} + ipk\rho^{-1}\frac{\partial\tilde{v}}{\partial\rho} - ip(k+2)\rho^{-2}\tilde{v} = 0$$

$$\frac{\partial^2\tilde{v}}{\partial\rho^2} + \rho^{-1}\frac{\partial\tilde{v}}{\partial\rho} - [1 + (k+1)p^2]\rho^{-2}\tilde{v} + ipk\rho^{-1}\frac{\partial\tilde{u}}{\partial\rho} + ip(k+2)\rho^{-2}\tilde{u} = 0 \ ,$$

$$[4.28]$$

$$\tilde{\sigma}_\rho = (k+1)\frac{\partial\tilde{u}}{\partial\rho} + (k-1)\rho^{-1}(\tilde{u} + ip\tilde{v})$$

$$\tilde{\sigma}_{\rho\theta} = \rho^{-1}(ip\tilde{u} - \tilde{v}) + \frac{\partial\tilde{v}}{\partial\rho} \ , \qquad\qquad [4.29]$$

$$\tilde{\sigma}_{\rho} = -\tilde{q} \equiv \epsilon_p (2\pi)^{-1} \int_{-\pi}^{\pi} q(\theta) e^{i p \theta} d\theta \qquad (\rho = 1)$$

$$\tilde{\sigma}_{\rho}^{*} = \tilde{q}/\epsilon \qquad (\rho = \epsilon)$$

$$\tilde{\sigma}_{\rho\theta} = 0 \qquad (\rho = 1)$$

$$\tilde{\sigma}_{\rho\theta}^{*} = 0 \qquad (\rho = \epsilon)$$

$$\tilde{u}^{*} = \tilde{u}, \ \tilde{v}^{*} = \tilde{v}, \ \tilde{\sigma}_{\rho}^{*} = \tilde{\sigma}_{\rho}, \ \tilde{\sigma}_{\rho\theta}^{*} = \tilde{\sigma}_{\rho\theta} \qquad (\rho = \delta) \quad .$$

$$[4.30]$$

of course the super $\sim$ denotes the transformed quantity.

Solving the two sets of equations, [4.28] and [4.29], one set for $\delta < \rho \leq 1$ and the other for $\epsilon \leq \rho \leq \delta$, and letting $\epsilon \to 0$, the following results are obtained:

$$u(1,\theta) = u_0 + 2 \sum_{p=1}^{\infty} u_p \cos p\theta \quad , \qquad [4.31]$$

where

$$u_0 = \tilde{q}_0 \frac{\mu + \mu^* k^* + \delta^2 (\mu k - \mu^* k^*)}{2[uk(1 - \delta^2) + \mu^* k^* (k + \delta^2)]} \quad ,$$

$$u_1 = \tilde{q}_1 \frac{(2 - k)J_1 - \delta^2 J_2 + C}{\delta(k+1)[\mu k(k^* + 2)(1 - \delta^4) + \mu^* k^* (2\delta^4 + k + k\delta^4)]} \quad ,$$

$$u_p = \frac{\tilde{q}_p}{\Delta} \left\{ \frac{pk-2}{k(p+1)} \left[J_3 + (p+1)\delta^2 - p \right] \right.$$

$$+ \frac{1}{p+1} \left[J_4 - J_3 + \delta^2(p+1)(p-2) - p^2 - 1 \right]$$

$$+ \frac{1}{p-1} \left[-J_4 - (p-1)J_3 + \delta p(p-1) + p^2 \right]$$

$$\left. + \frac{pk+2}{k(p-1)} \left[J_4 - \delta^2(p-1) + p \right] \right\} \quad ,$$

$$J_1 = \mu \left[(-k*+2) - (2k*+k*k - 2k)\delta^2 \right] + \mu*k*(1+k\delta^2) \right] \quad ,$$

$$J_2 = \mu k \left[(kk* - 2k* - 2k) - (k*+2)\delta^2 \right] - \mu*k*(k^2 - k\delta^2 - 2\delta^2) \right] \quad ,$$

$$J_3 = \frac{\delta^{2(1-p)}\left[\mu k + \mu*(k+2) \right]}{(\mu - \mu*)k} \quad ,$$

$$J_4 = \frac{\delta^{2(p+1)}\left[\mu k(k*+2) - \mu*k*(k+2) \right]}{\mu k(k*+2) - \mu*k*k} \quad ,$$

$$\Delta = 4\left[(1+J_3)(J_4 - 1) + (\delta^2 - 1)(p^2 - 1) \right] \quad ,$$

$$\tilde{q} = \sum_{p=1}^{\infty} \tilde{q}_p \quad ,$$

and C is an arbitrary constant. Application of [4.31] is shown in a later chapter inasmuch as numerical calculations have to be resorted to.

H. A LAYERED ELASTIC SYSTEM UNDER A MOVING LOAD

Figure 4.7 differs from Figure 4.1 in two respects. First, there is a layer of elastic material which is characterized by distinct values of λ^*, μ^*, and ρ^* in general. Second, the composite is moving with a constant velocity V. For low values of V, this model is also akin to that shown in Figure 4.6. In this problem, the effect of inertia on the shear stress at the interface and surface, normal displacement are sought. Only the quasi-stationary solution (i.e., standing wave problem) is examined, however.

Now, starting from [1.1] without the body forces and [1.15] without the thermal effects, the plane strain equations of motion in terms of displacement are

$$(\alpha^2 - M^2) \frac{\partial^2 u}{\partial \xi^2} + \frac{\partial^2 u}{\partial \eta^2} + (\alpha^2 - 1) \frac{\partial^2 v}{\partial \xi \partial \eta} = 0$$

$$(1 - M^2) \frac{\partial^2 v}{\partial \xi^2} + \alpha^2 \frac{\partial^2 v}{\partial \eta^2} + (\alpha^2 - 1) \frac{\partial^2 u}{\partial \xi \partial \eta} = 0 \quad , [4.32]$$

$$\sigma_{\xi\xi} = \alpha^2 \frac{\partial u}{\partial \xi} + (\alpha^2 - 2) \frac{\partial v}{\partial \eta}$$

$$\sigma_{\xi\eta} = \frac{\partial u}{\partial \eta} + \frac{\partial v}{\partial \xi}$$

$$\sigma_{\eta\eta} = \alpha^2 \frac{\partial v}{\partial \eta} + (\alpha^2 - 2) \frac{\partial u}{\partial \xi} \quad , \qquad [4.33]$$

where $\xi \equiv x_1/H$, $\eta \equiv x_2/H$, $h \equiv H/l$, $\sigma_{\xi\xi} \equiv h\sigma_{11}/\mu$, $\sigma_{\eta\eta} \equiv h\sigma_{22}/\mu$, $\sigma_{\xi\eta} \equiv h\sigma_{12}/\mu$, $u \equiv u_1/l$, $v \equiv u_2/l$, a Mach number $M \equiv V/C_2$, $C_1 \equiv [(\lambda + 2\mu)/\rho]^{1/2}$, $C_2 \equiv (\mu/\rho)^{1/2}$, and $\alpha \equiv C_1/C_2$.

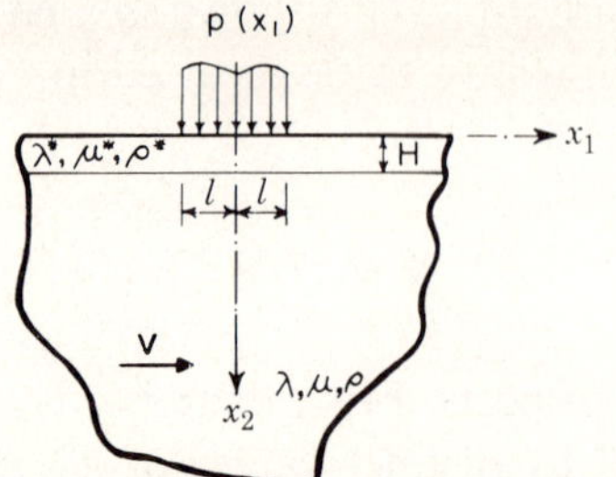

Figure 4.7

The above apply to the substrate. In the surface layer, [4.32] and [4.33] have the counterpart:

$$(\beta^2 - M^2)\frac{\partial^2 u}{\partial \xi^2} + r^2 \frac{\partial^2 u}{\partial \eta^2} + (\beta^2 - r^2)\frac{\partial^2 v}{\partial \xi \partial \eta} = 0$$

$$(r^2 - M^2)\frac{\partial^2 v}{\partial \xi^2} + \beta^2 \frac{\partial^2 v}{\partial \eta^2} + (\beta^2 - r^2)\frac{\partial^2 u}{\partial \xi \partial \eta} = 0 \quad , \quad [4.34]$$

$$\sigma_{\xi\xi} = \delta[\beta^2 \frac{\partial u}{\partial \xi} + (\beta^2 - 2r^2)\frac{\partial v}{\partial \eta}]$$

$$\sigma_{\xi\eta} = r^2 \delta(\frac{\partial u}{\partial \eta} + \frac{\partial v}{\partial \xi})$$

$$\sigma_{\eta\eta} = \delta[\beta^2 \frac{\partial v}{\partial \eta} + (\beta^2 - 2r^2)\frac{\partial u}{\partial \xi}] \quad , \quad [4.35]$$

where $C_1^* \equiv [(\lambda^* + 2\mu^*)/\rho^*]^{1/2}$, $C_2^* \equiv (\mu^*/\rho^*)^{1/2}$, $\beta = C_1^*/C_2$, $r \equiv C_2^*/C_2$, and $\delta \equiv \rho^*/\rho$.

The boundary conditions are

$$\sigma_{\xi\eta} = 0 \qquad\qquad (\xi = \xi , \eta = 0) \quad , \qquad [4.36]$$

$$\sigma_{\eta\eta} = \begin{cases} 0 & (1/h < |\xi| < \infty , \eta = 0) \\ q(\xi) & (0 \leq |\xi| \leq 1/h , \eta = 0), \quad [4.37] \end{cases}$$

$$\sigma_{\xi\xi} \, , \, \sigma_{\xi\eta} \, , \, \sigma_{\eta\eta} \, , \, u,v \to 0 \quad (0 \leq \eta \leq 1 \, , \, |\xi| \to \infty) \, , \quad [4.38]$$

$$\sigma_{\xi\xi} \, , \, \sigma_{\xi\eta} \, , \, \sigma_{\eta\eta} \, , \, u,v \to 0 \quad ([\xi^2 + \eta^2]^{1/2} \to \infty) \, , \quad [4.39]$$

and

$$\sigma_{\xi\eta} \, , \, \sigma_{\eta\eta} \, , \, u,v \text{ continuous} \quad (\xi = \xi \, , \, \eta = 1) \, . \quad [4.40]$$

Using Fourier transform [2.5], whose inverse is given by [2.12], the sets 4.32 and 4.34 are reduced to total differential equations; the transform parameter is s. The solutions of these are straightforward though matching the boundary condition is tedious. In transformed variables (i.e., a super $\sim$ over the variables already defined) the solutions are[43] as follows.

For subsonic regime for both layer and substrates: $M^2/r^2 < 1$, $M^2/\beta^2 < 1$. Letting

$$f,g = (1 - M^2/\alpha^2)^{1/2} \, , \qquad (1 - M^2)^{1/2} \, ,$$

$$j,k = (1 - M^2/\beta^2)^{1/2} \, , \qquad (1 - M^2/r^2)^{1/2} \, ,$$

$$\tilde{\sigma}_{\xi\eta}(s,1) = 2ae^{-f} + b(1 + g^2)e^{-g} \, , \quad [4.41]$$

$$\tilde{v}(s,0) = a^* + e^* \, , \quad [4.42]$$

$$\text{where for } s > 0 \text{ and } \tilde{q} \equiv \pi^{-1/2} \int_{-1/h}^{1/h} qe^{is\xi} \, d\xi \, ,$$

$$a \ = -\tilde{q}\,e^{fs}\,a_{12}/\delta r^2 \Delta \, ,$$

$$b \ = \ \tilde{q}\,e^{gs}\,a_{11}/\delta r^2 \Delta \, ,$$

$$a_{11} = 2(1+2L_2)c_j - \frac{2jL_1}{f} s_j + \frac{(1+L_1)(1+k^2)}{fk} s_k - 2L_2(1+k^2)c_k \quad ,$$

$$a_{12} = 2(1+L_1)c_j - 4gjL_2 s_j + \frac{(1+2L_2)(1+k^2)g}{k} s_k - L_1(1+k^2)c_k \quad ,$$

$$s_j = \sinh(js) \, , \quad c_j = \cosh(js) \, , \quad s_k = \sinh(ks) \, , \quad c_k = \cosh(ks) \, ,$$

$$L_1 = [(1+g^2)/\delta r^2 - 2]/(1-k^2) \quad ,$$

$$L_2 = [1/\delta r^2 - 1]/(1-k^2) \quad ,$$

$$\begin{aligned}
\Delta = 2 \Big\{ &[4gL_2(1+2L_2)(1+k^2) - 2L_1(1+L_1)(1+k^2)/f] \\[4pt]
&+ [2kL_1(1+2L_2) - 4kL_2(1+L_1) + (1+L_1)gL_2(1+k^2)^2/fk \\[4pt]
&- gL_1(1+2L_2)(1+k^2)^2/2kf]c_j s_k + [-2(1+2L_2)^2 g \\[4pt]
&- 2L_2{}^2 g(1+k^2)^2 + 2(1+L_1)^2/f + L_1{}^2(1+k^2)^2/2f]c_j c_k \\[4pt]
&+ [-2jkL_1{}^2/f - (1+L_1)^2(1+k^2)^2/2jkf + 8gjkL_2{}^2 \\[4pt]
&+ (1+k^2)(1+2L_2{}^2)^2 g/2kj]s_j s_k \\[4pt]
&+ [2jgL_1(1+2L_2)/f + (1+L_1)L_2(1+k^2)^2/j - 4gjL_2(1+L_1)/f \\[4pt]
&- L_1(1+2L_2)(1+k^2)^2/2j]s_j c_k \Big\} \quad ,
\end{aligned}$$

$$a^* = \frac{a e^{-fs}}{(1-k^2)} \left\{ \left(2k^2 - \frac{1+g^2}{\delta r^2}\right) \frac{s_j}{f} - \left[\frac{2}{\delta r^2} - (1+k^2) \right] \frac{c_j}{j} \right\}$$

$$+ \frac{a e^{-gs}}{1-k^2} \left\{ 2g\left(k^2 - \frac{1}{\delta r^2}\right) - \left[\frac{1+g^2}{\delta r^2} - (1+k^2) \right] \right\} \quad ,$$

$$e^* = \frac{a\,e^{-fs}}{(1-k^2)} \left\{ \left[\frac{1+g^2}{\delta r^2} - (1+k^2) \right] \frac{ks_k}{j} - 2(1 - \frac{1}{\delta r^2})c_k \right\}$$

$$+ \frac{b\,e^{-gs}}{1-k^2} \left\{ \left[\frac{2}{\delta r^2} - (1+k^2) \right] kgs_k - (2 - \frac{1+g^2}{\delta r^2})c_k \right\} \quad .$$

The above solutions are of the form

$$[\tilde{q}_1(s) + iq_2(s)]F(s,1) \quad .$$

For $s < 0$, replacing s by n, the solutions are of the form

$$[\tilde{q}_1(n) - iq_2(n)]F(n,1) \quad .$$

Again, application of [4.41] and [4.42] is shown in a later chapter.

I. STRESSES AT THE INTERFACE OF A LAYERED ELASTIC SYSTEM EXHIBITING COUPLE-STRESSES

Application of the balance of moment of momentum in classical stress analysis for the continuum leads to [1.2] which says that the stress tensor is symmetric (i.e., $\tau_{ij} = \tau_{ji}$). For polar materials, which exhibit couple-stresses, the stress tensor is asymmetric in general (i.e., [1.2] is not applicable). In this section, only the classical couple-stress[44,45] theory is outlined. Materials so described are known as Cosserat materials. It should be noted that there is a great flourish of the development of couple-stress theory recently.[46-48] Figure 4.8 shows, in two dimensions, components of couple-stresses, μ_{13} and μ_{23}, with x_3 being perpendicular to the plane of the paper. In the classical couple-stress theory [1.1], in the absence of inertia forces, is

$$\sigma_{ij,i} + \rho f_j = 0 \quad . \tag{4.43}$$

Balance of the moment of momentum leads to

$$\mu_{ij,i} + \epsilon_{ji\ell}\tau_{i\ell} = 0 \quad , \tag{4.44}$$

where μ_{ij} is the couple-stress tensor.

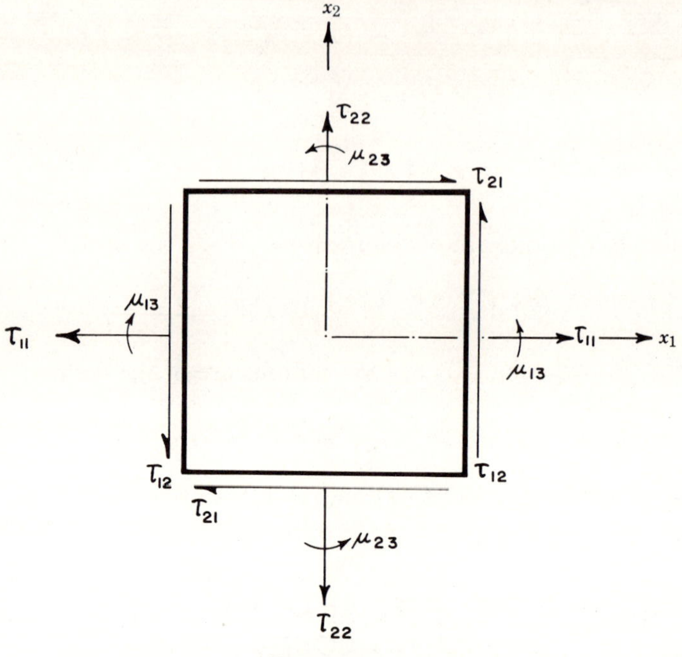

Figure 4.8

The strain-displacement relationships remain as those shown in [4.1].
The curvature-displacement relationships are

$$k_{ij} = \frac{1}{2}\,\epsilon_{i\ell k}\,\mu_{k,\ell j} \quad , \qquad [4.45]$$

where k_{ij} is the curvature-twist tensor. The constitutive relationship is
[1.15] for isothermal conditions; σ_{ij} in [1.15] should be replaced by σ^s_{ij},
the symmetrical part of σ_{ij}. The deviatoric components of μ_{ij} (i.e., μ^0_{ij})
are related to the curvature-twist tensor through

$$\mu^0_{ij} = 4(\eta\,k_{ij} + \eta'\,k_{ji}) \quad , \qquad [4.46]$$

where η and η' are couple-stress material constants. Figure 4.9 shows
a problem similar to that shown in Figure 4.6. The materials are distinct for the two regions and they obey the couple-stress theory shown
above. In view of the complexity of the general solution[49,50] and the
algebra, only the results of a special case are shown below. That is, a
uniform normal stress σ is distributed over an arc of 2Ω. For the case of

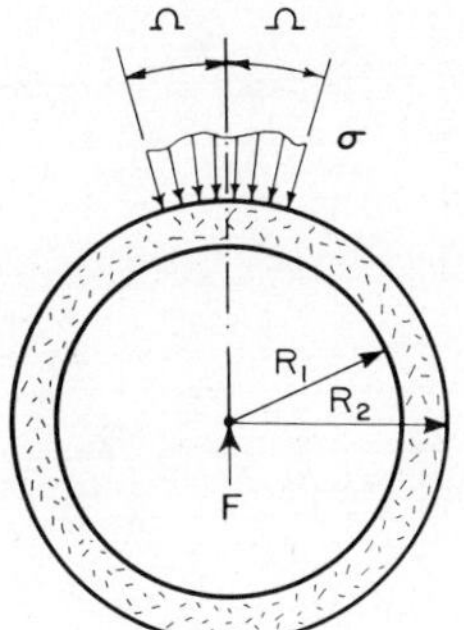

Figure 4.9

Al_2O_3 + Cr on steel, numerical values have been worked out. Figure 4.10 shows a plot of $\hat{\tau}_{\theta r}$ versus ξ. The $\hat{\tau}_{\theta r}$ is a dimensionless measure of $\tau_{\theta r}$ (i.e., $\tau_{\theta r}/\sigma$), and ξ is a dimensionless measure of radius (i.e., r/R_2); $\xi = 0.8$ is the interface. Note the discontinuity at the interface (this data is for $\theta/\Omega = 0.5$). This effect should play a role in the bonding strength of materials.

J.   THE QUARTER-SPACE UNDER EDGE LOAD

In this section and the next, two examples involving edge effect are discussed. This section is devoted to that of a quarter-space under edge load as shown in Figure 4.11. Using plane strain (or plane stress by a suitable adjustment of constant, as in Section 4.A), the vertical displacement on the surface due to $p(x_1)$ for $0 \leq x_1 \leq l$ is sought. In what follows,

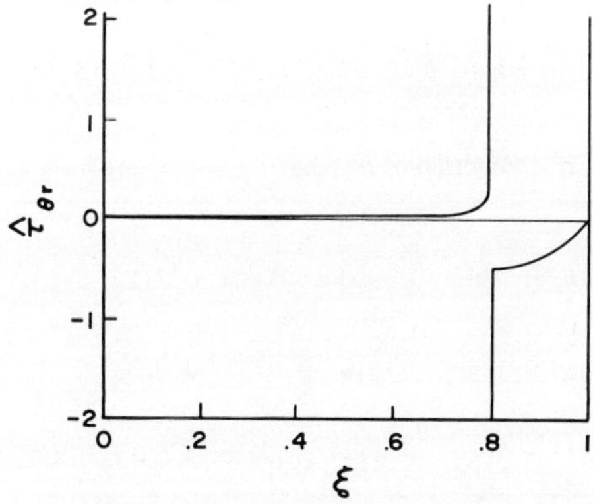

Figure 4.10

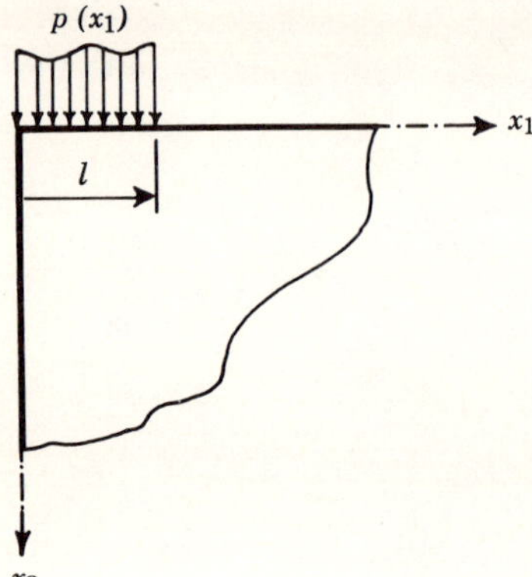

Figure 4.11

the variables used in Section 4.A prevail. To find the surface displacement, consider first the concentrated load $q(\xi)d\alpha$ on the quarter plane ($\xi > 0$, $\eta > 0$). It has been shown that the loading picture shown in Figure 4.12a is equivalent to that shown in Figure 4.12b, where dF and $qd\alpha$ are applied on $\xi > 0$ while dG is applied on $\eta > 0$. The dF and dG are functions of q and $\alpha \equiv a/l$.[51] Also dF and dG are dimensionless as $qd\alpha$.

Note, since dF does not produce any displacement at the plane of symmetry, $\eta = 0$, dG does not enter the particular facet of the problem at hand. In view of the above and [4.13] the solution sought is

$$v(\xi,0) = \int_0^1 q(\alpha)\left[K(\xi,\alpha) + \int_{-\infty}^{\infty} dF(\xi',\alpha)K(\xi,\xi')d\xi'\right]d\alpha \quad ,$$

$$[4.47]$$

where $\alpha dF/qd\alpha$ are tabulated in Table 1 and $K(\xi, \alpha)$ is as defined in [4.14].

Table 1. $\alpha dF/qd\alpha$ as a Function of $ln(\xi/\alpha)$

$-ln(\xi/\alpha)$	5.0	3.0	2.0	1.4	1.0	0.8	0.6	
$-\alpha dF/qd\alpha$	0.5000	0.4999	0.4804	0.4353	0.3820	0.3483	0.3110	

$ln(\xi/\alpha)$	-0.4	0.2	0	0.2	0.4	0.6	0.8	
$\alpha dF/qd\alpha$	0.2713	0.2309	0.1915	0.1548	0.1219	0.0937	0.0703	

$ln(\xi/\alpha)$	1.0	1.2	1.4	1.6	1.8	2.0	3.0	4.0
$\alpha dF/qd\alpha$	0.0517	0.0373	0.0265	0.0185	0.0128	0.0088	0.0012	0.0002

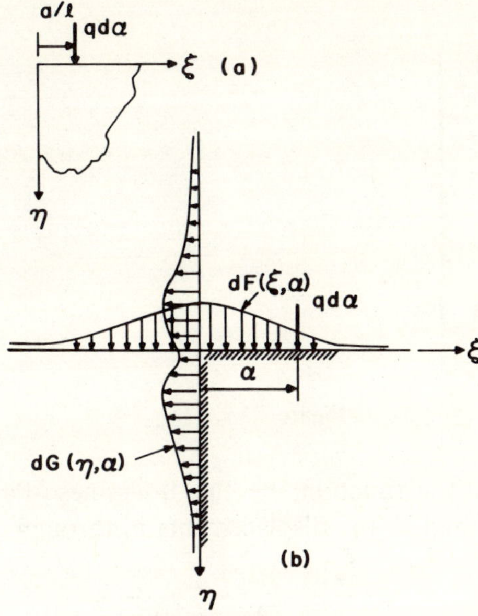

Figure 4.12

Similar treatment can be made of the quarter-space in plane strain or quarter-plane with tangential load.[51] The three-dimensional quarter-space problem is handled in an analogous manner.[52,53] For the quarter-space, two recent papers are of special interest.[54]

K. THE SEMIINFINITE CIRCULAR CYLINDER

Figure 4.13 shows the cross-section through the axis of a semiinfinite circular cylinder with a normal loading which is axisymmetric but otherwise arbitrary. There is no shear traction on this end. Also the side is free of tractions. In this problem the surface displacement is sought. To this end, the solution to a fundamental problem will be obtained in tabular form, that is, the problem with a ring normal load while everything else is the same. The ring will have small but finite width. Again, is convenient to use the displacement formulation of the equilibrium equations ([1.1] without inertia and body forces together with [1.15] without the thermal elastic term and [4.1]). Let Ψ_0, a scalar function,

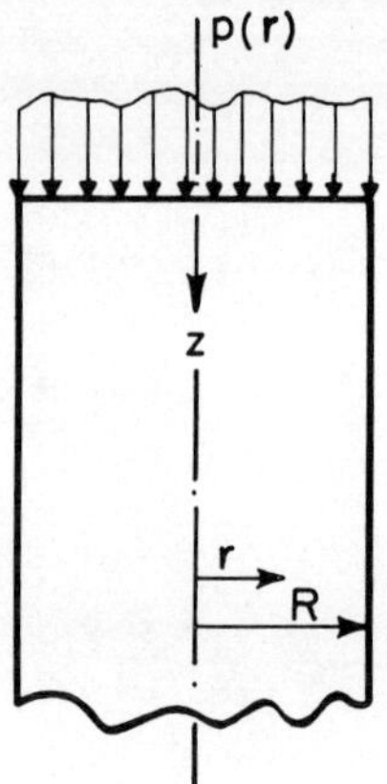

Figure 4.13

and Ψ_i, a vector function, be the Boussinesq-Papkovitch potentials which are related to the displacements u_i through

$$u_i = \frac{4(m-1)}{m} \psi_i - (\psi_0 - x_j \psi_j)_{,i} \quad , \qquad [4.48]$$

where $m = 1/\nu$. For the axisymmetric case, it has been shown that two of the four scalar functions suffice.[55] Using polar coordinates and selecting Ψ_0 and Ψ_r to be nonzero, [4.48] leads to

$$u = \frac{4(m-1)}{m} B_\rho - \frac{\partial}{\partial \rho}(\rho B_\rho + B_0)$$

$$w = -\frac{\partial}{\partial \zeta}(\rho B_\rho + B_0) \quad , \qquad [4.49]$$

where $u \equiv u_r/R$, $w \equiv u_3/R$, $\rho \equiv r/R$, $\zeta \equiv z/R$, $B_0 \equiv \Psi_0/R$, and $B_\rho \equiv \Psi_r/R$. The B_0 and B_ρ satisfy, respectively,

$$\frac{\partial^2 B_\rho}{\partial \rho^2} + \rho^{-1}\frac{\partial B_\rho}{\partial \rho} - \rho^{-2} B_\rho + \frac{\partial^2 B_\rho}{\partial \zeta^2} = 0 \quad , \qquad [4.50]$$

$$\frac{\partial^2 B_0}{\partial \rho^2} + \rho^{-1}\frac{\partial B_0}{\partial \rho} + \frac{\partial^2 B_0}{\partial \zeta^2} = 0 \quad . \qquad [4.51]$$

Letting $q(\rho) \equiv p/\mu$, $\sigma_\zeta \equiv \sigma_z/\mu$, $\sigma_{\rho\zeta} \equiv \sigma_{rz}/\mu$, $\sigma_\rho \equiv \sigma_r/\mu$, and $\sigma_\phi \equiv \sigma_\theta/r$, the boundary conditions are

$$\sigma_\zeta = q(\rho) \qquad\qquad (\zeta = 0)$$

$$\sigma_{\rho\phi} = \sigma_{\rho\zeta} = 0 \qquad\qquad (\zeta = 0) \quad , \qquad\qquad [4.52]$$

$$\sigma_\rho = \sigma_{\rho\zeta} = 0 \qquad\qquad (\rho = 1) \quad , \qquad\qquad [4.53]$$

$$u, \ w, \ \sigma_\rho, \ \sigma_\phi, \ \sigma_\rho, \ \sigma_{\rho\zeta} \to 0 \qquad\qquad [(\rho^2 + \zeta^2)^{1/2} \to \infty] \quad .$$

$$[4.54]$$

The above is supplemented by an equilibrium equation:

$$\int q(\rho)\rho \ d\rho = 0 \quad . \qquad\qquad [4.55]$$

Define the fundamental problem as one in which all but the first of [4.52] remain the same as above while $q(\rho)$ is replaced by $\delta(a)$, the Dirac Delta function. Therefore in place of [4.52],

$$\sigma_\rho = \delta(\rho - a) \qquad\qquad (\zeta = 0)$$

$$\sigma_{\rho\phi} = \sigma_{\rho\zeta} = 0 \qquad\qquad (\zeta = 0) \quad . \qquad\qquad [4.56]$$

Once the solution of the fundamental problem is found, that is,

$$w(\rho, 0; a) \quad , \qquad\qquad [4.57]$$

the solution of w for the general problem is merely

$$2\pi \int_0^1 q(a) w(\rho, 0; a) a \ da \quad . \qquad\qquad [4.58]$$

Now, returning to the fundamental problem, with transformed quantities denoted by a super $\sim$, [4.50] and [4.51] become, respectively,

$$\frac{d^2 \tilde{B}_\rho}{d\rho^2} + \rho^{-1} \frac{d\tilde{B}_\rho}{d\rho} - (s^2 + \rho^{-2})\tilde{B}_\rho = 0 \quad , \qquad [4.59]$$

$$\frac{d^2 \tilde{B}_0}{d\rho^2} + \rho^{-1} \frac{d\tilde{B}_0}{d\rho} - s^2 \tilde{B}_0 = 0 \quad . \qquad [4.60]$$

Equations 4.59 and 4.60 possess, respectively, the solutions

$$B_\rho = C\, I_1(s\rho) \quad , \qquad [4.61]$$

$$B_0 = D\, I_0(s\rho) \quad , \qquad [4.62]$$

where due consideration has been given to [4.54].

Using [4.61] and [4.62] through the transforms of [4.49] and those of stress-displacement relationships, expressions of the transformed stresses $\tilde{\sigma}_\rho$, $\tilde{\sigma}_{\rho\zeta}$, and $\tilde{\sigma}_\zeta$ are found. Enforcement of [4.53] leads to two homogeneous equations for C and D. These in turn, require the vanishing of the determinant of the coefficients for nontrivial solution and a relationship between C and D

$$s^3[I_0^{\,3}(s) - I_1^{\,2}(s)] - \frac{2(m-1)s}{m} I_1^{\,2}(s) = 0 \quad , \qquad [4.63]$$

$$D = [\frac{2(m-1)}{ms_t} - \lambda_t]C \quad , \qquad [4.64]$$

where s_t are the zeros of [4.61] and $\lambda_t = I_1(s_t)/I_0(s_t)$.

Executing the inverse transform on $\tilde{\sigma}_\zeta$ and $\tilde{\sigma}_{\rho\zeta}$ and enforcing [4.52] and [4.55],

$$F(\rho) = 2a - \delta(\rho-a) = \sum_{t=1}^{\infty} M_t[\alpha_t^{\,r}(\rho) + \alpha_i^{\,i}(\rho)] + N_t[\alpha_t^{\,r}(\rho) - \alpha_i^{\,i}(\rho)]$$

$$0 = \sum_{t=1}^{\infty} M_t[\beta_t^{\,r}(\rho) + \beta_t^{\,i}(\rho)] + N_t[\beta_t^{\,r}(\rho) - \beta_t^{\,i}(\rho)] \quad , \qquad [4.65]$$

where $\alpha_t^{\,r}$ and $\alpha_t^{\,i}$ are, respectively, the real and imaginary parts of

$$\frac{2}{\pi\, I_1(s_t)} \left[(2 - s_t\lambda_t)I_0(s_t\,\rho) + \rho\, s_t\, I_1(s_t\rho) \right] \quad ,$$

and correspondingly $\beta_t^{\,r}$ and $\beta_t^{\,i}$ are those of

$$\frac{s_t}{I_1(s_t)} \left[\rho\, I_0(s_t\,\rho) - \lambda_t\, I_1(s_t\,\rho) \right] \quad .$$

Equation 4.65 should be solved approximately for M_t's and N_t's numerically. In particular, [4.65] is satisfied through the following scheme:[56] Define

$$\gamma = \int_0^1 \left\{ [F(\rho) - (\alpha_t^{\,r} + i\,\alpha_t^{\,i})]^2 + (\beta_t^{\,r} + i\,\beta_t^{\,i})^2 \right\} \quad .$$

It is required that $\partial\gamma/\partial M_t = 0$ and $\partial\gamma/\partial N_t = 0$.

Numerically [4.57] is far more cumbersome than [4.58], if $q(a)$ is slowly varying, for a given degree of accuracy. This is because of the condition 4.56. On the other hand, it is precisely the nature of [4.56] which will make any tabulation of [4.57] generally meaningful and useful. In what follows, $w^*(\rho, 0; a)$ instead of $w(\rho, 0; a)$ is sought numerically, where

$$w^* = w - w_h \quad , \qquad [4.66]$$

and w_h is that which corresponds to the half-space as given by [4.20]. It reads, with the appropriate change of notations to ones used in the present section,

$$w_h(\rho,a) = \frac{2(m-1)}{m} K(k)\left(1 + \frac{\rho}{a}\right)^{-1} , \qquad [4.67]$$

where $k = 2\sqrt{\rho/a}\,[1 + (\rho/a)]^{-1}$. To obtain w^*, a two-step procedure is used, the second of which makes use of the procedure outlined above and culminated in [4.65] with $\sigma_\zeta = q^*(\rho, a)$ in place of the condition shown as the first of [4.56]. Since $q^*(\rho, a)$ is slowly varying the procedure is expected to yield accurate results without excessive machine labor. The $q^*(a)$ is now found as the first of two steps mentioned earlier.

Now $q^*(\rho, a)$ is expressible as

$$q^*(\rho,a) = 2\int_0^\infty \sigma_\zeta^{(ah)}(1,\zeta)K_n(\rho,0;1,\zeta)\,d\zeta$$

$$+ 2\int_0^\infty \sigma_{\rho\zeta}^{(ah)}(1,\zeta)K_t(\rho,0;1,\zeta)\,d\zeta , \qquad [4.68]$$

where $\sigma^{(ah)}(1, \zeta)$ and $\sigma_{\rho\zeta}^{(ah)}(1, \zeta)$ are the normal and shear stresses, respectively, generated for $\rho = 1$ and $\zeta = \zeta$ due to a ring load of unit dimensionless intensity at radius a for the half-space. Also $K_n(\rho, 0; 1, \zeta)$ and $K_t(\rho, 0; 1, \zeta)$ are the normal stresses at $\rho = \rho$ and $\zeta = 0$ due to a pair of band (width b shown in Figure 4.14) normal and shear load, respectively, at $\rho = 1$ and $\pm\zeta$. In the case of shear load, the sense of the pair is such that there is no net axial force. The intensity of the forces over each band per unit arc length is that of unity; $\sigma_\zeta^{(ah)}(1, \zeta)$ and $\sigma_{\rho\zeta}^{(ah)}(1, \zeta)$ are found by the method used in Section 4.D. Also $K_n(\rho, 0; 1, \zeta)$ and $K_t(\rho, 0; 1, \zeta)$ are found by the Fourier transform method and earlier in this section.[57]

In obtaining the displacement on a half-space due to a ring of pressure, a width of 2ϵ is used for computing the results shown in Table 2. These are to be added to the half-space solution, [4.20]. The correction terms are considered to be due to a ring of forces which has been averaged over some thickness Δr to represent a ring of pressure. For example, if $\epsilon = 0.01$, then all of the ring sizes in the table are used.

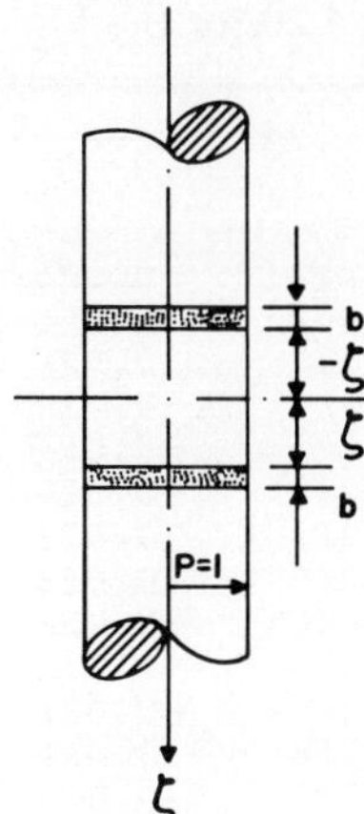

Figure 4.14

However, if $\epsilon = 0.02$, then the average radii for the rings of pressure would be $a = 0.01$ (center disc), 0,04, 0.08, . . . , 0.20, 0.24, 0.28, . . . , 0.98. Hence only the correction terms associated with rings of forces at 0.2, 0.24, 0.28, and so on would be used. The correction terms must also be converted from the equivalent pressure associated with averaging the ring of force over a thickness Δr, to a unit pressure. This is done by merely multiplying the correction terms by 2ϵ before adding them to the half-space solution.

The physical setup of the tabulation is as follows: the headings across the top of the pages refer to the radius of the load ring; the numbers in the far left-hand column refer to the radius for which the displacement is given in the subsequent column on that page.

Table 2. Correction Functions to Equation 4.20 for the Finite Cylinder

	A = .20	A = .22	A = .24	A = .26	A = .28
R = 0.0	−0.017	−0.019	−0.022	−0.025	−0.028
R = 0.02	−0.017	−0.019	−0.022	−0.025	−0.028
R = 0.04	−0.017	−0.019	−0.022	−0.025	−0.028
R = 0.06	−0.017	−0.019	−0.022	−0.024	−0.027
R = 0.08	−0.017	−0.019	−0.021	−0.024	−0.027
R = 0.10	−0.016	−0.019	−0.021	−0.024	−0.027
R = 0.12	−0.016	−0.018	−0.021	−0.023	−0.026
R = 0.14	−0.016	−0.018	−0.020	−0.023	−0.026
R = 0.16	−0.015	−0.017	−0.020	−0.022	−0.025
R = 0.18	−0.015	−0.017	−0.019	−0.022	−0.025
R = 0.20	−0.014	−0.017	−0.019	−0.021	−0.024
R = 0.22	−0.014	−0.016	−0.018	−0.021	−0.023
R = 0.24	−0.013	−0.015	−0.017	−0.020	−0.022
R = 0.26	−0.013	−0.015	−0.017	−0.019	−0.021
R = 0.28	−0.012	−0.014	−0.016	−0.018	−0.021
R = 0.30	−0.012	−0.013	−0.015	−0.017	−0.020
R = 0.32	−0.011	−0.013	−0.014	−0.016	−0.019
R = 0.34	−0.010	−0.012	−0.014	−0.015	−0.018
R = 0.36	−0.010	−0.011	−0.013	−0.015	−0.017
R = 0.38	−0.009	−0.010	−0.012	−0.013	−0.015
R = 0.40	−0.008	−0.009	−0.011	−0.012	−0.014
R = 0.42	−0.007	−0.009	−0.010	−0.011	−0.013
R = 0.44	−0.007	−0.008	−0.009	−0.010	−0.012
R = 0.46	−0.006	−0.007	−0.008	−0.009	−0.010
R = 0.48	−0.005	−0.006	−0.007	−0.008	−0.009
R = 0.50	−0.004	−0.005	−0.006	−0.006	−0.008
R = 0.52	−0.003	−0.004	−0.004	−0.005	−0.006
R = 0.54	−0.002	−0.003	−0.003	−0.004	−0.005
R = 0.56	−0.001	−0.002	−0.002	−0.003	−0.003
R = 0.58	−0.000	−0.000	−0.001	−0.001	−0.002
R = 0.60	0.001	0.001	0.000	0.000	−0.000
R = 0.62	0.002	0.002	0.002	0.002	0.001
R = 0.64	0.002	0.003	0.003	0.003	0.003
R = 0.66	0.003	0.004	0.004	0.004	0.004
R = 0.68	0.004	0.005	0.005	0.005	0.006
R = 0.70	0.005	0.006	0.006	0.007	0.007
R = 0.72	0.006	0.006	0.007	0.008	0.009
R = 0.74	0.006	0.007	0.008	0.009	0.010
R = 0.76	0.007	0.008	0.009	0.010	0.011
R = 0.78	0.007	0.008	0.009	0.010	0.011
R = 0.80	0.008	0.009	0.010	0.011	0.012
R = 0.82	0.008	0.009	0.010	0.011	0.012
R = 0.84	0.007	0.008	0.010	0.011	0.012
R = 0.86	0.007	0.008	0.009	0.010	0.012
R = 0.88	0.006	0.007	0.008	0.010	0.011
R = 0.90	0.005	0.006	0.007	0.008	0.010
R = 0.92	0.004	0.005	0.005	0.007	0.008
R = 0.94	0.002	0.003	0.003	0.004	0.006
R = 0.96	0.000	0.000	0.001	0.002	0.003
R = 0.98	0.000	0.000	0.000	0.000	0.000
R = 1.00	0.000	0.000	0.000	0.000	0.000

(Continued)

[108]

Table 2 (*Continued*)

	A = .30	A = .32	A = .34	A = .36	A = .38
R = 0.0	-0.031	-0.034	-0.038	-0.043	-0.047
R = 0.02	-0.031	-0.034	-0.038	-0.043	-0.047
R = 0.04	-0.031	-0.034	-0.038	-0.042	-0.047
R = 0.06	-0.030	-0.034	-0.038	-0.042	-0.047
R = 0.08	-0.030	-0.034	-0.038	-0.042	-0.046
R = 0.10	-0.030	-0.033	-0.037	-0.041	-0.046
R = 0.12	-0.029	-0.033	-0.037	-0.041	-0.045
R = 0.14	-0.029	-0.032	-0.036	-0.040	-0.044
R = 0.16	-0.028	-0.032	-0.035	-0.039	-0.044
R = 0.18	-0.028	-0.031	-0.034	-0.038	-0.043
R = 0.20	-0.027	-0.030	-0.033	-0.037	-0.042
R = 0.22	-0.026	-0.029	-0.033	-0.036	-0.040
R = 0.24	-0.025	-0.028	-0.031	-0.035	-0.039
R = 0.26	-0.024	-0.027	-0.030	-0.034	-0.038
R = 0.28	-0.023	-0.026	-0.029	-0.033	-0.036
R = 0.30	-0.022	-0.025	-0.028	-0.031	-0.035
R = 0.32	-0.021	-0.024	-0.027	-0.030	-0.033
R = 0.34	-0.020	-0.022	-0.025	-0.028	-0.032
R = 0.36	-0.019	-0.021	-0.024	-0.027	-0.030
R = 0.38	-0.017	-0.020	-0.022	-0.025	-0.028
R = 0.40	-0.016	-0.018	-0.021	-0.023	-0.026
R = 0.42	-0.015	-0.017	-0.019	-0.022	-0.024
R = 0.44	-0.013	-0.015	-0.017	-0.020	-0.022
R = 0.46	-0.012	-0.014	-0.016	-0.018	-0.020
R = 0.48	-0.010	-0.012	-0.014	-0.016	-0.018
R = 0.50	-0.009	-0.010	-0.012	-0.014	-0.016
R = 0.52	-0.007	-0.008	-0.010	-0.011	-0.013
R = 0.54	-0.006	-0.007	-0.008	-0.009	-0.011
R = 0.56	-0.004	-0.005	-0.006	-0.007	-0.008
R = 0.58	-0.002	-0.003	-0.004	-0.005	-0.006
R = 0.60	-0.000	-0.001	-0.002	-0.002	-0.003
R = 0.62	0.001	0.001	0.001	0.000	-0.000
R = 0.64	0.003	0.003	0.003	0.002	0.002
R = 0.66	0.005	0.005	0.005	0.005	0.005
R = 0.68	0.006	0.007	0.007	0.007	0.007
R = 0.70	0.008	0.008	0.009	0.009	0.010
R = 0.72	0.009	0.010	0.011	0.012	0.012
R = 0.74	0.011	0.012	0.013	0.014	0.015
R = 0.76	0.012	0.013	0.014	0.015	0.017
R = 0.78	0.013	0.014	0.015	0.017	0.019
R = 0.80	0.013	0.015	0.017	0.018	0.020
R = 0.82	0.014	0.015	0.017	0.019	0.021
R = 0.84	0.014	0.016	0.018	0.020	0.022
R = 0.86	0.014	0.015	0.017	0.020	0.022
R = 0.88	0.013	0.015	0.017	0.019	0.021
R = 0.90	0.011	0.013	0.015	0.017	0.020
R = 0.92	0.009	0.011	0.013	0.015	0.018
R = 0.94	0.007	0.008	0.010	0.012	0.015
R = 0.96	0.004	0.005	0.007	0.009	0.011
R = 0.98	0.001	0.002	0.004	0.005	0.008
R = 1.00	0.000	0.000	0.001	0.002	0.005

(Continued)

	A = .40	A = .42	A = .44	A = .46	A = .48
R = 0.0	−0.052	−0.058	−0.064	−0.071	−0.079
R = 0.02	−0.052	−0.058	−0.064	−0.071	−0.078
R = 0.04	−0.052	−0.058	−0.064	−0.071	−0.078
R = 0.06	−0.052	−0.057	−0.064	−0.070	−0.078
R = 0.08	−0.051	−0.057	−0.063	−0.070	−0.077
R = 0.10	−0.051	−0.056	−0.062	−0.069	−0.076
R = 0.12	−0.050	−0.056	−0.062	−0.068	−0.075
R = 0.14	−0.049	−0.055	−0.061	−0.067	−0.074
R = 0.16	−0.048	−0.054	−0.060	−0.066	−0.073
R = 0.18	−0.047	−0.053	−0.058	−0.065	−0.072
R = 0.20	−0.046	−0.051	−0.057	−0.063	−0.070
R = 0.22	−0.045	−0.050	−0.056	−0.062	−0.068
R = 0.24	−0.044	−0.049	−0.054	−0.060	−0.067
R = 0.26	−0.042	−0.047	−0.052	−0.058	−0.065
R = 0.28	−0.041	−0.045	−0.051	−0.056	−0.063
R = 0.30	−0.039	−0.044	−0.049	−0.054	−0.060
R = 0.32	−0.037	−0.042	−0.047	−0.052	−0.058
R = 0.34	−0.036	−0.040	−0.045	−0.050	−0.055
R = 0.36	−0.034	−0.038	−0.042	−0.047	−0.053
R = 0.38	−0.032	−0.036	−0.040	−0.045	−0.050
R = 0.40	−0.030	−0.033	−0.038	−0.042	−0.047
R = 0.42	−0.028	−0.031	−0.035	−0.039	−0.044
R = 0.44	−0.025	−0.029	−0.032	−0.037	−0.041
R = 0.46	−0.023	−0.026	−0.030	−0.033	−0.038
R = 0.48	−0.021	−0.023	−0.027	−0.030	−0.034
R = 0.50	−0.018	−0.021	−0.024	−0.027	−0.031
R = 0.52	−0.015	−0.018	−0.020	−0.023	−0.027
R = 0.54	−0.013	−0.015	−0.017	−0.020	−0.023
R = 0.56	−0.010	−0.012	−0.014	−0.016	−0.019
R = 0.58	−0.007	−0.009	−0.010	−0.012	−0.014
R = 0.60	−0.004	−0.005	−0.007	−0.008	−0.010
R = 0.62	−0.001	−0.002	−0.003	−0.004	−0.006
R = 0.64	0.002	0.001	0.000	−0.000	−0.001
R = 0.66	0.005	0.004	0.004	0.004	0.003
R = 0.68	0.008	0.008	0.008	0.008	0.008
R = 0.70	0.010	0.011	0.012	0.012	0.012
R = 0.72	0.013	0.014	0.015	0.016	0.017
R = 0.74	0.016	0.017	0.018	0.020	0.021
R = 0.76	0.018	0.020	0.022	0.023	0.025
R = 0.78	0.020	0.022	0.024	0.027	0.029
R = 0.80	0.022	0.025	0.027	0.030	0.033
R = 0.82	0.024	0.026	0.029	0.032	0.035
R = 0.84	0.024	0.027	0.030	0.034	0.038
R = 0.86	0.025	0.028	0.031	0.035	0.039
R = 0.88	0.024	0.028	0.031	0.035	0.040
R = 0.90	0.023	0.026	0.030	0.034	0.039
R = 0.92	0.021	0.024	0.028	0.032	0.037
R = 0.94	0.018	0.021	0.025	0.029	0.034
R = 0.96	0.014	0.017	0.021	0.025	0.030
R = 0.98	0.010	0.013	0.017	0.021	0.026
R = 1.00	0.007	0.010	0.014	0.018	0.023

(Continued)

Table 2 (*Continued*)

	A = .50	A = .52	A = .54	A = .56	A = .58
R = 0.0	-0.087	-0.096	-0.106	-0.116	-0.128
R = 0.02	-0.087	-0.096	-0.106	-0.116	-0.128
R = 0.04	-0.086	-0.095	-0.105	-0.116	-0.128
R = 0.06	-0.086	-0.095	-0.105	-0.115	-0.127
R = 0.08	-0.085	-0.094	-0.104	-0.115	-0.127
R = 0.10	-0.084	-0.093	-0.103	-0.114	-0.125
R = 0.12	-0.083	-0.092	-0.102	-0.112	-0.124
R = 0.14	-0.082	-0.091	-0.101	-0.111	-0.123
R = 0.16	-0.081	-0.090	-0.099	-0.109	-0.121
R = 0.18	-0.079	-0.088	-0.097	-0.107	-0.119
R = 0.20	-0.078	-0.086	-0.095	-0.105	-0.117
R = 0.22	-0.076	-0.084	-0.093	-0.103	-0.114
R = 0.24	-0.074	-0.082	-0.091	-0.101	-0.111
R = 0.26	-0.072	-0.080	-0.088	-0.098	-0.109
R = 0.28	-0.069	-0.077	-0.086	-0.095	-0.106
R = 0.30	-0.067	-0.075	-0.083	-0.092	-0.102
R = 0.32	-0.065	-0.072	-0.080	-0.089	-0.099
R = 0.34	-0.062	-0.069	-0.077	-0.086	-0.095
R = 0.36	-0.059	-0.066	-0.074	-0.082	-0.091
R = 0.38	-0.056	-0.063	-0.070	-0.078	-0.087
R = 0.40	-0.053	-0.059	-0.067	-0.074	-0.083
R = 0.42	-0.050	-0.056	-0.063	-0.070	-0.079
R = 0.44	-0.046	-0.052	-0.059	-0.066	-0.074
R = 0.46	-0.043	-0.048	-0.054	-0.061	-0.069
R = 0.48	-0.039	-0.044	-0.050	-0.056	-0.063
R = 0.50	-0.035	-0.040	-0.045	-0.051	-0.057
R = 0.52	-0.030	-0.035	-0.040	-0.045	-0.051
R = 0.54	-0.026	-0.030	-0.035	-0.040	-0.045
R = 0.56	-0.022	-0.025	-0.029	-0.034	-0.039
R = 0.58	-0.017	-0.020	-0.024	-0.027	-0.032
R = 0.60	-0.012	-0.015	-0.018	-0.021	-0.025
R = 0.62	-0.007	-0.009	-0.012	-0.014	-0.017
R = 0.64	-0.003	-0.004	-0.006	-0.008	-0.010
R = 0.66	0.003	0.002	0.001	-0.001	-0.002
R = 0.68	0.008	0.007	0.007	0.006	0.006
R = 0.70	0.013	0.013	0.013	0.014	0.014
R = 0.72	0.018	0.019	0.020	0.021	0.022
R = 0.74	0.023	0.024	0.026	0.028	0.030
R = 0.76	0.028	0.030	0.032	0.035	0.038
R = 0.78	0.032	0.035	0.038	0.041	0.045
R = 0.80	0.036	0.039	0.043	0.048	0.052
R = 0.82	0.039	0.043	0.048	0.053	0.059
R = 0.84	0.042	0.047	0.052	0.058	0.064
R = 0.86	0.044	0.049	0.055	0.062	0.069
R = 0.88	0.045	0.051	0.057	0.064	0.072
R = 0.90	0.044	0.051	0.057	0.065	0.074
R = 0.92	0.043	0.049	0.056	0.064	0.073
R = 0.94	0.040	0.046	0.054	0.062	0.071
R = 0.96	0.036	0.042	0.050	0.058	0.068
R = 0.98	0.031	0.038	0.046	0.054	0.064
R = 1.00	0.028	0.035	0.043	0.051	0.062

(*Continued*)

Table 2 (*Continued*)

	A = .60	A = .62	A = .64	A = .66	A = .68
R = 0.0	-0.142	-0.156	-0.172	-0.189	-0.208
R = 0.02	-0.141	-0.156	-0.172	-0.189	-0.208
R = 0.04	-0.141	-0.155	-0.171	-0.189	-0.208
R = 0.06	-0.140	-0.155	-0.170	-0.188	-0.207
R = 0.08	-0.139	-0.154	-0.169	-0.187	-0.206
R = 0.10	-0.138	-0.153	-0.168	-0.185	-0.204
R = 0.12	-0.137	-0.151	-0.167	-0.184	-0.203
R = 0.14	-0.135	-0.149	-0.165	-0.182	-0.200
R = 0.16	-0.133	-0.147	-0.162	-0.179	-0.198
R = 0.18	-0.131	-0.145	-0.160	-0.177	-0.195
R = 0.20	-0.129	-0.142	-0.157	-0.174	-0.192
R = 0.22	-0.126	-0.140	-0.154	-0.171	-0.189
R = 0.24	-0.123	-0.137	-0.151	-0.167	-0.185
R = 0.26	-0.120	-0.133	-0.148	-0.163	-0.181
R = 0.28	-0.117	-0.130	-0.144	-0.159	-0.177
R = 0.30	-0.114	-0.126	-0.140	-0.155	-0.172
R = 0.32	-0.110	-0.122	-0.136	-0.151	-0.167
R = 0.34	-0.106	-0.118	-0.131	-0.146	-0.162
R = 0.36	-0.102	-0.113	-0.126	-0.141	-0.157
R = 0.38	-0.097	-0.109	-0.121	-0.135	-0.151
R = 0.40	-0.093	-0.104	-0.116	-0.129	-0.144
R = 0.42	-0.088	-0.098	-0.110	-0.123	-0.138
R = 0.44	-0.083	-0.093	-0.104	-0.116	-0.131
R = 0.46	-0.077	-0.087	-0.097	-0.109	-0.123
R = 0.48	-0.071	-0.080	-0.090	-0.102	-0.115
R = 0.50	-0.065	-0.073	-0.083	-0.094	-0.106
R = 0.52	-0.058	-0.066	-0.075	-0.085	-0.096
R = 0.54	-0.052	-0.059	-0.067	-0.076	-0.087
R = 0.56	-0.044	-0.051	-0.058	-0.067	-0.076
R = 0.58	-0.037	-0.043	-0.049	-0.057	-0.065
R = 0.60	-0.029	-0.034	-0.040	-0.046	-0.054
R = 0.62	-0.021	-0.025	-0.030	-0.035	-0.042
R = 0.64	-0.013	-0.016	-0.020	-0.024	-0.030
R = 0.66	-0.004	-0.007	-0.009	-0.013	-0.017
R = 0.68	0.005	0.003	0.002	-0.001	-0.003
R = 0.70	0.014	0.013	0.013	0.012	0.011
R = 0.72	0.023	0.023	0.024	0.025	0.025
R = 0.74	0.032	0.034	0.036	0.038	0.040
R = 0.76	0.040	0.044	0.047	0.050	0.054
R = 0.78	0.049	0.053	0.058	0.063	0.069
R = 0.80	0.057	0.063	0.069	0.076	0.083
R = 0.82	0.065	0.072	0.079	0.087	0.096
R = 0.84	0.071	0.079	0.088	0.098	0.109
R = 0.86	0.077	0.086	0.096	0.108	0.120
R = 0.88	0.081	0.091	0.103	0.115	0.130
R = 0.90	0.084	0.094	0.107	0.121	0.137
R = 0.92	0.084	0.095	0.109	0.124	0.141
R = 0.94	0.082	0.094	0.108	0.124	0.142
R = 0.96	0.079	0.091	0.106	0.122	0.141
R = 0.98	0.075	0.088	0.103	0.120	0.139
R = 1.00	0.073	0.087	0.102	0.120	0.140

(*Continued*)

[112]

Table 2 (Continued)

	A = .70	A = .72	A = .74	A = .76	A = .78
R = 0.0	−0.229	−0.253	−0.278	−0.307	−0.338
R = 0.02	−0.229	−0.253	−0.278	−0.306	−0.338
R = 0.04	−0.229	−0.252	−0.278	−0.306	−0.337
R = 0.06	−0.228	−0.251	−0.277	−0.305	−0.336
R = 0.08	−0.227	−0.250	−0.275	−0.303	−0.335
R = 0.10	−0.225	−0.248	−0.274	−0.302	−0.333
R = 0.12	−0.223	−0.246	−0.272	−0.299	−0.330
R = 0.14	−0.221	−0.244	−0.269	−0.297	−0.328
R = 0.16	−0.218	−0.241	−0.266	−0.294	−0.324
R = 0.18	−0.215	−0.238	−0.263	−0.290	−0.321
R = 0.20	−0.212	−0.234	−0.259	−0.286	−0.317
R = 0.22	−0.208	−0.231	−0.255	−0.282	−0.312
R = 0.24	−0.205	−0.226	−0.251	−0.277	−0.307
R = 0.26	−0.200	−0.222	−0.246	−0.272	−0.302
R = 0.28	−0.196	−0.217	−0.241	−0.267	−0.296
R = 0.30	−0.191	−0.212	−0.235	−0.261	−0.290
R = 0.32	−0.186	−0.206	−0.229	−0.255	−0.283
R = 0.34	−0.180	−0.200	−0.223	−0.248	−0.276
R = 0.36	−0.174	−0.194	−0.216	−0.241	−0.268
R = 0.38	−0.168	−0.187	−0.209	−0.233	−0.260
R = 0.40	−0.161	−0.180	−0.201	−0.225	−0.251
R = 0.42	−0.154	−0.172	−0.193	−0.216	−0.242
R = 0.44	−0.146	−0.164	−0.184	−0.206	−0.232
R = 0.46	−0.138	−0.155	−0.174	−0.196	−0.220
R = 0.48	−0.129	−0.145	−0.164	−0.185	−0.208
R = 0.50	−0.119	−0.135	−0.152	−0.172	−0.195
R = 0.52	−0.109	−0.124	−0.140	−0.159	−0.181
R = 0.54	−0.099	−0.112	−0.128	−0.145	−0.166
R = 0.56	−0.087	−0.100	−0.114	−0.130	−0.149
R = 0.58	−0.075	−0.087	−0.100	−0.115	−0.132
R = 0.60	−0.063	−0.073	−0.084	−0.098	−0.113
R = 0.62	−0.049	−0.058	−0.068	−0.080	−0.094
R = 0.64	−0.036	−0.043	−0.052	−0.062	−0.073
R = 0.66	−0.021	−0.027	−0.034	−0.042	−0.052
R = 0.68	−0.006	−0.010	−0.015	−0.021	−0.028
R = 0.70	0.009	0.007	0.004	0.001	−0.004
R = 0.72	0.025	0.025	0.025	0.023	0.021
R = 0.74	0.042	0.043	0.045	0.047	0.048
R = 0.76	0.058	0.062	0.066	0.070	0.075
R = 0.78	0.074	0.081	0.088	0.095	0.102
R = 0.80	0.091	0.099	0.109	0.119	0.130
R = 0.82	0.106	0.118	0.130	0.143	0.158
R = 0.84	0.121	0.135	0.150	0.167	0.186
R = 0.86	0.135	0.151	0.169	0.189	0.211
R = 0.88	0.146	0.164	0.185	0.208	0.235
R = 0.90	0.154	0.175	0.198	0.224	0.254
R = 0.92	0.160	0.182	0.207	0.236	0.269
R = 0.94	0.162	0.186	0.213	0.244	0.279
R = 0.96	0.162	0.187	0.215	0.248	0.287
R = 0.98	0.162	0.188	0.218	0.253	0.293
R = 1.00	0.164	0.191	0.223	0.260	0.304

(Continued)

Table 2 (*Continued*)

	A = .80	A = .82	A = .84	A = .86	A = .88
R = 0.0	−0.372	−0.411	−0.452	−0.498	−0.547
R = 0.02	−0.372	−0.411	−0.452	−0.498	−0.547
R = 0.04	−0.371	−0.410	−0.452	−0.497	−0.547
R = 0.06	−0.370	−0.409	−0.451	−0.496	−0.546
R = 0.08	−0.369	−0.407	−0.449	−0.494	−0.544
R = 0.10	−0.367	−0.405	−0.447	−0.492	−0.542
R = 0.12	−0.364	−0.403	−0.444	−0.490	−0.540
R = 0.14	−0.362	−0.400	−0.441	−0.487	−0.537
R = 0.16	−0.358	−0.396	−0.438	−0.483	−0.533
R = 0.18	−0.354	−0.392	−0.434	−0.479	−0.529
R = 0.20	−0.350	−0.388	−0.429	−0.474	−0.524
R = 0.22	−0.345	−0.383	−0.424	−0.469	−0.519
R = 0.24	−0.340	−0.377	−0.418	−0.463	−0.513
R = 0.26	−0.334	−0.372	−0.412	−0.457	−0.507
R = 0.28	−0.328	−0.365	−0.406	−0.450	−0.500
R = 0.30	−0.322	−0.359	−0.399	−0.443	−0.493
R = 0.32	−0.315	−0.351	−0.391	−0.436	−0.485
R = 0.34	−0.307	−0.343	−0.383	−0.427	−0.477
R = 0.36	−0.299	−0.335	−0.374	−0.418	−0.468
R = 0.38	−0.291	−0.326	−0.365	−0.408	−0.458
R = 0.40	−0.281	−0.316	−0.354	−0.398	−0.447
R = 0.42	−0.271	−0.305	−0.343	−0.386	−0.435
R = 0.44	−0.260	−0.293	−0.331	−0.373	−0.421
R = 0.46	−0.248	−0.280	−0.317	−0.358	−0.406
R = 0.48	−0.235	−0.266	−0.302	−0.342	−0.389
R = 0.50	−0.221	−0.251	−0.285	−0.324	−0.370
R = 0.52	−0.205	−0.234	−0.266	−0.305	−0.349
R = 0.54	−0.189	−0.216	−0.247	−0.284	−0.327
R = 0.56	−0.171	−0.196	−0.226	−0.261	−0.302
R = 0.58	−0.152	−0.175	−0.203	−0.236	−0.274
R = 0.60	−0.132	−0.153	−0.179	−0.209	−0.245
R = 0.62	−0.110	−0.129	−0.152	−0.180	−0.213
R = 0.64	−0.087	−0.104	−0.125	−0.149	−0.179
R = 0.66	−0.063	−0.077	−0.095	−0.116	−0.143
R = 0.68	−0.037	−0.049	−0.063	−0.080	−0.103
R = 0.70	−0.010	−0.018	−0.029	−0.042	−0.060
R = 0.72	0.018	0.014	0.007	−0.002	−0.015
R = 0.74	0.048	0.047	0.045	0.041	0.033
R = 0.76	0.079	0.082	0.085	0.086	0.085
R = 0.78	0.110	0.118	0.126	0.133	0.139
R = 0.80	0.142	0.155	0.169	0.183	0.196
R = 0.82	0.175	0.193	0.212	0.233	0.255
R = 0.84	0.207	0.230	0.256	0.285	0.316
R = 0.86	0.237	0.266	0.299	0.335	0.376
R = 0.88	0.265	0.299	0.338	0.383	0.433
R = 0.90	0.288	0.328	0.373	0.426	0.486
R = 0.92	0.307	0.351	0.403	0.462	0.532
R = 0.94	0.321	0.369	0.426	0.493	0.572
R = 0.96	0.331	0.384	0.446	0.520	0.608
R = 0.98	0.341	0.398	0.465	0.546	0.644
R = 1.00	0.355	0.417	0.490	0.578	0.686

(*Continued*)

[114]

Table 2 (Continued)

	A = .90	A = .92	A = .94	A = .96	A = .98
R = 0.0	-0.598	-0.654	-0.715	-0.765	-0.800
R = 0.02	-0.598	-0.654	-0.714	-0.765	-0.800
R = 0.04	-0.597	-0.654	-0.714	-0.765	-0.799
R = 0.06	-0.597	-0.654	-0.714	-0.764	-0.799
R = 0.08	-0.596	-0.653	-0.713	-0.764	-0.798
R = 0.10	-0.594	-0.652	-0.713	-0.764	-0.798
R = 0.12	-0.592	-0.650	-0.712	-0.763	-0.797
R = 0.14	-0.589	-0.647	-0.710	-0.763	-0.797
R = 0.16	-0.585	-0.644	-0.708	-0.762	-0.796
R = 0.18	-0.581	-0.641	-0.705	-0.760	-0.796
R = 0.20	-0.577	-0.636	-0.701	-0.758	-0.795
R = 0.22	-0.572	-0.631	-0.697	-0.755	-0.795
R = 0.24	-0.566	-0.626	-0.692	-0.752	-0.795
R = 0.26	-0.560	-0.621	-0.688	-0.750	-0.794
R = 0.28	-0.554	-0.615	-0.683	-0.748	-0.794
R = 0.30	-0.547	-0.609	-0.679	-0.746	-0.793
R = 0.32	-0.540	-0.602	-0.674	-0.745	-0.793
R = 0.34	-0.532	-0.595	-0.668	-0.744	-0.792
R = 0.36	-0.523	-0.587	-0.662	-0.742	-0.792
R = 0.38	-0.513	-0.578	-0.655	-0.740	-0.791
R = 0.40	-0.502	-0.568	-0.646	-0.736	-0.791
R = 0.42	-0.490	-0.556	-0.636	-0.730	-0.791
R = 0.44	-0.476	-0.543	-0.624	-0.722	-0.790
R = 0.46	-0.461	-0.527	-0.610	-0.712	-0.790
R = 0.48	-0.443	-0.510	-0.592	-0.699	-0.789
R = 0.50	-0.423	-0.489	-0.572	-0.682	-0.789
R = 0.52	-0.401	-0.466	-0.549	-0.663	-0.788
R = 0.54	-0.377	-0.441	-0.524	-0.640	-0.788
R = 0.56	-0.351	-0.412	-0.494	-0.614	-0.787
R = 0.58	-0.321	-0.381	-0.461	-0.583	-0.787
R = 0.60	-0.290	-0.347	-0.425	-0.547	-0.786
R = 0.62	-0.255	-0.309	-0.384	-0.506	-0.785
R = 0.64	-0.218	-0.268	-0.339	-0.459	-0.769
R = 0.66	-0.177	-0.223	-0.289	-0.406	-0.743
R = 0.68	-0.133	-0.174	-0.235	-0.346	-0.702
R = 0.70	-0.085	-0.120	-0.175	-0.279	-0.645
R = 0.72	-0.034	-0.063	-0.109	-0.204	-0.569
R = 0.74	0.021	-0.000	-0.037	-0.120	-0.476
R = 0.76	0.079	0.067	0.041	-0.027	-0.365
R = 0.78	0.142	0.140	0.126	0.075	-0.236
R = 0.80	0.208	0.216	0.216	0.184	-0.094
R = 0.82	0.277	0.298	0.313	0.304	0.066
R = 0.84	0.348	0.383	0.416	0.432	0.243
R = 0.86	0.420	0.469	0.521	0.567	0.439
R = 0.88	0.489	0.554	0.629	0.710	0.659
R = 0.90	0.555	0.637	0.736	0.860	0.909
R = 0.92	0.614	0.714	0.842	1.016	1.193
R = 0.94	0.667	0.787	0.944	1.177	1.510
R = 0.96	0.717	0.856	1.047	1.343	1.852
R = 0.98	0.766	0.926	1.149	1.509	2.194
R = 1.00	0.822	1.001	1.255	1.674	2.510

CHAPTER 5

THERMOELASTICITY

The subject of thermoelasticity has been pursued by many eminent investigators and mostly their contributions have been summarized in several recent books.[58–61] The materials in this chapter are largely those not found in the books mentioned.

A. SEMIINFINITE SOLID

Figure 5.1 shows a semiinfinite elastic solid with a rectangular heat source. The remaining surfaces are insulated. The surface displacement normal to the surface is sought. As before, the solution is expressed as an integral, the kernel of which is the fundamental solution applicable to any geometry over which an arbitrary source of $q(x_1, x_2)$ is given. Equation 1.1, in the absence of inertia forces and together with [1.15], is expressible as

$$ v_{i,kk} + (\delta^2 - 1)v_{k,ki} = b\,\phi_{,i} \,, \qquad [5.1] $$

where $v_i \equiv u_i/l$, $\delta^2 - 1 \equiv (\lambda + \mu)/\mu$, $b \equiv (3\lambda + 2\mu)\alpha q_0 l/K$, $\phi \equiv KT/q_0 L$, $\beta \equiv a/l$, $p \equiv q/q_0$, and $(,i)$ denotes differentiation with respect to

[116]

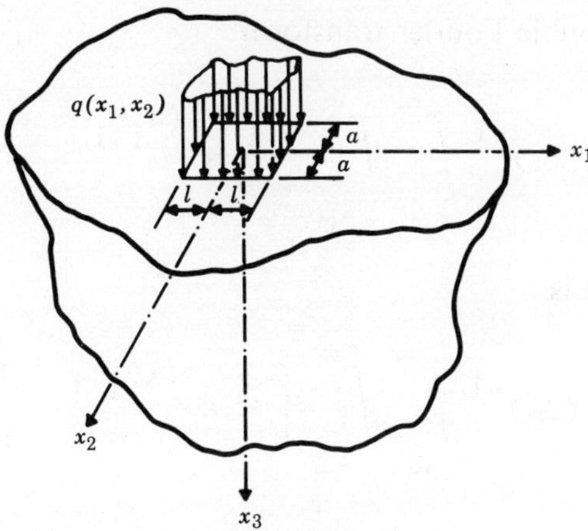

Figure 5.1

$\xi_i \equiv x_i/l$. The stress-displacement relationships are

$$\tau_{ij} = v_{i,j} + v_{j,i} + [(\delta^2 - 2)v_{k,k} - b\phi]\delta_{ij} \quad , \quad [5.2]$$

where $\tau_{ij} \equiv \sigma_{ij}/\mu$, while ϕ satisfies

$$\phi_{,kk} = 0 \quad . \quad\quad [5.3]$$

The boundary conditions are

$$\tau_{12} = \tau_{13} = \tau_{33} = 0 \qquad (\xi_3 = 0) \quad , \quad\quad\quad [5.4]$$

$$-\frac{\partial\phi}{\partial\xi_3} = \begin{cases} p(\xi_1,\xi_2) & (|\xi_1| \leq 1,\ |\xi_2| \leq \beta,\ \xi_3 = 0) \\ 0 & (\text{elsewhere on } \xi_3 = 0) \quad , \end{cases} \quad [5.5]$$

$$\tau_{ij},\ v_i,\ \phi \to 0 \qquad [(\xi_1^2 + \xi_2^2 + \xi_3^2)^{1/2} \to \infty] \quad . \quad [5.6]$$

Taking the double Fourier transform

$$\mathcal{F}\{\} = (2\pi)^{-1} \int_{-\infty}^{\infty} \int_{-\infty}^{\infty} \{\} \, e^{i(a_1\xi_1 + a_2\xi_2)} \, d\xi_1 d\xi_2 \quad , \tag{5.7}$$

whose inverse is

$$\mathcal{F}^{-1}\{\} = (2\pi)^{-1} \int_{-\infty}^{\infty} \int_{-\infty}^{\infty} \{\} \, e^{-i(a_1\xi_1 + a_2\xi_2)} \, da_1 da_2 \quad , \tag{5.8}$$

and denoting the transformed quantities by a super $\sim$, [5.1] through [5.6] become, respectively,

$$\frac{d^2\tilde{v}_1}{d\xi_3^2} - (\delta^2 a_1^2 + a_2^2)\tilde{v}_1 - (\delta^2 - 1)a_1 a_2 \tilde{v}_2 - i(\delta^2 - 1)a_1 a_3 \frac{d\tilde{v}_3}{d\xi_3} = -iba_1$$

$$\frac{d^2\tilde{v}_2}{d\xi_3^2} - (a_1^2 + \delta^2 a_2^2)\tilde{v}_2 - (\delta^2 - 1)a_1 a_2 \tilde{v}_1 - i(\delta^2 - 1)a_1 a_3 \frac{d\tilde{v}_3}{d\xi_3} = iba_2$$

$$\frac{d^2\tilde{v}_3}{d\xi_3^2} - a^2\tilde{v}_3 - i(\delta^2 - 1)a_3^2 \frac{d\tilde{v}_3}{d\xi_3} - (\delta^2-1)a_1 a_3\tilde{v}_1 - (\delta^2-1)a_2 a_3\tilde{v}_2 = -iba_3\tilde{\phi} \tag{5.9}$$

$$\tilde{\tau}_{13} = \frac{d\tilde{v}_1}{d\xi_3} - i a_1 \tilde{v}_3$$

$$\tilde{\tau}_{23} = \frac{d\tilde{v}_2}{d\xi_3} - i a \tilde{v}_3$$

$$\tilde{\tau}_{33} = \delta^2 \frac{d\tilde{v}_3}{d\xi_3} - (\delta^2-2)(-i a_1 \tilde{v}_1 - i a_2 \tilde{v}_2 - b\tilde{\phi}) \quad , \tag{5.10}$$

$$\frac{d^2\tilde{\phi}}{d\xi_3^2} - a^2\tilde{\phi} = 0 \quad , \tag{5.11}$$

$$\tilde{\tau}_{12} = \tilde{\tau}_{23} = \tilde{\tau}_{33} = 0 \qquad (\xi_3 = 0) \quad , \quad [5.12]$$

$$-\frac{d\tilde{\phi}}{d\xi_3} = \tilde{p} \equiv (2\pi)^{-1} \int_{-1}^{1} \int_{-\beta}^{\beta} p(\xi_1,\xi_2) e^{i(a_1\xi_1 + a_2\xi_2)} d\xi_1 d\xi_2 \qquad (\xi_3 = 0) \quad ,$$

$$[5.13]$$

$$\tilde{\tau}_{ij} \, , \, \tilde{v}_i \, , \, \tilde{\phi} \to 0 \qquad (\xi_3 \to \infty) \quad , \qquad [5.14]$$

where $a^2 = a_1{}^2 + a_2{}^2$.

Equations 5.9 and 5.11, subject to [5.14], lead to the solution

$$\tilde{\phi} = C e^{-a\xi_3}$$

$$\tilde{v}_1 = (A_1 + D a_1 \xi_3) e^{-a\xi_3}$$

$$\tilde{v}_2 = (A_2 + D a_2 \xi_3) e^{-a\xi_3}$$

$$\tilde{v}_3 = (A_3 - i a D \xi_3) e^{-a\xi_3} \quad , \qquad [5.15]$$

where $D = [(1 - \delta^2)/(r + \delta^2)a] \{a_1 A_1 + a_2 A_2 - iaA_3 - [ibC/(\delta^2 - 1)]\}$.
Enforcement of [5.12] and [5.13] on [5.15] leads to $C = \tilde{p}/a$.

$$A_1 = \frac{i b \tilde{q} a_1}{2(\delta^2 - 1)a^2} \, , \quad A_2 = \frac{i b q a_2}{2(\delta^2 - 1)a^2} \, , \quad A_3 = -\frac{b \tilde{q}}{2(\delta^2 - 1)a^2} \, . \quad \text{Thus}$$

$$\tilde{v}_3 = -\frac{b \tilde{q}}{2(\delta^2 - 1)a^2} e^{-a\xi_3} \quad ,$$

and, upon application of [5.8], for $\xi_3 = 0$,

$$v_3(\xi_1,\xi_2,0) = \int_{-1}^{1} \int_{-\beta}^{\beta} p(\xi_1',\xi_2') K(\xi_1,\xi_2;\xi_1',\xi_2') d\xi_1' d\xi_2' \quad ,$$

$$[5.16]$$

where

$$K(\xi_1,\xi_2;\xi_1{}',\xi_2{}') = -\frac{b}{8\pi^2(\delta^2-1)} \int_{-\infty}^{\infty}\int_{-\infty}^{\infty} \frac{1}{a^2} e^{-i[a_1(\xi_1-\xi_1{}')+a_2(\xi_2-\xi_2{}')]} \, da_1 da_2$$

$$= \frac{b}{4\pi(\delta^2-1)} \ln\sqrt{(\xi_1-\xi_1{}')^2 + (\xi_2-\xi_2{}')^2} \quad . \qquad [5.17]$$

Again, [5.17] is the fundamental solution and, as such, the area integral in [5.16] may be arbitrary.

B. EFFECT OF INERTIA

Equations 5.1 in Section 5.A are the equations of equilibrium and therefore are ones in which inertia is neglected. Figure 5.2 represents a simple model in which the flat surface of a semi-infinite solid has a prescribed temperature,

$$T_0(t) = \begin{cases} T_0 t/t_0 & (0 \le t \le t_0) \\ T_0 & (t_0 \le t \le \infty) \end{cases} , \qquad [5.18]$$

while the initial temperature is zero. For this example[62] the surface displacement is sought where the inertia effect is considered. Define the following variables, all except the first two of which are dimensionless: the speed of propagation of dilatational waves $C \equiv \sqrt{(\lambda + \mu)/\rho} \equiv \sqrt{[2(1 - \nu)]\mu/[(1 - 2\mu)\rho]}$, $a \equiv \kappa/C$, $\xi \equiv x_1/a$, $\tau \equiv \kappa t/a^2$, $\phi \equiv T/T_0$, $u \equiv [(1 - \nu)u_1]/[(1 + \nu)\alpha T_0 a]$, $\tau_{11} \equiv [(1 - 2\nu)\sigma_{11}]/[2(1 + \nu)\alpha T_0\mu]$ and $\tau_{22} \equiv [(1 - 2\nu)(1 - \nu)\sigma_{22}]/[2(1 + \nu)\alpha T_0\mu]$.

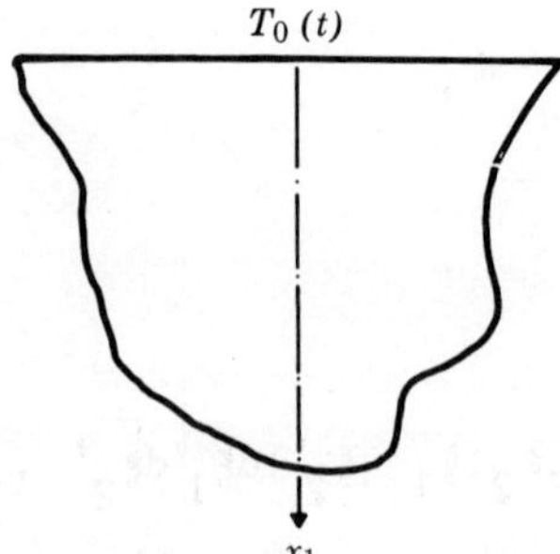

Figure 5.2

The heat equation in one dimension, the initial condition and boundary conditions are

$$\frac{\partial^2 \phi}{\partial \xi^2} = \frac{\partial \phi}{\partial \tau} \quad , \qquad\qquad\qquad\qquad [5.19]$$

$$\phi = 0 \qquad\qquad (\xi = \xi, \ \tau = 0) \quad , \qquad [5.20]$$

$$\phi = \phi(\tau) \equiv \begin{cases} \tau/\tau_0 & (\xi = 0, \ 0 \le \tau \le \tau_0) \\ 1 & (\xi = 0, \ \tau_0 \le \tau < \infty) \end{cases} \quad , \quad [5.21]$$

$$\phi \to 0 \qquad\qquad (\xi = \infty) \quad . \qquad\qquad [5.22]$$

The equation of motion in one dimension is

$$\frac{\partial^2 u}{\partial \xi^2} = \frac{\partial^2 u}{\partial \tau^2} + \frac{\partial \phi}{\partial \xi} \quad , \qquad\qquad [5.23]$$

with the stress-displacement relationships

$$\tau_{11} = \frac{\partial u}{\partial \xi} - \phi$$

$$\tau_{22} = \nu \frac{\partial u}{\partial \xi} - (1-\nu)\phi \quad , \qquad\qquad [5.24]$$

the initial conditions

$$u = \frac{\partial u}{\partial \tau} = 0 \qquad (\xi = \xi, \ \tau = 0) \quad , \quad [5.25]$$

and the boundary conditions

$$\tau_{11} = 0 \qquad\qquad (\xi = 0, \ \tau = \tau) \quad , \quad [5.26]$$

$$u, \ \tau_{11}, \ \tau_{22} \to 0 \qquad (\xi \to \infty) \quad . \qquad [5.27]$$

As an artifice, examine the subproblem in which the following is replacing [5.21] of the above set:[63]

$$\phi = h(\tau) \quad , \tag{5.28}$$

where $h(\tau)$ is the Heaviside step function.

The Laplace transform [2.90] of the set of equations for the subproblem are

$$\frac{d^2\tilde{\phi}}{d\xi^2} - is\tilde{\phi} = 0 \quad , \tag{5.29}$$

$$\tilde{\phi} = \frac{1}{s} \qquad (\xi = 0) \quad , \tag{5.30}$$

$$\tilde{\phi} = 0 \qquad (\xi \to \infty) \quad , \tag{5.31}$$

$$\frac{d^2\tilde{u}}{d\xi^2} - s^2\tilde{u} = \frac{d\tilde{\phi}}{d\xi} \quad , \tag{5.32}$$

$$\tilde{\tau}_{11} = \frac{d\tilde{u}}{d\xi} - \tilde{\phi} \quad , \tag{5.33}$$

$$\tilde{\tau}_{11} = 0 \qquad (\xi = 0) \quad , \tag{5.34}$$

$$\tilde{\tau}_{11}, \tilde{\tau}_{22}, \tilde{u} \to 0 \qquad (\xi \to \infty) \quad , \tag{5.35}$$

where the super $\sim$ denotes transformed quantities.

Equations 5.29 through 5.35 possess a solution, which upon inversion [2.91], yields for the surface

$$u_s(\tau) = h(\tau)(1-e^\tau) - 2\sqrt{\frac{\tau}{\pi}} + \frac{e^\tau}{2}[\mathrm{erf}\, c(-\sqrt{\tau}) - \mathrm{erf}\, c(\sqrt{\tau})] \quad .$$

$$\tag{5.36}$$

By convolution the actual problem (i.e., that which fulfills [5.21] can be obtained from [5.36])

$$u(\tau) = \int_0^\tau \frac{d\phi(\lambda)}{d\lambda} \, u_s(\tau - \lambda) d\lambda$$

or

$$u(\tau) = u'(\tau) \equiv \frac{h(\tau)}{\tau_0} \left[1 + \tau - e^\tau \operatorname{erf} c \sqrt{\tau} - \frac{2}{3}(3 + 2\tau) \sqrt{\frac{\tau}{\pi}} \right]$$

$$(0 \leq \tau \leq \tau_0) \ , \quad [5.37]$$

$$u(\tau) = u'(\tau) - u'(\tau - \tau_0) \qquad (\tau_0 < \tau < \infty) \ . \qquad [5.38]$$

It is noted in [5.37] and [5.38] that $u(\tau)$ for $\tau_0 = 0$ is zero. In other words, the quasistatic solution gives no surface displacement.

C.   EFFECT OF THERMOELASTIC COUPLING

Section 5.A above involved the thermoelastic equations in an uncoupled fashion. Section 5.B above discussed the effect of inertia. An examination of [1.1] in the absence of body forces for simplicity, [1.15] [1.19] in the absence of heat generation for simplicity, and [4.1] shows that the thermoelastic equations are actually coupled (i.e., they have to be solved simultaneously). These equations are

$$T_{,ii} = \kappa^{-1} \dot{T} + \eta \, \dot{u}_{i,i} \ , \qquad [5.39]$$

where $\eta = (3\lambda + 2\mu)T_0/K$ and

$$(\lambda + \mu) u_{k,ki} + \mu \, u_{i,kk} = \rho \, \ddot{u}_i + (3\lambda + 2\mu)\alpha \, T_{,i} \ . \quad [5.40]$$

Of course, [5.39] implies isotropic heat conduction.

Defining $\Omega_i = \epsilon_{ijk}u_{k,j}$ (i.e., the curl of $\vec{u}$), [5.40] becomes

$$\Omega_{i,kk} = C_2^2 \ddot{\Omega}_i \ , \qquad [5.41]$$

where $C_2^2 = \mu/\rho$. The C_2 is the speed of propagation of shear waves.

Note, thermal effects do not enter in the case of shear waves since there is no associated volume change. On the other hand, defining $e = u_{k,k}$ (i.e., divergence of $\vec{u}$), [5.40] becomes

$$e_{,kk} = C_1^2 \ddot{e} + m\,T_{,kk} \quad , \qquad [5.42]$$

where $C_1^2 = (\lambda + 2\mu)/\rho$ and $m = (3\lambda + 2\mu)\alpha/(\lambda + 2\mu)$. The C_1 is the speed of propagation of dilatational waves.

Consider now the example of a harmonic plane wave.[64] For this, the following solutions to [5.41] and [5.42] are assumed:

$$e = e^{\circ}\,\exp(i\,\omega\,t + k\,x_i\,n_i)$$

$$T = T^{\circ}\,\exp(i\,\omega\,t + k\,x_i\,n_i) \quad , \qquad [5.43]$$

where n_i represents the ith component of the unit vector. The phase velocity of the waves C is given by

$$C = \omega/\,\mathcal{I}m\,(k) \quad , \qquad [5.44]$$

and the attenuation ϑ is given by

$$\vartheta = \mathcal{R}e\,(k) \quad . \qquad [5.45]$$

Equations 5.43 into 5.39 and 5.42 lead to

$$k^4 + k^2[\sigma^2 - q(1+\epsilon)] - \sigma^2 q = 0 \quad , \qquad [5.46]$$

where $q = i\omega/\kappa$, $\sigma^2 = \omega^2/C_1^2$, and $\epsilon = \eta\kappa M$.

The roots of [5.46] are $\pm k_1$ and $\pm k_2$, where

$$k_{1,2}^2 = \frac{C_1^2}{2\kappa^2}\,[-\lambda^2 + i\,\lambda(1+\epsilon) \pm \Delta] \quad , \qquad [5.47]$$

$$\Delta = \left\{ \lambda^2[\lambda^2 - (1+\epsilon)^2] + 2i\,\lambda^3(1-\epsilon) \right\}^{1/2} \quad .$$

Here $\lambda = \omega/\omega^*$ and $\omega^* = C_1^2/\kappa$. For $\epsilon = 0$, the case of the uncoupled problem, $k_1 = \sqrt{(q/2)}\,(1 + i)$ and $k_2 = i\sigma$ correspond to a pure thermal wave and elastic wave, respectively. For $\epsilon \neq 0$, k_1 and k_2 correspond to a modified thermal wave and elastic wave, respectively, for the coupled case. Equation 5.47 should be studied numerically in general. However, for $\lambda \gg 1$ (i.e., corresponding to high frequencies)

$$k_1 = \pm \sqrt{\frac{q}{2}} \; [\,(1 - \frac{\epsilon}{2\lambda}) + i(1 + \frac{\epsilon}{2\lambda})\,]$$

$$k_2 = \pm \sigma(\frac{\epsilon}{2} + i\,\lambda) \quad , \qquad\qquad [5.48]$$

and for $\lambda \ll 1$ (i.e., corresponding to low frequencies)

$$k_1 = \pm \sqrt{\frac{q}{2}} \; (\frac{1+\epsilon}{2})^{1/2} \left\{ [\,1 - \frac{\lambda \epsilon}{2(1+\epsilon)^2}\,] + i[\,1 + \frac{\lambda \epsilon}{2(1+\epsilon)^2}\,] \right\}$$

$$k_2 = \pm \sigma \, [\, \frac{\lambda^2 \epsilon}{2(1+\epsilon)^{5/2}} + \frac{i\lambda}{(1+\epsilon)^{1/2}} \,] \quad . \qquad\qquad [5.49]$$

D. CONVECTIVE HALF-SPACE WITH MOVING HEAT SOURCE—TWO-DIMENSIONAL CASE

Now, having seen two examples on the effect of inertia and coupling of the generalized heat equation with the equations of motion, respectively, discussions now center on the uncoupled theory of thermoelasticity as in Section 5.A. For this section, the problem of Section 2.A is solved considering the body to be in elastic and thermal strain.[65] For the thermal part, see Figure 2.1; [2.2] through [2.4] define the problem in dimensionless form. Mechanically, using the notations in Section 4.A where applicable and $\beta = \alpha q_0 l / K$, the equations of equilibrium, the stress-displacement relations and the boundary conditions are, respectively,

$$(r+2) \frac{\partial^2 u}{\partial \xi^2} + \frac{\partial^2 u}{\partial \eta^2} + (r+1) \frac{\partial^2 v}{\partial \xi \partial \eta} = (3r+2)\beta \frac{\partial \phi}{\partial \xi}$$

$$\frac{\partial^2 v}{\partial \xi^2} + (r+2) \frac{\partial^2 v}{\partial \eta^2} + (r+1) \frac{\partial^2 u}{\partial \xi \partial \eta} = (3r+2)\beta \frac{\partial \phi}{\partial \eta} \quad , [5.50]$$

$$\sigma_{\xi} = (r+2)\,\frac{\partial u}{\partial \xi} + r\,\frac{\partial v}{\partial \eta} - (3r+2)\beta\,\phi$$

$$\sigma_{\eta} = r\,\frac{\partial u}{\partial \xi} + (r+2)\,\frac{\partial v}{\partial \eta} - (3r+2)\beta\,\phi$$

$$\sigma_{\xi\eta} = \frac{\partial u}{\partial \eta} + \frac{\partial v}{\partial \xi}\;, \qquad\qquad [5.51]$$

$$\sigma_{\xi\eta} = \sigma_{\eta} = 0 \quad (\xi = \xi,\ \eta = 0)\;, \qquad\qquad [5.52]$$

$$\sigma_{\xi}\;,\ \sigma_{\eta}\;,\ \sigma_{\xi\eta}\;,\ u,\ v \to 0 \quad [(\xi^{2} + \eta^{2})^{1/2} \to \infty]\;. \quad [5.53]$$

Returning to Section 2.A and [2.2], which is the heat equation, the case for $R \equiv Vl/\kappa \gg 1$ is considered. For this case $\partial^{2}\phi/\partial\xi^{2}$ may be neglected and [2.2] becomes

$$\frac{\partial^{2}\phi}{\partial\eta^{2}} = R\,\frac{\partial\phi}{\partial\xi}\;, \qquad\qquad [5.54]$$

subject to the boundary conditions

$$\frac{\partial\phi}{\partial\eta} = \begin{cases} -P(\xi) & (0 \le |\xi| \le 1,\ \eta = 0) \\[2ex] H\phi & (1 < |\xi| < \infty,\ \eta = 0) \end{cases}\;, \quad [5.55]$$

$$\phi \to 0 \qquad [(\xi^{2} + \eta^{2})^{1/2} \to \infty]\;. \qquad\qquad [5.56]$$

Taking the Fourier transform [2.5] and denoting the transformed quantities by a superposed $\sim$, [5.50] through [5.56] become, with some grouping

$$\frac{\partial^2 \tilde{u}}{\partial \eta^2} - s^2(r+2)\tilde{u} - is(r+1)\frac{\partial \tilde{v}}{\partial \eta} = -is(3r+2)\beta\tilde{\phi}$$

$$(r+2)\frac{\partial^2 \tilde{v}}{\partial \eta^2} - s^2\tilde{v} - is(r+1)\frac{\partial \tilde{u}}{\partial \eta} = (3r+2)\frac{\partial \tilde{\phi}}{\partial \eta} \quad , \quad [5.57]$$

$$\tilde{\sigma}_\xi = -is(r+2)\tilde{u} + r\frac{\partial \tilde{v}}{\partial \eta} - (3r+2)\beta\tilde{\phi}$$

$$\tilde{\sigma}_\eta = -isr\tilde{u} + (r+2)\frac{\partial \tilde{v}}{\partial \eta} - (3r+2)\beta\tilde{\phi}$$

$$\tilde{\sigma}_{\xi\eta} = \frac{\partial \tilde{u}}{\partial \eta} - is\tilde{v} \quad , \qquad\qquad [5.58]$$

$$\frac{\partial^2 \tilde{\phi}}{\partial \eta^2} = -isR\tilde{\phi} \quad , \qquad\qquad [5.59]$$

$$\tilde{\sigma}_{\xi\eta} = \tilde{\sigma}_\eta = 0 \qquad (\eta = 0) \quad , \qquad [5.60]$$

$$-\frac{\partial \phi}{\partial \eta} = \tilde{P} \equiv \pi^{-1/2} \int_{-\infty}^{\infty} P^*(\xi)e^{is\xi}d\xi \qquad (\eta = 0) \quad , \quad [5.61]$$

$$\tilde{\sigma}_\xi \, , \, \tilde{\sigma}_\eta \, , \, \tilde{\sigma}_{\xi\eta} \, , \, \tilde{u}, \, \tilde{v}, \, \tilde{\phi} \to 0 \qquad (\eta \to \infty) \quad , \qquad [5.62]$$

where

$$P^*(\xi) \equiv \begin{cases} P(\xi) & (0 \le |\xi| \le 1) \\ -H\phi(\xi,0) & (1 < |\xi| < \infty) \end{cases} \quad . [5.63]$$

First, the heat equation is solved. Let $\tilde{\phi} = \tilde{\phi}_1 + i\tilde{\phi}_2$ and $\tilde{P} = \tilde{P}_1 + i\tilde{P}_2$, where $\tilde{\phi}_1$, $\tilde{\phi}_2$, $\tilde{P}_1$, and $\tilde{P}_2$ are real. Equations 5.59, 5.61, and 5.62 lead to, for $s > 0$,

$$\tilde{\phi}_1 = (2sR)^{-1/2} e^{-\sqrt{\frac{sR}{2\eta}}} [(\tilde{P}_1 - \tilde{P}_2)\cos\sqrt{\frac{sR}{2\eta}} - (\tilde{P}_1 + \tilde{P}_2)\sin\sqrt{\frac{sR}{2\eta}}]$$

$$\tilde{\phi}_2 = (2sR)^{-1/2} e^{-\sqrt{\frac{sR}{2\eta}}} [(\tilde{P}_1 + \tilde{P}_2)\cos\sqrt{\frac{sR}{2\eta}} + (\tilde{P}_1 - \tilde{P}_2)\sin\sqrt{\frac{sR}{2\eta}}] \quad .$$

$$[5.64]$$

For $s < 0$, [5.64] may be used, provided s is replaced by $n \equiv -s$ and a negative sign is added to $\tilde{\phi}_2$.

Proceed now to the solution of the displacement equations. From [5.57]

$$i s \tilde{u} = s^{-2}(r+1)^{-1}\frac{\partial^3 \tilde{v}}{\partial\eta^3} + r(r+1)^{-1}\frac{\partial \tilde{v}}{\partial\eta} - s^{-2}(r+1)m\frac{\partial^2 \tilde{\phi}}{\partial\eta^2} - m\tilde{\phi} \quad ,$$

$$[5.65]$$

and

$$\frac{\partial^4 \tilde{v}}{\partial\eta^4} - 2s^2\frac{\partial^2 \tilde{v}}{\partial\eta^2} + s^4\tilde{v} = m\frac{\partial^3 \tilde{\phi}}{\partial\eta^3} - s^2 m\tilde{\phi} \quad , \quad [5.66]$$

where

$$m = \frac{(3r+2)\beta}{r+2} \quad .$$

With [5.64] in [5.66], the latter may be solved and the complimentary solution is

$$\tilde{v}_c = (A_1 + i A_2)e^{-s\eta} + (B_1 + i B_2)\eta e^{-s\eta} \quad , \quad [5.67]$$

where A_1, A_2, B_1, and B_2 are constants of integration, and condition 5.62 has been taken into consideration without the benefit of the details concerning the coefficients.

For the particular integral,

$$\tilde{v}_p = m(s^2 + R^2)^{-1} e^{-\sqrt{\frac{sR}{2\eta}}} \left\{ [\tilde{P}_1 + (\frac{R}{s})\tilde{P}_2] \right.$$

$$+ i[-(\frac{R}{s})\tilde{P}_1 + \tilde{P}_2] \right\} \cos\sqrt{\frac{sR}{2\eta}}$$

$$+ \left\{ [(\frac{R}{s})\tilde{P}_1 - \tilde{P}_2] + i[\tilde{P}_1 + (\frac{R}{s})\tilde{P}_2] \right\} \sin\sqrt{\frac{sR}{2\eta}} \quad .$$

$$[5.68]$$

Enforcement of [5.60] leads to

$$A_1 = (r+1)^{-1} s^{-1} \left\{ [s - (r+2)\sqrt{\frac{s}{2R}}(R+s)]\tilde{q}_1 \right.$$

$$+ [R - (r+2)\sqrt{\frac{s}{2R}}(R-s)]\tilde{q}_2 \right\} \quad ,$$

$$A_2 = (r+1)^{-1} s^{-1} \left\{ [-R + (r+2)\sqrt{\frac{s}{2R}}(R-s)]\tilde{q}_1 \right.$$

$$+ [s - (r+2)\sqrt{\frac{s}{2R}}(R+s)]\tilde{q}_2 \right\} \quad ,$$

$$B_1 = [s - \sqrt{\frac{s}{2R}}(R+s)]\tilde{q}_1 + [R - \sqrt{\frac{s}{2R}}(R-s)]\tilde{q}_2 \quad ,$$

$$B_2 = [-R + \sqrt{\frac{s}{2R}}(R-s)]\tilde{q}_1 + [s - \sqrt{\frac{s}{2R}}(R+s)]\tilde{q}_2 \quad ,$$

$$[5.69]$$

where $\tilde{q}_1 = m\tilde{P}_1/(s^2 + R^2)$ and $\tilde{q}_2 = m\tilde{P}_2/(s^2 + R^2)$.

For $s < 0$, with $n \equiv -s$, A_1 and B_1 take on the same form as in [5.69] with n replacing s; A_2 and B_2 take on the same form as in [5.69] with n replacing s and multiplied by a negative sign.

Inversion [2.12] of [5.64] leads to

$$\phi_0 = \pi^{-1}\sqrt{\frac{2}{R}} \int_{-\infty}^{\infty} P^*(\xi')d\xi' \int_0^{\infty} [\cos(\xi-\xi')s + \sin(\xi-\xi')s]\frac{ds}{\sqrt{s}} \quad , \qquad [5.70]$$

where ϕ_0 denoted ϕ for $\eta \to 0$. Equation 5.70 may be shown to be

$$\phi_0(\xi) = 2(\pi R)^{-1/2} \int_{-\infty}^{\infty} P^*(\xi')K_\phi^*(\xi,\xi')d\xi' \quad , \qquad [5.71]$$

where

$$K_\phi^* = \begin{cases} (\xi - \xi')^{1/2} & [(\xi - \xi') \geq 0] \\ 0 & [(\xi - \xi') < 0] \end{cases} \quad . \quad [5.72]$$

Recalling the definition of $P^*(\xi)$ in [5.63],

$$\phi_0(\xi) = \begin{cases} 0 & (-\infty \leq \xi < -1) \\ 2(\pi R)^{-1/2} \int_{-1}^{\xi} P(\xi')(\xi-\xi')^{-1/2}d\xi' & (-1 \leq \xi \leq 1) \\ 2(\pi R)^{-1/2} \int_{-1}^{1} P(\xi')(\xi-\xi')^{-1/2}d\xi' - H\int_1^{\xi} \phi_0(\xi')(\xi-\xi')^{1/2}d\xi' & \end{cases}$$

$$(1 < \xi \leq \infty) \quad . \qquad [5.73]$$

Coming now to the inverse transform of the displacement component $(v = v_1 + v_2)$, which may be written as

$$\tilde{v}_1(s,0) = m(r+2)(r+1)^{-1}s^{-1}(s^2+R^2)^{-1}$$

$$\cdot \left\{ [s - \sqrt{\frac{s}{2R}}(R+s)]\tilde{P}_1 + [R - \sqrt{\frac{s}{2R}}(R-s)]\tilde{P}_2 \right\} \qquad s \geq 0$$

$$\tilde{v}_2(s,0) = m(r+2)(r+1)^{-1}s^{-1}(s^2+R^2)^{-1}$$

$$\cdot \left\{ [-R + \sqrt{\frac{s}{2R}}(R-s)]\tilde{P}_1 + [s - \sqrt{\frac{s}{2R}}(R+s)]\tilde{P}_2 \right\} \qquad s \geq 0 \quad . \quad [5.74]$$

For $s < 0$, s in the first of [5.74] and in the negative of the second of [5.74] should be replaced by n. As in the treatment of ϕ_0, $v_0(\xi)$ which is $v(\xi, \eta)$ for $\eta \to 0$ may be shown to be[64]

$$v_0(\xi) = m R^{-1} (r+2)(r+1)^{-1} \int_{-1}^{\infty} P^*(\xi')K^*(\xi,\xi')d\xi' \quad , \qquad [5.75]$$

where

$$K^* = \begin{cases} -2 & [(\xi-\xi') \geq 0] \\ -2e^{-R(\xi-\xi')} \operatorname{erf} c \sqrt{R(\xi-\xi')} & [(\xi-\xi') \leq 0] \end{cases} . \qquad [5.76]$$

Equation 5.76 is the asymptotic form of [2.14]. A problem for constant temperature boundary condition over a finite length and impervious surface outside the heat source has been solved.[66]

E.　ELASTIC HALF-SPACE WITH MOVING HEAT SOURCE—THREE-DIMENSIONAL CASE

Figure 5.3 shows the three-dimensional counterpart of the problem in Section 5.D and shown in Figure 2.1 except that no convection is to take place on the surface $x_3 = 0$. Using the same notation as before

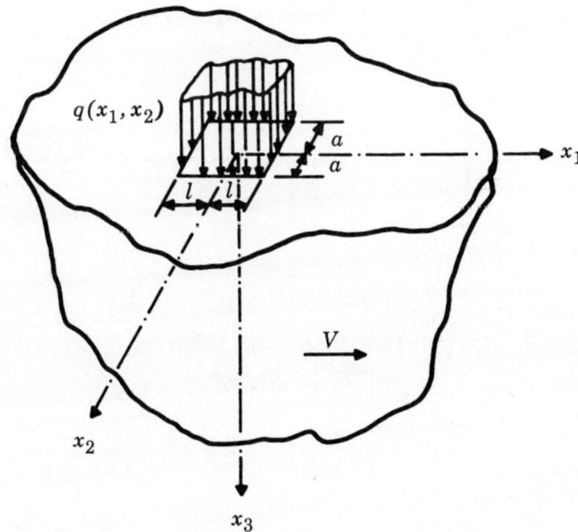

Figure 5.3

(e.g., $w \equiv u_3/l$, $p \equiv q/q_0$, where l is some characteristic length), the displacement at the surface in dimensionless form is expressible as

$$w(\xi,\eta,0) = \int_A p(\xi',\eta')K(\xi,\eta;\xi',\eta')d\xi'\,d\eta' \quad , \qquad [5.77]$$

where the integration is over that area A over which q is applied. As such $K(\xi, \eta; \xi', \eta')$ is the fundamental solution. The K may be found directly as in Section 5.D. It may also be found in parts, making use of formulae already derived. As a first part, consider the problem of a moving, point heat-source of intensity q_0 in an infinite space. A procedure, using transform method, leads to the dimensionless temperature solution

$$2\pi^{-1}\,\frac{e^{-R[r - (\xi-\xi')]}}{r} \quad , \qquad [5.78]$$

where $R \equiv Vl/\kappa$, $r \equiv [(\xi - \xi')^2 + (\eta - \eta')^2 + (\zeta - \zeta')^2]^{1/2}$, (ξ, η, ζ) is the point in question while (ξ', η', ζ') is the point where q_0 is located, $\xi \equiv x_1/l$, $\eta \equiv x_2/l$, $\zeta \equiv x_3/l$, $\sigma_{\xi\zeta} \equiv \sigma_{13}/\mu$, $\sigma_{\eta\zeta} \equiv \sigma_{23}/\mu$, and $\sigma_{\zeta\zeta} \equiv \sigma_{33}/\mu$. See [2.19] for the surface value of [5.78]. Also, the thermoelastic solution yields $w = \sigma_{\xi\zeta} = \sigma_{\eta\zeta} = 0$ at ζ while

$$\sigma_{\zeta\zeta}(\xi,\eta,0;\xi',\eta') = -\frac{M}{Rr[r-(\xi-\xi')]}\Bigg[e^{-R[r-(\xi-\xi')]}\left\{1 - \frac{(\eta-\eta')^2}{r^2}[1-Rr+\frac{r}{r-(\xi-\xi')}]\right.$$

$$\left. - \frac{r-(\xi-\xi')}{r}[\frac{\xi-\xi'}{r} + R(r-(\xi-\xi'))]\right\}$$

$$- \left\{1 - \frac{(\eta-\eta')^2}{r^2}[1 + \frac{r}{r-(\xi-\xi')}] + \frac{(\xi-\xi')[r-(\xi-\xi')]}{r^2}\right\}\Bigg] \quad ,$$

$$[5.79]$$

where $M \equiv [\beta(3\lambda + 2\mu)]/[4\pi(\lambda + 2\mu)]$ and $\beta \equiv q_0\alpha l/\kappa$.

As a second part, recalling the Boussinesq problem, [4.15] and [4.16], $K(\xi, \eta; \xi', \eta')$ in [5.77] is simply

$$K(\xi,\eta;\xi',\eta') = \int_{-\infty}^{\infty}\int_{-\infty}^{\infty} \sigma(\xi'',\eta'';\xi',\eta')K_1(\xi,\eta;\xi'',\eta'')d\xi''d\eta'' \quad ,$$

$$[5.80]$$

where K_1 is given by [4.16].

CHAPTER 6

VISCOELASTICITY

A. MODELS

Materials which exhibit time-dependent behavior are termed visco-elastic materials. The emphasis here is in the more solid-like materials (the more fluid-like ones are sometimes called elasticoviscous materials). Typical examples of the mechanical behavior which show time dependencies are creep and stress-relaxation. For isotropic and homogeneous elastic material [1.15] gives the constitutive relationships

$$\sigma_{ij} = 2\mu\, \epsilon_{ij} + (\lambda e - \gamma T)\delta_{ij} \; , \qquad [6.1]$$

where $\gamma = (3\lambda + 2\mu)\alpha$ and $e = \epsilon_{kk}$.

Letting compressibility $k \equiv \lambda + 2\mu$, $s \equiv \sigma_{kk}/3$, and $e \equiv \epsilon_{kk}/3$, [6.1] may be replaced by

$$s_{ij} = 2\mu\, e_{ij} \; , \qquad s = k e - \gamma T \; , \qquad [6.2]$$

where $s_{ij} = \sigma_{ij} - \tfrac{1}{3}s\delta_{ij}$, $e_{ij} = \epsilon_{ij} - \tfrac{1}{3}e\delta_{ij}$. For the simple Hookean material s_{ij} is linearly proportional to e_{ij}. The simplest relationship between the stresses and strains in a viscoelastic material is derived by adding to Hooke's law a term which represents Newton's viscosity. Thus

$$s_{ij} = 2\mu(e_{ij} + \vartheta\, \dot{e}_{ij}) \; , \qquad [6.3]$$

[133]

where $\vartheta = \eta/\mu$ is the retardation time and η is the Newtonian viscosity coefficient. For pure compression or tension, the body behaves elastically, that is,

$$s = ke - \gamma T \ . \qquad [6.4]$$

Equations 6.3 and 6.4 describe the Kelvin-Voigt model. Mechanically [6.3] amounts to an elastic spring and a dashpot in parallel. If they are arranged in series,

$$\dot{s}_{ij} + \frac{s_{ij}}{v} = 2\mu\, \dot{e}_{ij} \ , \qquad [6.5]$$

applies and $v = \mu/\eta$ is the relaxation time. Equations 6.5 and 6.4 are to be used together in the Maxwell model.

More general linear models may be written:

$$P_1 s_{ij} = P_2 e_{ij} \qquad [6.6]$$

$$P_3 s = P_4 (e - \alpha T) \ , \qquad [6.7]$$

where $P_i (i = 1, 2, 3, 4)$ are linear differential operators

$$P_i = \sum_{n=0}^{N_i} a_i^{(n)} \left(\frac{\partial}{\partial t} \right)^n \qquad (a_i^{(N_i)} \neq 0) \ . \quad [6.8]$$

Equation 6.6 with [6.8] are as shown by [1.28].

Alternatively, integral laws of Boltzmann and Biot may be convenient to use. They appear in continuous spectra

$$\sigma_{ij} = 2 \int_0^t a(t-\tau)\dot{\epsilon}_{ij}\, d\tau + \delta_{ij} \int_0^t \left\{ b(t-\tau)\dot{e} \right.$$

$$\left. - [3b(t-\tau) + 2a(t-\tau)] \alpha \dot{T} \right\} d\tau \ , \qquad [6.9]$$

where the super dot refers to differentiation with respect to τ.

B. ELASTICITY-VISCOELASTICITY ANALOGY

Since the constitutive relationships for viscoelastic bodies are time-dependent, the stresses also vary with time, for slow variations of the stresses with time, the problem may be regarded as quasistatic, the inertia terms being neglected in the equations of motion. For equilibrium,

$$\sigma_{ij,j} = 0 \quad , \qquad\qquad [6.10]$$

and, using the strain-displacement relationship,

$$\epsilon_{ij} = \frac{1}{2} (u_{i,j} + u_{j,i}) \quad ,$$

[6.9] becomes

$$\int_0^t \left\{ a(t-\tau)\dot{u}_{i,kk} + [b(t-\tau) + a(t-\tau)]\dot{u}_{k,ki} \right\} d\tau$$

$$= a_t \int_0^t [3b(t-\tau) + 2a(t-\tau)]\dot{T}_{,i} \, d\tau \quad . \qquad [6.11]$$

Also, [6.6] and [6.7] may be put in the general form

$$P_1 P_3 \sigma_{ij} = P_2 P_3 \epsilon_{ij} + \frac{1}{3} \delta_{ij} [(P_1 P_4 - P_2 P_3)e - P_1 P_4 \alpha T] \quad .$$

$$[6.12]$$

Taking the Laplace transform (with parameters of [6.11] and [6.12] and assuming that for $t \leq 0$ the viscoelastic body is in the natural state (i.e., $u_i = \dot{u}_i = \ldots = \partial^n u_i/\partial t^n = 0$ for $x_i = x_i$ and $t = 0$), [6.11] and [6.12] become

$$\bar{\mu}(s) \bar{u}_{i,kk} + [\bar{\lambda}(s) + \bar{\mu}(s)]\bar{u}_{k,ki} = \bar{\gamma}(s) \bar{T}_{,i} \qquad [6.13]$$

where

$$\bar{\gamma}(s) = [2\bar{\mu}(s) + 3\bar{\lambda}(s)]\alpha$$

$$\bar{\mu}(s) = \frac{P_2(s)}{2P_1(s)} = s\,\bar{a}(s)$$

$$\bar{\lambda}(s) = \frac{P_1(s)\, P_4(s) - P_2(s)\, P_3(s)}{3P_1(s)\, P_3(s)} = s\,\bar{b}(s) \quad ,$$

and the super bar denotes the transformed quantities.

Laplace transform of [6.10] is

$$\bar{\sigma}_{ij} = 2\bar{\mu}\,\bar{\epsilon}_{ij} + (\bar{x}\,\bar{e} - \bar{T}\,\bar{\gamma})\delta_{ij} \quad . \qquad [6.14]$$

The boundary conditions are

$$\bar{\sigma}_{ij}\, n_j = 0 \qquad\qquad [6.15]$$

if the boundary is traction free and

$$\bar{u}_i = f_i(x_r) \qquad (r = 1,\, 2,\, 3) \quad , \qquad [6.16]$$

in the case of kinematic boundary conditions.

The counterpart of [6.13] for elastic material is

$$\bar{\mu}\,\bar{u}^{(0)}_{i,kk} + (\bar{\lambda} + \bar{\mu})\bar{u}^{(0)}_{k,ki} = \gamma\,\bar{T}_{,i} \quad , \qquad [6.17]$$

where the super (0) denotes quantities for the elastic material. It is to be noted that [6.17] and [6.13] correspond[67-70] in that these equations are the same provided the Lamé constants λ and μ are replaced by the Laplace transform of $a(t)$ and $b(t)$, each multiplied by s. Thus in solving a viscoelastic problem, for which the counterpart elastic problem has been solved, the solution may be obtained by obtaining the inverse Laplace transorm of the elastic solution with $\bar{\lambda}(s)$ and $\bar{\mu}(s)$ replacing λ and μ, respectively.

C. OTHER REPRESENTATIONS OF MECHANICAL PROPERTIES

In Sections 6.A and 6.B the mechanical properties were represented by the constants $a_i^{(n)}$ in [6.8] or a and b functions in [6.9]. Still other representations are used for easy application. First, often used are the integral representations in one dimension

$$\epsilon(t) = \int_0^t f(t-\tau)\,\frac{d\sigma(\tau)}{d\tau}\,d\tau \qquad [6.18]$$

or

$$\sigma(t) = \int_0^t E(t-\tau) \frac{de(\tau)}{d\tau} d\tau \quad , \qquad [6.19]$$

where f is the creep compliance and E is the relaxation modulus. These are most adaptable to empirical determination. Second, sometimes it is more convenient to determine the f and E functions by correlation with vibration experiments. Again, in one dimension,

$$\sigma = \sigma_0 \, e^{i\omega t} \quad , \quad \epsilon = \epsilon_0 \, e^{i\omega t} \qquad [6.20]$$

where [6.20] is represented in complex variables.

The complex modulus is defined as

$$E^*(\omega) = \frac{\sigma}{\epsilon} = \frac{\sigma_0}{\epsilon_0} = E_1(\omega) + i E_2(\omega) \quad , \quad [6.21]$$

and the complex compliance

$$f^*(\omega) = E^{-1}(\omega) \quad . \qquad [6.22]$$

These are related to P_1 and P_2, in [6.8] for instance, through

$$E^*(\omega) = \frac{\sigma}{\epsilon} = \frac{P_2(i\omega)}{P_1(i\omega)} \quad . \qquad [6.23]$$

This is so since $\partial/\partial t$ is equivalent to multiplication by $i\omega$ because of the $e^{i\omega t}$ time variation.

D.   INFLUENCE OF TEMPERATURE ON VISCOELASTIC BEHAVIOR

Equation 6.19 in three dimensions and deviatoric terms is[71]

$$s_{ij} = \int_0^t G_T(t-t') \frac{\partial e_{ij}(x,t')}{\partial t'} dt' \quad , \qquad [6.24]$$

where $x = (x_1, x_2, x_3)$ and G_T is the relaxation modulus for the stress deviators at temperature T.

Among the amorphous high polymers which satisfy approximately the linear viscoelastic laws at uniform temperature are a group which exhibit approximately a particularly simple property with change of temperature. This is a translational shift (no change in shape) of the relaxation modulus plotted against the logarithm of time at different uniform temperatures. This leads to an equivalent relation between temperature and *lnt*.

Let $E_r(lnt)$ be the relaxation modulus as a function of *lnt* at uniform temperature T. Accordingly,

$$E_T(\ell n\ t) = E_{T_0}[\ell n\ t + f(T)] \quad , \qquad [6.25]$$

where $f(T)$ is measured relative to some arbitrary temperature T_0. Figure 6.1 shows, for thermorheologically simple material, the shift $f(T)$ which is a positive increasing function for $T > T_0$. If $G_T(t)$ denotes the relaxation modulus as a function of time at uniform temperature T, so that $G_T(t) = E_T(lnt)$, and a shift factor $a(T)$ is defined by

$$a(T) = \exp[f(T)] \quad , \qquad [6.26]$$

then

$$G_T(t) = G_{T_0}(\xi) \quad , \qquad [6.27]$$

where ξ, the reduced time, is defined by

$$\xi = t \cdot a(T) \quad . \qquad [6.28]$$

For an increase in temperature above T_0, ξ is $> t$ since $a(T) > 1$.

Expressing the stress and strain as functions of x and pseudotime, the viscoelastic law becomes

$$s_{ij}(x,\xi) = \int_0^{\xi} G_T(\xi-\xi') \frac{\partial e_{ij}(x,\xi')}{\partial \xi'} d\xi' \quad . \qquad [6.29]$$

Next extension to include the general case of temperature field $T(x, t)$ will now be made. Toward this end, consider first that the temperature is dependent only on the space coordinates, $T(x)$. Assuming that the shift law applies to each particle with pseudotime which depends on its position, the stress-strain law again has the form [6.29]

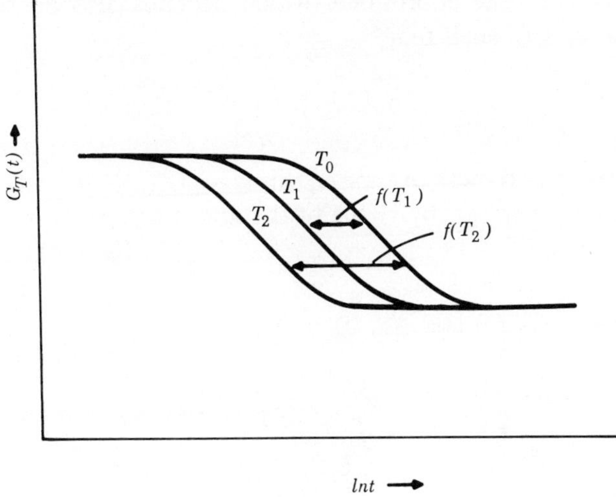

Figure 6.1

except now $\xi = ta[T(x)]$. It should, therefore, be noted that $s_{ij}(x, \xi)$ and $e_{ij}(x, \xi)$ similarly are mappings of the functions $s_{ij}(x, t)$ and $e_{ij}(x, t)$ from the (x, t) plane into the (x, ξ) plane and not the same functions with ξ replacing t.

To recapitulate, for the uniform temperature result, if the time taken for $G_{T_0}(t)$ to decrease from some value $G_{T_0}^{(1)}$ to some other value $G_{T_0}^{(2)}$ is δt, then the time for $G_T(t)$ to decrease from $G_T^{(1)}$ to $G_T^{(2)}$ is $\delta(t)/a(T)$. This implies that the molecular processes associated with the particular relaxation behavior (shear or dilatation) are accelerated by a constant factor $a(T)$ at a higher temperature. If now a temperature $T(x, t)$ is considered, each particle has a temperature varying with time. The assumption is made that these molecular processes, hence the rate of change of the relaxation modulus at each temperature level, are accelerated by the same factor $a(T)$, where T is the current temperature, independent of the temperature history. Letting $G^*(t)$ denote the relaxation modulus for a temperature field $T(x, t)$, accordingly dependent on the particle position, and deducing from the shift property that $G^*(t)$ decreases monotonically from the instantaneous modulus to the static modulus which are independent of temperature, there is some mapping of the time coordinate, which depends also on the particle position, $\xi = \xi(t)$, such that

$$G^*(t) = G_{T_0}(\xi) \quad . \qquad [6.30]$$

After a further increment of time δt, this gives $G^*(t + \delta t) = G_{T_0}(\xi + \delta\xi)$, where the increment $\delta\xi$, by virtue of the above assumption, satisfies

$$\delta\xi = a[T(x,t)]\delta t \quad . \qquad [6.31]$$

Since $G^*(0) = G_{T_0}(\xi)$ and $\xi(x, 0) = 0$,

$$\xi(x,t) = \int_0^t a[T(x,t')]dt' \quad . \qquad [6.32]$$

Finally, consider the response under the same temperature field to a unit applied strain at some instant t_0, and let $G(t - t_0)$ denote the relaxation modulus, then there will be some $\zeta(x, t)$ such that $\zeta(x, t_0) = 0$ and

$$G(t-t_0) = G_{T_0}(\zeta) \quad , \qquad [6.33]$$

where

$$d\zeta = a[T(x,t)]dt \quad ,$$

so that

$$\zeta(x,t) = \int_{t_0}^t a[T(x,t')]dt'$$

$$= \xi(x,t) - \xi(x_0,t) \quad . \qquad [6.34]$$

For small strain, the heat equation $T_{,ii} = \kappa^{-1}\,\partial T/\partial t$ is assumed to hold since small strains prevent the dissipation of appreciable mechanical energy to constitute a distributed heat source within the body, [6.29] applies. For the hydrostatic components,

$$\sigma_{ii}(x,\xi) = k\left\{\epsilon_{ii}(x,\xi) - 3\alpha[T(x,\xi) - T_0]\right\} \quad . \quad [6.35]$$

For a temperature dependent α, the expression $3\alpha(T - T_0)$ must be replaced by

$$\int_{T_0}^{T} \alpha(T')dT' \ .$$

But this can be written as

$$3\alpha_0 \, [T^*(x,\xi) - T_0]$$

so that [6.35] has the same form in terms of a pseudotemperature field $T^*(x, \xi)$, α_0 being a constant.

In the absence of body forces, the problem to solve in the quasi-static analysis is

$$\left. \frac{\partial \sigma_{ij}}{\partial x_j} \right|_t = 0 \qquad\qquad [6.36]$$

together with [6.29] and [6.35], subject to compatibility condition and boundary conditions. Note that [6.29] is in the (x, ξ) coordinates while [6.36] and the rest in (x, t) coordinates. Uniformity can be obtained by rewriting [6.36] as

$$\left. \frac{\partial \sigma_{ij}}{\partial x_j} \right|_\xi + \left. \frac{\partial \sigma_{ij}}{\partial \xi} \right|_x \left(\frac{\partial \xi}{\partial x_j} \right)_t = 0 \ , \qquad [6.37]$$

where $(\partial \xi / \partial x_j)_t$ is a known function of (x, t), hence of (x, ξ) because of [6.32]. In general, the viscoelastic analogy is no longer valid for the thermoviscoelastic problems in which temperature dependence of moduli is considered.

E.   EXAMPLE ON THE ELASTICITY-VISCOELASTICITY ANALOGY

In general quasistatic elastic solutions are time dependent, but many are independent of time. The latter category offers the simplest examples of elasticity-viscoelasticity analogy. Equation 4.16 gives the fundamental static or quasistatic solution of surface displacement for an elastic body. The Laplace transform [2.90] of this is

$$\widetilde{K}^{E} = (4\pi)^{-1}(r+2)(r+1)^{-1}[(\xi-\xi')^{2} + (\eta-\eta')^{2}]^{-1/2} s^{-1} \, ,$$

$$[6.38]$$

where r was defined as in Section 4.A.

Now, for a Maxwell material [6.5], the operators $P_i(i = 1, 2, 3, 4)$ in [6.6] and [6.7] are

$$P_1 = \frac{1}{2\mu} \frac{\partial}{\partial t} + \frac{1}{2\eta}$$

$$P_2 = \frac{\partial}{\partial t}$$

$$P_4/P_3 = 3\lambda + 2\mu = k \quad , \qquad [6.39]$$

where k is, of course, the compressibility. According to the definitions associated with [6.13],

$$\widetilde{\mu}(s) = \frac{P_2(s)}{2P_1(s)} = \frac{\mu s}{s + \mu/\eta}$$

and

$$\widetilde{\lambda}(s) = \frac{P_1(s)[P_4(s)/P_3(s)] - P_2(s)}{3P_1(s)}$$

$$= \frac{s + k\mu/3\eta}{s + \mu/\eta} \, . \qquad [6.40]$$

Note $k = 3\lambda + 2\mu = 3\widetilde{\lambda} + 2\widetilde{\mu}$.

According to the elasticity-viscoelasticity analogy, the Laplace transform of the viscoelasticity counterpart of $\widetilde{K}^E$ (i.e., $\widetilde{K}^V$) is obtained by replacing λ and μ by $\widetilde{\lambda}$ and $\widetilde{\mu}$, respectively, in [6.38]. Thus since $r = \lambda/\mu$, it is to be replaced by $\tilde{\lambda}/\tilde{\mu} = [\lambda s + k\mu/3]/\mu s$, and

$$\tilde{K}^V = (4\pi)^{-1} \frac{(\lambda + 2\mu)s + k\mu/3\eta}{s[(\lambda + \mu)s + k\mu/3\eta]} [(\xi-\xi')^2 + (\eta-\eta')^2]^{-1/2}$$

The inverse Laplace transform [2.91] yields

$$\tilde{K}^V(\xi,\eta,t;\xi',\eta') = (4\pi)^{-1}[(\xi-\xi')^2 + (\eta-\eta')^2]^{-1/2}$$

$$\cdot \left[1 + \frac{\mu}{\lambda + \mu} e^{-\frac{k\mu t}{3\eta(\lambda + \mu)}} \right] \cdot$$

$$[6.41]$$

For Kelvin-Voigt material,

$$P_1 = 1 \quad ,$$

$$P_2 = 2\mu + 2\eta \frac{\partial}{\partial t} \quad ,$$

$$P_4/P_3 = k \quad . \qquad [6.42]$$

With the above, the viscoelastic counterpart of [4.18] is

$$\tilde{K}^V(\xi,\eta,t;\xi',\eta') = (4\pi)^{-1}[(\xi-\xi')^2 + (\eta-\eta')^2]^{1/2}$$

$$\cdot \left[\frac{\lambda + 2\mu}{\lambda + \mu} + \frac{k}{\lambda + \mu} e^{-\frac{3(\lambda + \mu)t}{\eta}} \right] \cdot$$

$$[6.43]$$

F. QUASISTATIONARY RESPONSE OF A VISCOELASTIC HALF-SPACE TO MOVING LOADS

Carrying the example in Section 6.E a step further to conditions under which the speed of a moving load is such that the quasistatic assumption is no longer valid. Also, for a more general description of the visco-elastic material, the creep compliance in [6.18] will be used. In particular,

$$f(t) = \mu_d^{-1}\left[1 + \sum_n F_n(1 - e^{-t/\tau_n})\right] \quad , \quad [6.44]$$

where μ_d is the dynamic shear modulus, τ_n are the retardation times, and F_n are the strengths of retardation spectra. This is the moving counter-part of the model shown in Figure 4.2 for viscoelastic bodies for which the fundamental solutions for the surface displacement are given by [6.41] and [6.43], respectively, for the Maxwell and the Kelvin-Voigt models.

In the examples above, the elastic solution [4.16] was referred to. It may be shown[72] that $w(\xi, \eta, 0)$ satisfies

$$w(\xi,\eta,0) = -\frac{(1-\nu)}{\mu}\left.\Phi\right|_{\zeta=0}$$

and

$$p(\xi,\eta) = -\left.\frac{\partial\Phi}{\partial\zeta}\right|_{\zeta=0} \quad , \qquad [6.45]$$

where Φ satisfies $\Phi_{,ii} = 0$, ν is the Poisson ratio and μ is the shear modulus.

Now according to the associated equations of [6.13], the Laplace transform of μ is

$$\tilde{\mu}(s) = [s\,\tilde{f}(s)]^{-1} \quad , \qquad [6.46]$$

where f is given by [6.44]. The Laplace transform of [6.45] yields

$$\tilde{w}(\xi,\eta,0) = -\frac{(1-\nu)}{\mu}\left.\tilde{\Phi}\right|_{\zeta=0}$$

$$\tilde{p}(\xi,\eta) = -\left.\frac{\partial\tilde{\Phi}}{\partial\zeta}\right|_{\zeta=0} \quad , \qquad [6.47]$$

where Φ satisfies $\Phi_{,ii} = 0$.

According to the viscoelasticity-elasticity analogy, the Laplace transforms of the viscoelastic counterpart of [6.45] are

$$\tilde{w}^{V}(\xi,\eta,0,s) = -\frac{(1-\nu)}{\tilde{\mu}(s)}\,\tilde{\Phi}\Big|_{\zeta=0}$$

$$\tilde{p}(\xi,\eta) = -\frac{\partial\tilde{\Phi}}{\partial\zeta}\Big|_{\zeta=0} \quad , \qquad [6.48]$$

where $\tilde{\mu}(s)$ is as given by [6.46] and $\tilde{\nu}$ is given by

$$\tilde{\nu}(s) = \frac{\tilde{\lambda}(s)}{2[\tilde{\lambda}(s)+\tilde{\mu}(s)]} \quad . \qquad [6.49]$$

Assume that the material behaves similarly in both shear and dilation; then [6.49] defines the Poisson ratio, independent of s. This approximation is not unduly restrictive since a more refined treatment in this respect merely substitutes a more complicated function of time for the quantity $f(t)$ in the ensuing analysis.

With $\nu = $ constant and the help of the Faltung theorem, [6.48] may be formally inverted in the form

$$w(\xi,\eta,0,t) = -(1-\nu)\,\Psi(\xi,\eta,0,t)$$

$$p(\xi,\eta) = -\frac{\partial\Phi}{\partial\zeta}\Big|_{\zeta=0} \quad , \qquad [6.50]$$

where Φ is the Laplace inversion of $\tilde{\Phi}$, and

$$\Psi(\xi,\eta,\zeta,t) = \int_{0}^{t} f(t-t')\frac{\partial}{\partial t'}\Phi(\xi,\eta,\zeta,t')dt' \quad . \qquad [6.51]$$

Restrict the analysis to only quasistationary state, that is, deformation field which appears stationary to an observer traveling with constant velocity V. Let V be in the direction of positive x_1. As such $x_1 - Vt$

gives the moving coordinate for the loading. Thus

$$w(\xi,\eta,0) = -(1-\nu)\,\Psi(\xi,\eta,0)$$

$$p(\xi,\eta) = -\frac{\partial\Phi}{\partial\zeta}\bigg|_{\zeta=0}$$

$$\Psi = -\int_0^\infty f(s/V)\,\frac{\partial}{\partial s}\Phi(\xi+s,\eta,\zeta)\,ds \quad , \qquad [6.52]$$

where Φ and Ψ are both harmonic functions.

Again, a solution sought is of the form

$$w(\xi,\eta,0) = \int_A q(\xi',\eta')\,K(\xi,\eta;\xi',\eta')\,d\xi'\,d\eta' \quad , \qquad [6.53]$$

where $q \equiv p/\mu_d$ and K is the fundamental solution. Now working with the Laplace equation, whose solution is harmonic, and double Fourier transform [5.7] it is found that for

$$\frac{\partial\Phi^{(f)}}{\partial\zeta}\bigg|_{\zeta=0} = \mu_d\,\delta(\xi)\,\delta(\eta) \quad , \qquad [6.54]$$

and the appropriate boundary conditions, $\Phi^{(f)}$ must satisfy

$$\Phi^{(f)} = \int_{-\infty}^\infty \int_{-\infty}^\infty A(a_1,a_2)e^{-i(a_1\xi+a_2\eta)}\,e^{-\sqrt{a_1^2+a_2^2}\,\zeta}\,da_1\,da_2 \quad , \qquad [6.55]$$

where $A(a_1,a_2) = -\mu_d/4\pi^2\sqrt{a_1^2+a_2^2}$. In the above, $\Phi^{(f)}$ refers to the fundamental solution, for example, [6.54] multiplied by $q(\xi,\eta) \equiv p(\xi,\eta)\mu_d$ yields the correct condition shown in the second of [6.52]. Accordingly, [6.52], [6.55], and [6.44] lead to

$$K(\xi,\eta) = -\frac{(1-\nu)}{\mu_d}\int_{-\infty}^\infty \int_{-\infty}^\infty \left(1 + \sum^n \frac{f_n}{1 + i\,a_1\,V\,\tau_n/\ell}\right)$$

$$\cdot (a_1^2+a_2^2)^{-1}\,e^{-i(a_1\xi+a_2\eta)}\,da_1\,da_2 \quad .$$

Upon integrating[71] and substituting $\xi - \xi'$ for ξ and $\eta - \eta'$ for η,

$$K(\xi,\eta;\xi',\eta') = \frac{1}{2\pi}\left\{\left[(\xi-\xi')^2 + (\eta-\eta')\right]^{-1/2}\right.$$

$$\left. + \sum_n f_n \int_0^\infty \frac{e^{-\rho}\,d\rho}{\left[(\xi-\xi') + V\tau_n\rho/\ell)^2 + (\eta-\eta')^2\right]^{1/2}}\right\}. \qquad [6.56]$$

The first term is the familiar result for an elastic solid with shear modulus μ_d. In the limit $V \to \infty$, the remaining terms vanish so that for sufficiently large velocities V, the viscoelastic material behaves as an elastic solid with shear modulus μ_d. On the other hand, for $V \to 0$, the material behaves elastically and the modulus μ_d is replaced by the static modulus $\mu_d/(1 + \sum_n f_n)$.

In two dimensions, by redefining the origin of displacement to dispose of an infinite constant (see [4.14]),

$$K(\xi,\xi') = -\frac{(1-v)}{\pi}\left[\ln|\xi-\xi'|\right.$$

$$\left. + \sum_n f_n \int_0^\infty \ln|\xi-\xi' + V\tau_n\rho/\ell|\,e^{-\rho}d\rho\right]. \qquad [6.57]$$

Other treatments of the subject, to varying degrees of approximation, for rolling problem have appeared earlier.[73-75]

G.   DEFORMATION OF A SOFT LAYER MATERIAL UNDER A MOVING LOAD

In Section 6.F, surface deformation of viscoelastic material was discussed. The ultimate goal of such solutions is in application. As such, the equations derived are not the easiest to apply, especially when the moving viscoelastic material is in the form of a layer, see Figure 6.2. On occasion, the physical problem may be such that the bulk of the information sought may be revealed by a simplified analysis. That is, where the material is very soft, a viscous analysis may suffice even though the material is actually viscoelastic. To this end such a viscous

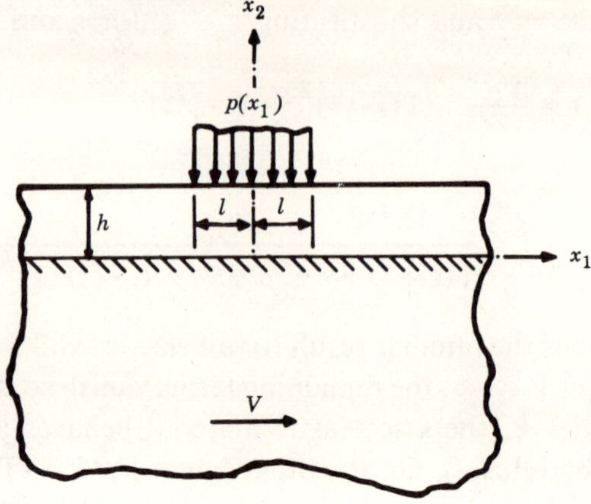

Figure 6.2

analysis is given.[76] In the subsequent section a more elaborate model is examined.

Let the surface profile under load at time t and position x_1 be $F_1(x_1, t)$, the undeformed surface being flat (i.e., $F_1(x_1, t) = 0$). Let $u_0(x_1)$, $v_0(x_1)$ be the velocity components in the x_1, x_2 directions, respectively, due to a fixed and unchanging load. The origin of x_1 is fixed with the load, the material as a whole moves to the right with a velocity V. At dt later, F_1 assumes a new value

$$F_1[x_1 + (V + u_0)dt, \ t + dt] \quad,$$

while the vertical velocity is given by the Eulerian expression

$$\frac{\partial F_1}{\partial t} + V \frac{\partial F_1}{\partial x_1} = v_0(x_1) \quad, \qquad [6.58]$$

where u_0 has been neglected in comparison with V.

The general solution to [6.58] is

$$F_1 = F(x_1 - Vt) + V^{-1} I(x_1) \quad, \qquad [6.59]$$

where F is arbitrary and

$$I(x_1) = \int_0^x v_0(\eta)d\eta \quad. \qquad [6.60]$$

Since $F_1(x, 0) = 0$, [6.59] determines F, and

$$F_1 = V^{-1}[I(x_1) - I(x_1 - Vt)] \quad . \qquad [6.61]$$

The term $I(x_1)$ due to penetration should be appreciable only at and near the loading zone. The term $I(x_1 - Vt)$ represents, of course, a wave traveling downstream with the material. At some time later it will no longer affect the profile in the neighborhood of the load.

The function $v_0(x_1)$ in [6.58] represents the vertical velocity of the surface in the slow motion of the viscous layer (i.e., $V = 0$) under the load $p(x_1)$. For small departures from the originally level surface the change of the profile need not be taken into account. The boundary conditions are

$$u = v = 0 \qquad\qquad (x_2 = 0) \quad , \qquad\qquad [6.62]$$

$$\sigma_{22} = \begin{cases} p(x_1) & (|x_1| \leq \ell,\ x_2 = h) \\ 0 & (|x_1| > \ell,\ x_2 = h) \end{cases} , \qquad [6.63]$$

$$\sigma_{12} = 0 \qquad\qquad (x_2 = h) \quad . \qquad\qquad [6.64]$$

The governing differential equation, for incompressible viscous fluid with constant viscosity μ, is obtained by using [1.23] in conjunction with equilibrium ([1.1] in the absence of body and inertia forces):

$$p_{,i} = \mu\, u_{i,jj} \qquad\qquad [6.65]$$

where

$$\sigma_{ij} = -p\,\delta_{ij} + \mu(u_{i,j} + u_{j,i}) \quad . \qquad [6.66]$$

Furthermore, under Stokes' assumption $p = 0$, [6.65] and [6.66] reduce to, respectively,

$$u_{i,jj} = 0 \qquad\qquad [6.67]$$

and

$$\sigma_{ij} = \mu(u_{i,j} + u_{j,i}) \quad . \qquad [6.68]$$

Taking the Fourier transform [2.5] of the set [6.62], [6.63], [6.64], [6.67], and [6.68], solving the transformed equations and taking the inverse transform [2.12], the following expression results for the surface displacement $v_0(x_1)$:

$$v_0^*(\xi) = \int_{-1}^{1} p(\xi') \, K(\xi, \xi') \, d\xi' \,, \qquad [6.69]$$

where $v_0^*(x_1) \equiv \pi v_0(x_1)/l$, $q \equiv p/p_0$ ($p_0 = $ constant), $\xi \equiv x_1/l$, $H \equiv h/l$, and

$$K(\xi, \xi') = \int_{-\infty}^{\infty} \frac{\tanh SH \; e^{is(\xi - \xi')} \, dS}{S} \,. \qquad [6.70]$$

H. VISCOELASTIC LAYER ON AN ELASTIC HALF-SPACE

Figure 6.3 shows the mechanical model which involves a viscoelastic layer on a half-space.[77] The case of plane strain is considered. The (x_1', x_2') are coordinates fixed in the composite solid while the moving load $w(x_2)$ is attached to (x_1, x_2). The load moves with a constant velocity V. The physical system consists of two parts: domain I $(0 \leq x_1' = x_1 \leq H; \; |x_2'| = |x_2| < \infty)$ is the viscoelastic layer in the undeformed state. Domain II $(x_1' = x_1 > H; \; |x_2'| = |x_2| < \infty)$ is the elastic base in the undeformed state. The viscoelastic material is charac-

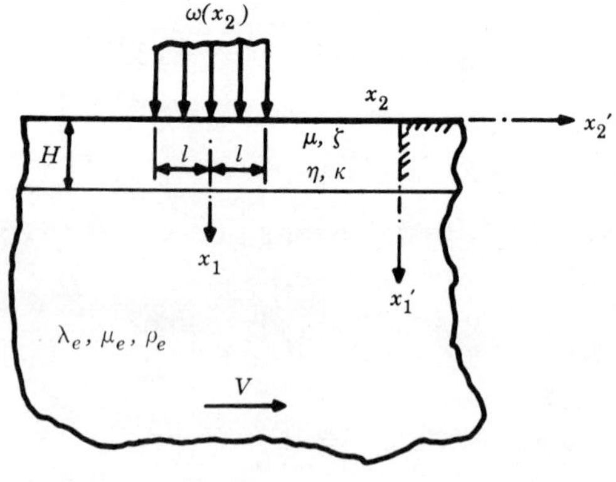

Figure 6.3

terized by μ, ζ, η, and κ; whereas λ_e, μ_e (the Lamé constants), and ρ_e (the mass density) characterize the elastic material.

The relationships between the (x_1, x_2) and (x'_1, x'_2) systems are

$$x_1 = x'_1$$

$$x_2 = x'_2 + Vt \; , \qquad\qquad [6.71]$$

with t denoting time. Moreover, in the (x_1, x_2) system, the material derivative $D(\ )/Dt$ is given by

$$\frac{D(\ )}{Dt} = \frac{\partial(\ )}{\partial t} + \frac{\partial(\ )}{\partial x'_2} \frac{\partial x'_2}{\partial t} = \frac{\partial(\ )}{\partial t} + V \frac{\partial(\ )}{\partial x_2} \; .$$

For the quasi-stationary state, $\partial(\ )/\partial t = 0$, and

$$\frac{D(\ )}{Dt} = V \frac{\partial(\ )}{\partial x_2} \; . \qquad\qquad [6.72]$$

The equation of equilibrium $\sigma_{\alpha\beta,\beta} = 0$ applies; $a, \beta = 1, 2$ refer now to the moving coordinates.

The constitutive equations for a material, elastic in bulk and represented by standard linear model in shear, are expressed in differential form as

$$P_1 \epsilon_{\alpha\beta} = P_2 \epsilon_{\alpha\beta} + (\kappa P_1 - \frac{P_2}{3}) \delta_{\alpha\beta} \epsilon_{kk} \; , \qquad [6.73]$$

where $\alpha, \beta = 1, 2$ referred to (x'_1, x'_2) system; $P_1 = (1/2\mu)(D/Dt) + (1/2\eta)$ and $P_2 = (D/Dt) + \zeta$ are the differential time operators with μ, η, and ζ the constants characteristic of the postulated viscoelastic material; κ is the bulk modulus. On expansion, [6.73] with the aid of [6.72] yields

$$(V \frac{\partial}{\partial x_2} + \frac{\mu}{\eta})(\sigma_{11}, \sigma_{22}) = [V(\kappa + \frac{4}{3}\mu)\frac{\partial}{\partial x_2} + (\frac{\kappa\mu}{\eta} + \frac{4}{3}\mu\zeta)](\frac{\partial u_1}{\partial x_1}, \frac{\partial u_2}{\partial x_2})$$

$$+ [V(\kappa - \frac{2}{3}\mu)\frac{\partial}{\partial x_2} + (\frac{\kappa\mu}{\eta} - \frac{2}{3}\mu\zeta)](\frac{\partial u_2}{\partial x_2}, \frac{\partial u_1}{\partial x_1}) \; ,$$

$$(V \frac{\partial}{\partial x_2} + \frac{\mu}{\eta}) \sigma_{12} = (\mu V \frac{\partial}{\partial x_2} + \mu \zeta)(\frac{\partial u_1}{\partial x_2} + \frac{\partial u_2}{\partial x_1}) \; . \qquad [6.74]$$

The following nondimensional quantities are now introduced:

$$\xi \equiv x_1/H \ , \quad \rho \equiv x_2/H \ , \quad h \equiv H/\ell \ , \quad u \equiv u_1/\ell \ , \quad v \equiv u_2/\ell \ ,$$

$$\mu \equiv \mu/k \ , \quad \bar{s} \equiv \mu/\eta\zeta \ , \quad \bar{V} \equiv V/\ell\zeta \ , \quad \bar{r} \equiv V/H\zeta \ , \quad \sigma_{\xi\xi} \equiv h\sigma_{11}/\mu e \ ,$$

$$\sigma_{\rho\rho} \equiv h\sigma_{22}/\mu e \ , \quad \sigma_{\xi\rho} \equiv h\sigma_{12}/\mu e \ , \quad q \equiv hw/\mu e \ .$$

Equilibrium can now be expressed as

$$\frac{\partial \sigma_{\xi\xi}}{\partial \xi} + \frac{\partial \sigma_{\xi\rho}}{\partial \rho} = 0 \ , \quad \frac{\partial \sigma_{\xi\rho}}{\partial \xi} + \frac{\partial \sigma_{\rho\rho}}{\partial \rho} = 0 \ . \qquad [6.75]$$

It is further assumed here that $\sigma_{\xi\xi}$, $\sigma_{\rho\rho}$, $\sigma_{\xi\rho}$, u, and v are continuous functions of ξ and ρ. Substituting [6.74] into [6.75],

$$(1 + \tfrac{4}{3}\bar{\mu})\,\bar{r}\,\frac{\partial^3 u}{\partial\rho\partial\xi^2} + (\bar{s} + \tfrac{4}{3}\bar{\mu})\,\frac{\partial^2 u}{\partial\xi^2} + (1 + \tfrac{1}{3}\bar{\mu})\,\bar{r}\,\frac{\partial^3 u}{\partial\xi\partial\rho^2}$$

$$+ (\bar{s} + \tfrac{1}{3}\bar{\mu})\,\frac{\partial^2 v}{\partial\xi\partial\rho} + \bar{\mu}\,\frac{\partial^2 u}{\partial\rho^2} + \bar{\mu}\,\bar{r}\,\frac{\partial^3 u}{\partial\rho^3} = 0 \ ,$$

$$\bar{\mu}\,\frac{\partial^2 v}{\partial\xi^2} + \bar{\mu}\,\bar{r}\,\frac{\partial^3 v}{\partial\rho\partial\xi^2} + (\bar{s} + \tfrac{1}{3}\bar{\mu})\,\frac{\partial^2 u}{\partial\xi\partial\rho} + (1 + \tfrac{1}{3}\bar{\mu})\,\bar{r}\,\frac{\partial^3 u}{\partial\xi\partial\rho^2}$$

$$+ (1 + \tfrac{4}{3}\bar{\mu})\,\bar{r}\,\frac{\partial^3 v}{\partial\rho^3} - (\bar{s} + \tfrac{4}{3}\bar{\mu})\,\frac{\partial^2 u}{\partial\rho^2} = 0 \ . \qquad [6.76]$$

For the case of plane strain elasticity for domain II, the equations of equilibrium in terms of displacements are

$$\alpha^2\,\frac{\partial^2 u}{\partial\xi^2} + \frac{\partial^2 u}{\partial\rho^2} + (\alpha^2 - 1)\,\frac{\partial^2 u}{\partial\xi\partial\rho} = 0 \ ,$$

$$(\alpha^2 - 1)\,\frac{\partial^2 u}{\partial\xi\partial\rho} + \frac{\partial^2 u}{\partial\xi^2} + \alpha^2\,\frac{\partial^2 v}{\partial\rho^2} = 0 \ ,$$

$$\text{where } \alpha^2 = c_1/c_2, \quad c_1 = \left(\frac{\lambda_e + \mu_e}{\rho_e}\right)^{1/2} \quad \text{and} \quad c_2 = \left(\frac{\mu_e}{\rho_e}\right)^{1/2} .$$

$$[6.77]$$

The boundary conditions are:

$$\sigma_{\xi\xi}^{I} = \begin{cases} -q(\rho) & (|\rho| \leq 1, \ \xi = 0) \\ 0 & (1 < |\rho| < \infty, \ \xi = 0) \end{cases} ,$$

$$\sigma_{\xi\rho}^{I} = 0 \qquad (|\rho| < \infty, \ \xi = 0) ,$$

$$\left. \begin{aligned} \sigma_{\xi\xi}^{I} &= \sigma_{\xi\xi}^{II} \\ \sigma_{\xi\rho}^{I} &= \sigma_{\xi\rho}^{II} \\ u^{I} &= u^{II} \\ v^{I} &= v^{II} \end{aligned} \right\} \qquad (|\rho| < \infty, \ \xi = 1) . \qquad [6.78]$$

The regularity conditions are $\sigma_{\xi\xi}^{I}$, $\sigma_{\xi\rho}^{I}$, u^{I}, $v^{I} \to 0$ for $0 \leq \xi \leq 1$ and $(\xi^2 + \rho^2)^{1/2} \to \infty$ and $\sigma_{\xi\xi}^{II}$, $\sigma_{\xi\rho}^{II}$, u^{II}, $v^{II} \to 0$ for $|\xi| > 1$ as $(\xi^2 + \rho^2)^{1/2} \to \infty$.

Applying the Fourier transform [2.5] to [6.76] and [6.77] respectively.

$$[(\bar{s} + \frac{4}{3} \bar{\mu}) + i p \bar{r}(1 + \frac{4}{3} \bar{\mu})] \frac{d^2 \bar{u}^{I}}{d\xi^2} + [-p^2(1 + \frac{1}{3} \bar{\mu})\bar{r}$$

$$-i p(\bar{s} + \frac{1}{3} \bar{\mu})] \frac{d\bar{v}^{I}}{d\xi} + [-\bar{\mu} p^2 + i p^3 \bar{\mu} \bar{r}] \bar{u}^{I} = 0 ,$$

$$(\bar{\mu} - i p \bar{\mu} \bar{r}) \frac{d^2 \bar{v}^I}{d\xi^2} + [-p^2 \bar{r} (1 + \frac{1}{3} \bar{\mu}) - i p (\bar{s} + \frac{1}{3} \bar{\mu})] \frac{d\bar{u}^I}{d\xi}$$

$$+[-p^2 (\bar{s} + \frac{4}{3} \bar{\mu}) + i p^3 \bar{r} (1 + \frac{4}{3} \bar{\mu})]\bar{v}^I = 0 \quad ,$$

$$[6.79]$$

and

$$\alpha^2 \frac{d^2 \bar{u}^{II}}{d\xi^2} - p^2 \bar{u}^{II} + i p (\alpha^2 - 1) \frac{d\bar{v}^{II}}{d\xi} = 0 \quad ,$$

$$- i p(\alpha^2 - 1) \frac{d\bar{u}^{II}}{d\xi} - p^2 \alpha^2 \bar{v}^{II} + \frac{d^2 \bar{v}^{II}}{d\xi^2} = 0 \quad .$$

$$[6.80]$$

Both sets of equations in [6.79] and [6.80] are coupled, linear differential equations. Their solutions are known. After a rather lengthy set of algebraic operations, as a result of satisfying the boundary conditions

Table 3. Material Constants for Steel

$$\rho_e = 487.296 \text{ lb/ft}^3$$
$$\nu = 0.3 \text{ (Poisson's Ratio)}$$
$$\mu_e = 11.53 \times 10^6 \text{ psi}$$
$$\lambda_e = [2\mu_e\nu/(1 - \nu)] = 17.295 \times 10^6 \text{ psi}$$
$$\alpha = [(\lambda_e + 2\mu_e)/\mu_e]^{1/2} = 1.8708$$

Table 4. Material Constants for PIB at 25°C

Bulk Modulus $\kappa = 58.012 \times 10^4$ psi

Frequency range (Hz)	μ (psi)	ζ (1/sec)	η (psi/sec)
30–300	241.73	701.374	0.0507
400–4000	4834.4	96267.99	0.0324

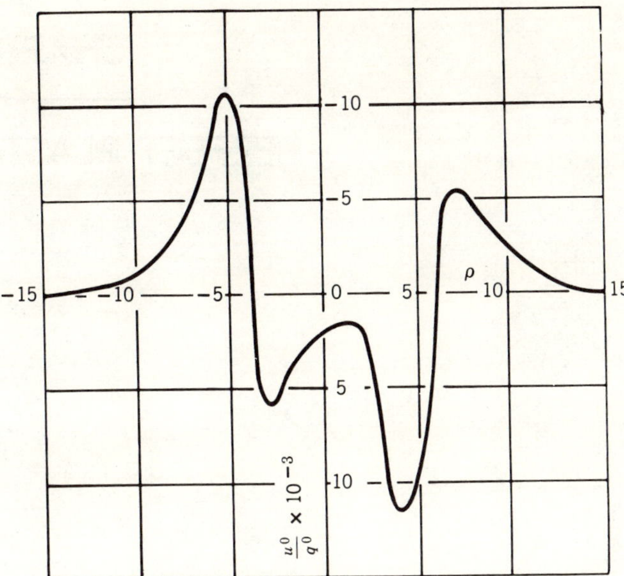

Figure 6.4

[6.78], six equations with complex coefficients appeared. It did not appear that these could be solved analytically.

A typical numerical result is shown on the surface displacement for the material constants given in Table 3 and Table 4.

Figure 6.4 shows the results for $q = q_0 = $ constant.

CHAPTER 7

PERFECT PLASTICITY

This chapter deals primarily with the slip-line theory and its application to statically determined problems which may be useful in surface phenomena. It is not the intent here to reproduce plasticity theory or methods of solution which are covered adequately elsewhere.[7,78,79]

A.   SLIP-LINE THEORY

The flow of plastic material is called plane if the velocity of flow is restricted to a plane, say the (x_1, x_2) plane. Thus the rate of strains

$$\dot{\epsilon}_{33} = \dot{\epsilon}_{23} = \dot{\epsilon}_{13} = 0 \quad . \qquad [7.1]$$

In rigid-perfect plasticity theory, the elastic strain is disregarded; therefore the total strain and the plastic strain are identical. Moreover, since the plastic mean normal strain is supposed to vanish, the total mean normal strain vanishes. That is, the material is incompressible. As such [1.26] is expressible as

$$s_{ij} = 2\mu\, \dot{\epsilon}_{ij} \quad . \qquad [7.2]$$

In view of [7.1] and [7.2],

$$s_{33} = s_{23} = s_{13} = 0 \quad , \qquad [7.3]$$

while the remaining components of stress deviation are independent of x_3, the direction perpendicular to the plane in question. In view of [7.1] and [7.3], ϵ_{12} may be denoted by γ and s_{12} by τ without causing confusion. By definition $s_{33} = \sigma_{33} - \frac{1}{3}(\sigma_{12} + \sigma_{22} + \sigma_{33})$. Since $s_{33} = 0$, by virtue of [7.3],

$$\sigma_{33} = \frac{1}{2}(\sigma_{11} + \sigma_{22}) \quad . \qquad [7.4]$$

Again in view of [7.3],

$$s = \frac{1}{2}(\sigma_{11} + \sigma_{22}) \quad . \qquad [7.5]$$

The third of the equations of equilibrium ([1.1] in the absence of inertia and body forces) shows σ_{33} is independent of x_3. In view of [7.4] and [7.5], this applies to s also. The normal components of the stress deviation are given by

$$s_{11} = \sigma_{11} - s = \frac{1}{2}(\sigma_{11} - \sigma_{22})$$

$$s_{22} = \sigma_{22} - s = -\frac{1}{2}(\sigma_{11} - \sigma_{22}) \qquad [7.6]$$

$$s_{33} = \frac{1}{2}(\sigma_{11} + \sigma_{22}) \quad .$$

In view of [7.6] and [7.3] both the Mises and Tresca yield conditions, [1.27] and [1.28], respectively, give

$$\frac{1}{4}(\sigma_{11} - \sigma_{22})^2 + \tau^2 - k^2 = 0 \quad . \qquad [7.7]$$

Figure 7.1 is a Mohr's circle showing how the plane state of stress at one orientation is related to those at another. The circle represents the locus of points, which have as coordinates the normal and shear stresses, for all orientations. The angle shown is twice that in the physical plane. For example (σ_{11}, τ) and $(\sigma_{22}, -\tau)$, representing the state of stress at the adjacent faces of a rectangular element, are 180° apart in Figure 7.1. Note the orientation from the reference surface to a given surface as well as the magnitude of the state of stress on this surface are readily obtainable. Thus the maximum and minimum stresses are represented by the left-most and right-most points of the circle. The directions are

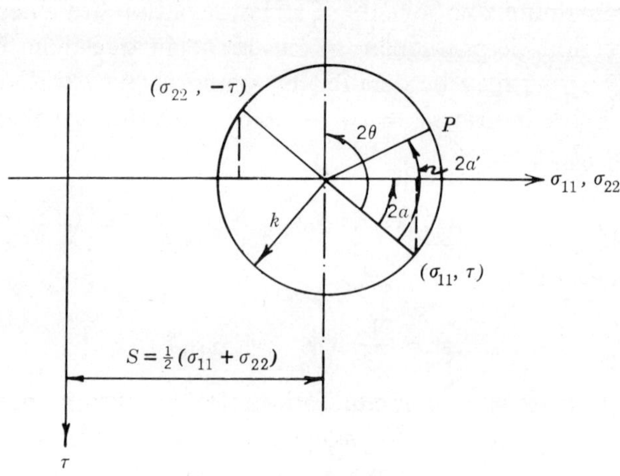

Figure 7.1

first and second principal directions. The maximum shear stresses are represented by the highest and lowest points on the circle. The directions are the first and second shear directions, the first being 45° counterclockwise from the first principal direction. If in addition, the radius is set at k, then [7.7] is satisfied. Thus the circle is also the yield circle. That is, the circle represents the locus of stress states satisfying the yield condition. Analytically, if $w \equiv s/2k$ and $2\theta = \tan^{-1}[2\tau/(\sigma_{11} - \sigma_{22})] + (\pi/2)$, then

$$\sigma_{11} = 2kw + k \sin 2\theta$$

$$\sigma_{22} = 2kw - k \sin 2\theta \qquad\qquad [7.8]$$

$$\tau = -k \cos 2\theta \quad .$$

Now the first two equations of equilibrium are

$$\frac{\partial \sigma_{11}}{\partial x_1} + \frac{\partial \tau}{\partial x_2} = 0 \; , \quad \frac{\partial \tau}{\partial x_1} + \frac{\partial \sigma_{22}}{\partial x_2} = 0 \quad . \quad [7.9]$$

Equations 7.9 and 7.7 form a set of three equations for the three unknowns σ_{11}, σ_{22}, and τ. Given the appropriate boundary conditions, these equations can be solved and the stress distribution determined in problems of plane plastic flow without further reference to the constitutive relationships. Such problems are called statically determinate. Using [7.8] in [7.9], which represent the yield condition and equilibrium,

$$\frac{\partial w}{\partial x_1} + \frac{\partial \theta}{\partial x_1} \cos 2\theta + \frac{\partial \theta}{\partial x_2} \sin 2\theta = 0$$

$$\frac{\partial w}{\partial x_2} + \frac{\partial \theta}{\partial x_1} \sin 2\theta - \frac{\partial \theta}{\partial x_2} \cos 2\theta = 0 \quad . \quad [7.10]$$

If the x_1- and x_2-axes are chosen along the first and second shear directions at a generic point, then $\theta = \pi/2$. Letting differentiations with respect to the first and second shear directions be denoted by d/ds_1 and d/ds_2, respectively, [7.10] assume the form

$$\frac{d}{ds_1}(w - \theta) = 0 \quad , \quad \frac{d}{ds_2}(w + \theta) = 0 \quad . \quad [7.11]$$

Thus $w - \theta$ is constant along the first shear line and $w + \theta$ along the second shear line.

In using the slip-line theory, it is useful to note the following theorems:

1. Hencky's theorem.[80]   The angle formed by the tangents of two fixed shear lines of one family at their points of intersection with a shear line of the second family does not depend on the choice of the intersecting shear line of the second family (as a corollary, if a family of shear lines contains a straight line, it consists entirely of straight lines).

2. Prandtl's theorem.[81]   Along a fixed shear line of one family, the center of curvature of the shear lines of the other family form an involute of the fixed shear line.

Given the above properties, the Hencky-Prandtl net[82] is the net formed by two orthogonal families of curves having the mentioned properties. To find the particular net of shear lines which is appropriate to the solution of a given problem, the boundary conditions must be used. In a statically determined problem of plane plastic flow, the normal and tangential tractions must be prescribed.

In the case of straight shear lines, $w - \theta = $ constant along the first and $w + \theta = $ constant along the second; w is constant along any straight shear line. Thus both w and θ are constant throughout a region where the shear line net is formed by two orthogonal families of parallel straight lines. Such a net constitutes a region of constant state. A region in which the shear lines of one and only one family are straight is called a fan. The centered fan is one in which the straight shear lines are concurrent.

Let the centered fan be the first shear lines. Then the second shear lines are concentric circles. Let the origin of the fan be the origin of a system of polar coordinates. Since θ = constant along each straight first shear line, θ also represents the angular coordinate. Denote the distance from the origin by r. Since $w - \theta$ must be constant along any first shear line, and $w + \theta$ along the second shear line, and since θ = constant along any first shear line, it follows that $w + \theta$ has a constant value, say c, throughout the fan. Thus

$$w = c - \theta$$

and w is independent of r. Also according to [7.8], σ_{11}, σ_{22}, and τ are independent of r. The components of stress with respect to the polar coordinates are

$$\sigma_r = \sigma_\theta = 2k(c - \theta) \quad , \quad \tau_{r\theta} = k \quad . \qquad [7.12]$$

The case of rigid-perfect plastic material obeying the Tresca yield condition and the associated flow rule is statically determinate when the Haar and von Karman hypothesis is valid.[83] This hypothesis stipulates that the circumferential stress is equal to one of the principal stresses in the meridional planes during plastic deformation. For axially symmetric situations, the comparable equations to [7.8] take the form:[84]

$$\sigma_r = 2kw - k \sin 2\theta$$

$$\sigma_z = 2kw + k \sin 2\theta$$

$$[7.13]$$

$$\sigma_{rz} = k \cos 2\theta$$

$$\sigma_\phi = 2kw + k \quad ,$$

where σ_r, σ_z, and σ_θ are the normal stress components in the polar coordinates (r, z, ϕ), σ_{rz} is the relevant shear stress for axially symmetric case, θ is the inclination of the α-lines to the r-axis, and w is as defined before. The slip-lines, which are lines of maximum shear stress in the (r, z) plane, will be used as an orthogonal system of curvilinear coordinates. These are the α- and β-lines with α associated with the algebraically greater principal stress.

Equations comparable to [7.10] take the form:

$$-2kw + 2k\ d\theta + k(\sin\ \theta + \cos\ \theta)ds_\alpha/r = 0 \text{ on } \alpha\text{-lines}$$

$$-2kw - 2k\ d\theta - k(\sin\ \theta + \cos\ \theta)ds_\beta/r = 0 \text{ on } \beta\text{-lines .}$$

$$[7.14]$$

It should be noted, though outside the scope of this monograph, the kinematically determinate cases are best handled with the aid of principal lines[85] which are the characteristics of the basic equations. Principal lines refer to directions of principal stresses and strain-rates.

B. STRESS FIELD IN A SEMIINFINITE SOLID UNDER LUBRICATED FLAT PUNCH

Figure 7.2 shows a semi-infinite rigid-perfect plastic body being loaded from above by a lubricated flat punch, which is assumed rigid. Lubrication infers zero shear traction. Of course, two-dimensional stress field is assumed; as such the extent of the contact is AB. The diagram shows the slip-line field[86] for incipient plastic flow which is shown by ACGFD. Regions ACG and AFD are right isosceles triangles. Region AGF is a

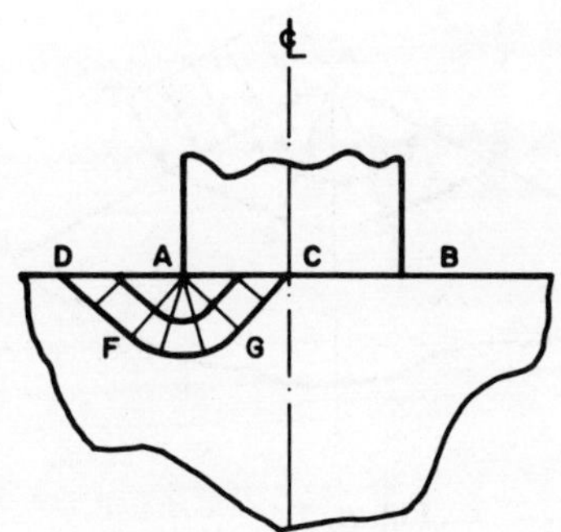

Figure 7.2

centered fan sector. Because of symmetry, only half of the slip-line field is shown. This pressure, p, is uniform along AB and

$$p = k(\pi + 2) \quad . \qquad [7.15]$$

It should be noted that elastic-plastic interface may be beyond DFGC into the solid. Immediately after incipient flow, the state of stress depicted in Figure 7.2 may not be correct and an alternate one, which gives the same pressure as in [7.15], may be more appropriate.[87]

C. STRESS FIELD IN A TRUNCATED WEDGE UNDER LUBRICATED FLAT PUNCH

Figure 7.3 shows a two-dimensional truncated wedge in contact with a lubricated flat punch which is assumed rigid. In other words, Section 7.B shows a special case of the one treated here. The slip-line field, similar to Figure 7.2, is ACGFD. Half the internal angle of the wedge is γ. The pressure at incipient flow is again constant and, for $0 < \gamma < \pi/2$,[86] is given by

$$p = 2k(1 + \gamma) \quad . \qquad [7.16]$$

D. STRESS FIELD IN A WEDGE UNDER LATERAL PRESSURE

The slip-line field of a wedge under lateral pressure is shown in Figure 7.4, ACGFD. Again regions ACG and AFD are right isosceles triangles

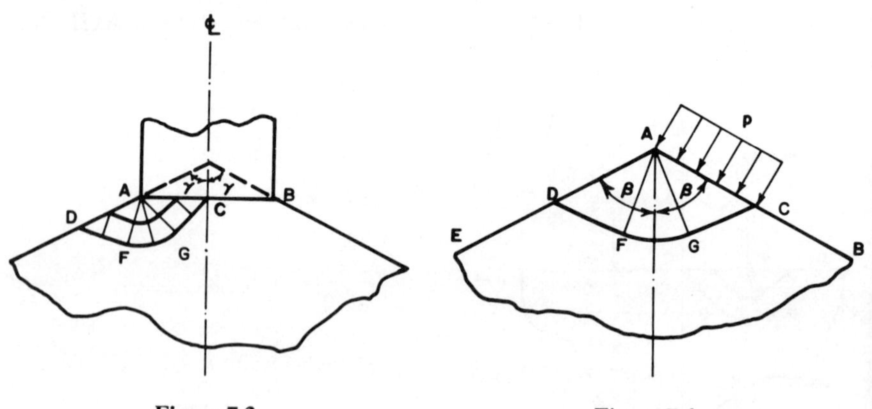

Figure 7.3   Figure 7.4

and region AGF is a centered fan sector. The semiinternal angle is β. The pressure at incipient flow, for $0 < \beta < \pi/2$, is

$$p = 2k(1 + 2\beta - \pi/2) \quad . \qquad [7.17]$$

E. COMPRESSION OF A WEDGE BY A FLAT DIE

The apex 0 of an infinite wedge of rigid-perfect plastic body with total internal angle, 2β, is compressed[79] symmetrically by a lubricated flat die, see Figure 7.5. The configuration preserves its geometrical similarity as the wedge is progressively flattened. The plastic-die interface at any moment is AA with B as the center of the die face. Although it can be inferred from the velocity boundary condition, let the total load, P, normal to AA be evenly distributed. Also let there be no shear tractions. Then the slip-line field discussed in Section 7.A takes the form of ABDEC with ABD and ACE being right isosceles triangles and ADE being formed by a fan net. Letting the internal angle be α, the contact

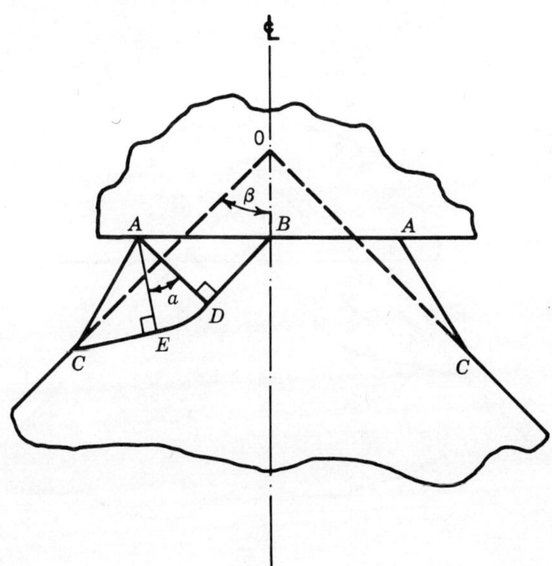

Figure 7.5

distance be *l* and the height of compression be *h*, the following are given by the slip-line theory:

$$P = 2k(1 + \alpha)\, \ell$$

$$\ell = 2d(1 + \sin\,\alpha)/\cos\,\alpha \quad , \qquad\qquad [7.18]$$

where

$$\tan\,\beta = (1 + \sin\,\alpha)^2/\cos\,\alpha(2 + \sin\,\alpha) \quad .$$

It may be shown that $\alpha \to 0$ as $\beta \to 26.6°$; this corresponds to the case where AC becomes vertical. Physically, this suggests that the very thin wedges are unstable. It should be noted that wedge deformation may well lead to an antisymmetric mode.[88]

In the case where the die is not frictionless,[89] Figure 7.6 shows the slip-line field which is similar to the one shown in Figure 7.5. The differences lie in the fact that Ψ and α are distinct from each other and θ is no longer 45°. Letting the coefficient of friction be μ and the corresponding traction due to friction be $\tau \equiv k \cos\theta$ with $s \equiv \tau/k$, the

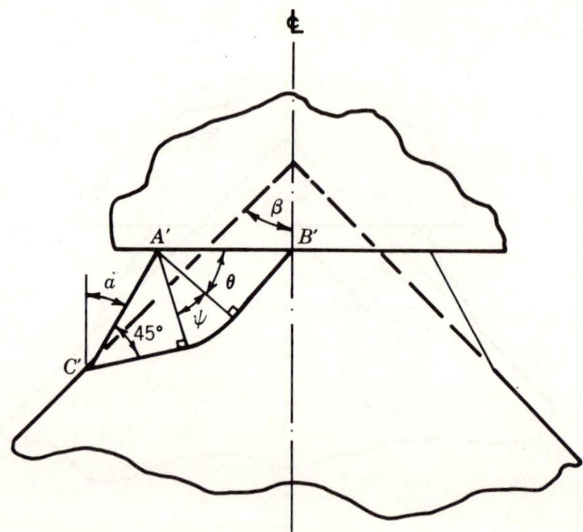

Figure 7.6

equations corresponding to [7.18] are

$$P = k\ell[1 + \pi/2 - \cos^{-1} s + (1 - s^2)^{1/2} + 2\alpha]$$

$$\ell = 2d[1 + (\sin \alpha)(1 + s_2)^{1/2}]/(\cos \alpha)(1 + s)^{1/2} \quad ,$$

$$[7.19]$$

where

$$\tan \beta = [1 + (\sin \alpha)(1 + s)^{1/2}]^2/(\cos \alpha)(1 + s)^{1/2}[2 + (\sin a)(1 + s_2)^{1/2}] \quad ,$$

$\Psi = \pi/2 - \theta - \alpha, 0 \leq s \leq 1$, and $\cos^{-1} s$ refer to principal value only. Of course, [7.19] degenerate into [7.18] for $s = 0$.

F. SLIDING OF A WEDGE UNDER A FLAT DIE UNDER LOAD

Figure 7.7 refers to a truncated wedge of internal angle 2β and width w in sliding condition. In this case, the equations comparable to [7.18] are

$$P = k\ell(1 + 2\psi + \sin 2\phi)$$

$$\ell/w = 2(1 + \beta)/(1 + 2\psi + \sin 2\phi) \qquad [7.20]$$

$$\ell/w = \left[(1 - 2H)^2 - 2H \tan \beta \left\{ G \right. \right.$$

$$\left. \left. + \sin \phi[\cos(\psi - \phi) - \sin(\psi - \phi)] \right\} \right]^{-1/2} \quad ,$$

where

$$G = - \left\{ \tan \phi[\cos(\psi - \phi) - \sin(\psi - \phi)] - \cos(\psi + \phi) - \sin(\psi + \phi) \right\} F$$

$$H = \left\{ \tan \beta[\cos(\psi + \phi) + \sin(\psi + \phi)] - \sin(\psi + \phi) + \cos(\psi + \phi) \right\} F$$

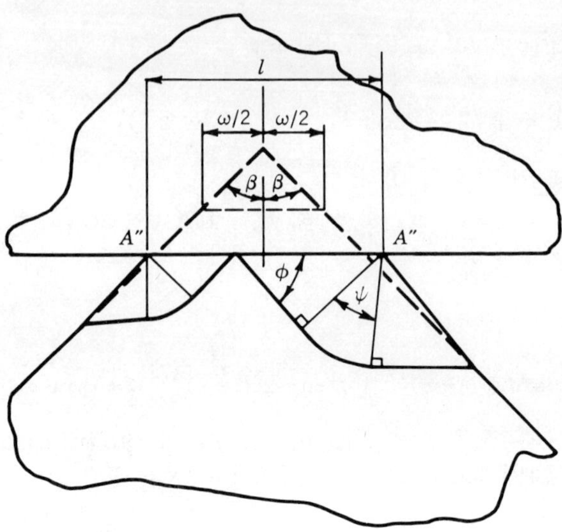

Figure 7.7

in which

$$F = \sin \phi \left\{ \tan \alpha [\cos(\psi - \phi) - \sin(\psi - \phi)] \right.$$

$$\left. - \cos(\psi - \phi) - \sin(\psi - \phi) \right\}$$

$$\cdot \left[\tan \phi \left\{ \tan \beta [\cos(\psi - \phi) - \sin(\psi - \phi)] \right. \right.$$

$$\left. - \cos(\psi - \phi) - \sin(\psi - \phi) \right\}$$

$$+ \tan \beta [\cos(\psi + \phi) + \sin(\psi + \phi)]$$

$$\left. + \cos(\psi + \phi) - \sin(\psi + \phi) \right]^{-1} .$$

In the above, shear traction $\tau = k \cos 2\phi$. For given values of β and ϕ, the last two equations in [7.20] must be solved numericaly for Ψ and l/w. This has been done[90] for $\beta = 45°$ and $\beta = 85°$.

G. INDENTATION OF A SEMIINFINITE SOLID BY A LUBRICATED WEDGE

A problem which is very similar to the flattening of a wedge by a die is shown in Figure 7.8. The slip-line field[91] covers ABDEC. This in turn consists of two right isosceles triangles and a fan net.

For a semiwedge angle β and in the absence of shear traction, which is the case in Figure 7.8,

$$p = k(1 + \alpha)$$

$$\cos(2\beta - \alpha) = \cos \alpha/(1 + \sin \alpha) \quad , \qquad [7.21]$$

where p is the pressure at the wedge interface. This pressure remains the same for all depths of indentation since there is a similarity and therefore a proportionality in the configurations for all depths of penetration.

H. INDENTATION OF A SEMIINFINITE SOLID BY A CYLINDER

Slip-line theory has been used[92] to solve problems involving the curve indenter and a semiinfinite solid with constant shear traction along the

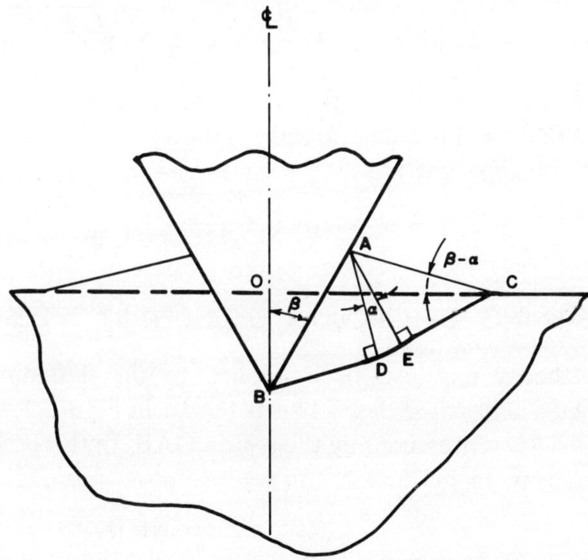

Figure 7.8

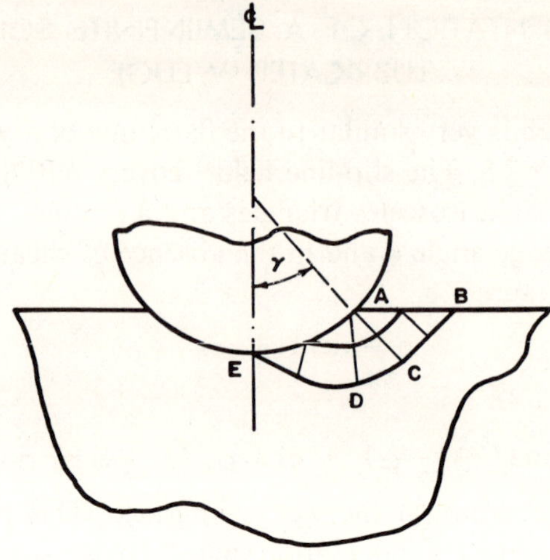

Figure 7.9

line of contact. An example is shown in Figure 7.9 for a cylindrical indenter. Half of the slip-line field is shown by the region ABCDE. The pressure normal to the surface for a general curved surface within the contact region is

$$p = 2k(\phi + \frac{\pi}{2} + \frac{1}{2} + \phi_0) + \sqrt{k^2 - \tau} \quad , \quad [7.22]$$

where ϕ is the angle between the tangent to the contour of the body and the axis AB, $\phi_0 = \frac{1}{2}$ arc sin τ/k and τ is the shear traction. For a circular cylinder with zero traction

$$p = 2k(1 + \frac{\pi}{2} - \phi) \quad . \quad\quad\quad [7.23]$$

I. FLATTENING OF CIRCULAR CYLINDER BY A LUBRICATED DIE

Slip-line theory has also been applied to the flattening of circular cylinder by a lubricated die.[92] This is shown in Figure 7.10. Half of the slip-line field is represented by the region OAB. In this case the pressure is nonuniform. In particular, for $\gamma = 15°$,

$$p = \begin{cases} 3.92k \text{ at point O} \\ \\ 4.62k \text{ at point A} \end{cases} \quad\quad [7.24]$$

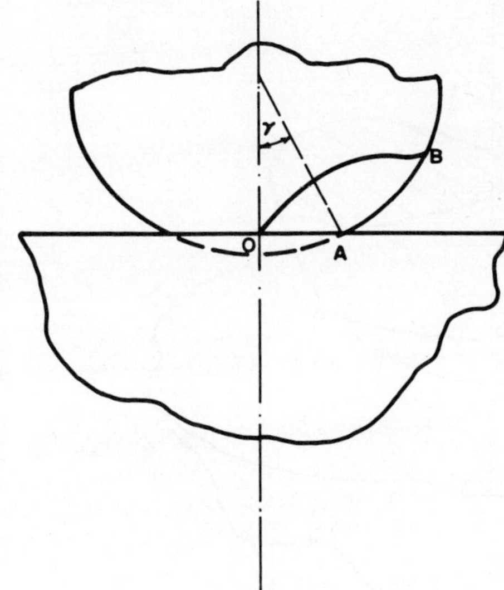

Figure 7.10

J. INDENTATION OF A SEMIINFINITE SOLID BY A LUBRICATED SPHERICAL DIE

Figure 7.11 shows a semiinfinite solid being indented by a lubricated spherical die. The slip-line field was computed by finite difference method.[93] Again, the normal pressure at the interface is nonuniform. As an example, for $r = 1063$ mm and $h = 78$ mm,

$$p = \begin{cases} 7.92k \text{ at point O} \\ \\ 4.38k \text{ at point A} \end{cases} \qquad [7.25]$$

K. INDENTATION OF A SEMIINFINITE SOLID BY THE END OF A LUBRICATED CYLINDER

The problem[84] concerns the indentation of the plane surface of a semi-infinite body of rigid-perfect plastic body by the end of a lubricated circular cylinder. Referring to Figure 7.12, the stress field in region ABC is generated by the stress free surface AB of the body. The line AC is an α-line and BC is a β-line. The field in the region OACD is determined by the slip-line AC and the condition of zero shearing stress on OA.

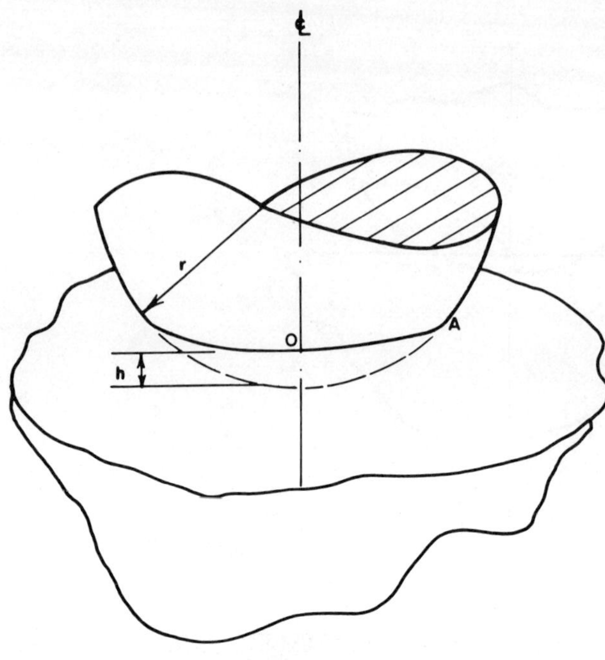

Figure 7.11

Point A is a singular point and, together with the characteristic AC, generates the fan-shaped region ACD of angle $\pi/2$. The region OAD is determined by the characteristic AD and the condition that the slip-lines meet OA at angles of $\pi/4$. The field of the region OACD is obtained by numerical integration.

The results on the pressure generated are shown in Figure 7.12 with an average value of $5.69k$ and OB is 1.58 times the distance OA.

L. INDENTATION OF A SEMIINFINITE SOLID BY A LUBRICATED TRUNCATED CONE

The analysis of the stress field associated with the indentation of a semi-infinite solid by a lubricated truncated cone is given in great detail.[94] For incipient deformation, where the flat surface remains flat, as shown in Figure 7.13, the contact pressure

$$p = 2k(1 + \gamma + \mu) \quad , \tag{7.26}$$

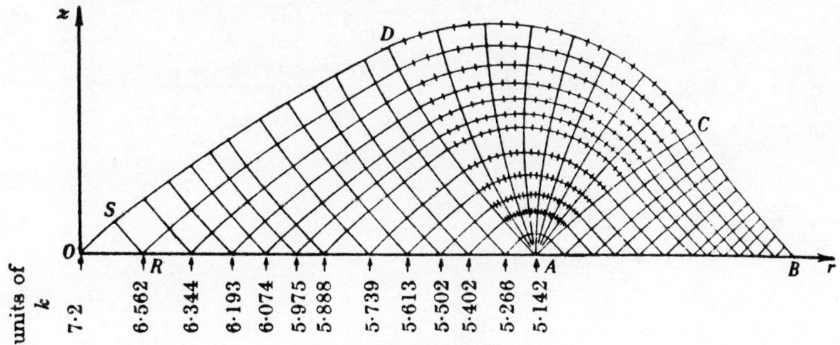

Figure 7.12

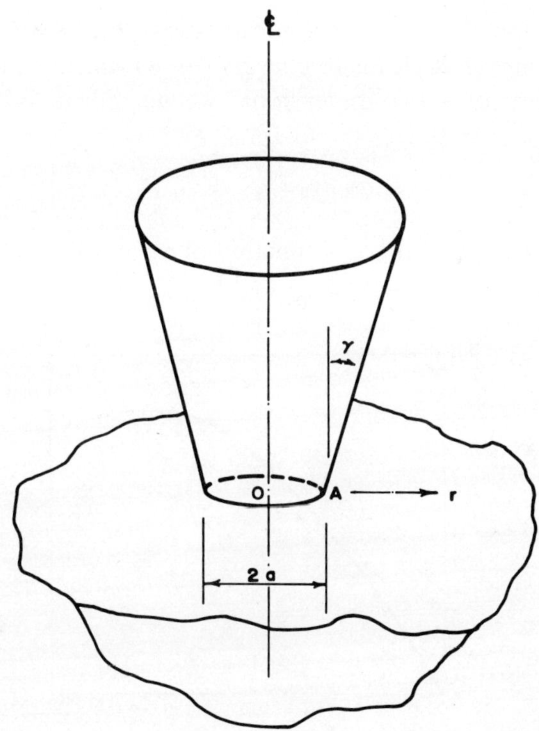

Figure 7.13

where

$$2\mu = [1 - \tan(\pi/4 - \gamma)]\,\ln\left[1 + \frac{\alpha\,\cos(\pi/4 - \gamma)}{1 - \alpha\,\cos(\pi/4 + \gamma)}\right]$$

$$+ \ln\frac{1 - \alpha\,\cos(\pi/4 + \gamma)}{1 - \alpha\,\cos\pi/4} + \gamma$$

$$- 2(1 - \alpha^2)^{-1/2}\left\{\arctan\left[\tan(\pi/8 + \gamma/2)\,(\tfrac{1 + \alpha}{1 - \alpha})^{1/2}\right]\right.$$

$$\left. - \arctan\left[\tan\pi/8\,(\tfrac{1 + \alpha}{1 - \alpha})^{1/2}\right]\right\}\;.$$

α is the direction of the principal normal stress. For $\alpha = 0$, which corresponds to $r = a$, [7.26] leads to $p = 2k(1 + \gamma)$ which is the solution for the indentation of a two-dimensional wedge. For $\alpha = \sqrt{2}/2$, which corresponds to $r = 0$, let

$$p = 2k(1 + \gamma + \mu_0)\;. \qquad\qquad [7.27]$$

μ_0 is shown in Figure 7.14 as a function of γ.

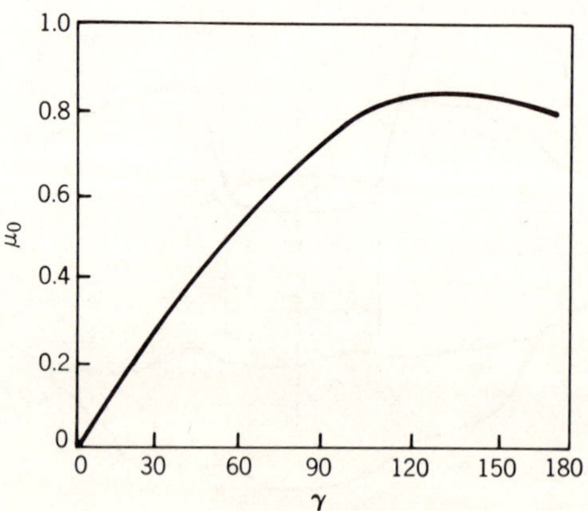

Figure 7.14

CHAPTER 8

ROUGH SURFACES

Surface roughness is of interest for various reasons and methods used, therefore, are dependent on the nature of technical fields concerned. For example, adsorption methods[95] are used for measuring the surface area of rough surfaces while other methods are designed to provide such information on surfaces as they may be needed for electromagnetic radiation,[96] heat transfer,[24] electrical contacts,[97] friction and wear,[98] and lubrication.[99,100] Actual methods, besides the adsorption method, for quantitative studies of the macrofeatures may be mechanical,[101,102] optical,[103] electromechanical,[104] or radiographic.[105] The advent of the scanning electron microscope should offer other possibilities, especially with respect to surfaces that are too soft for stylus incursions.

For the purposes of this chapter metal surfaces are inferred, although discussions are applicable to surfaces other than metal ones. One reason for emphasizing metal surfaces is the predominance of earlier studies and the continued use of metal surfaces.

A. BEARING AREA CURVES

As an example of metal surfaces, Figure 8.1 shows a standard norm developed by the Daimler-Benz Company in Germany. The roughness

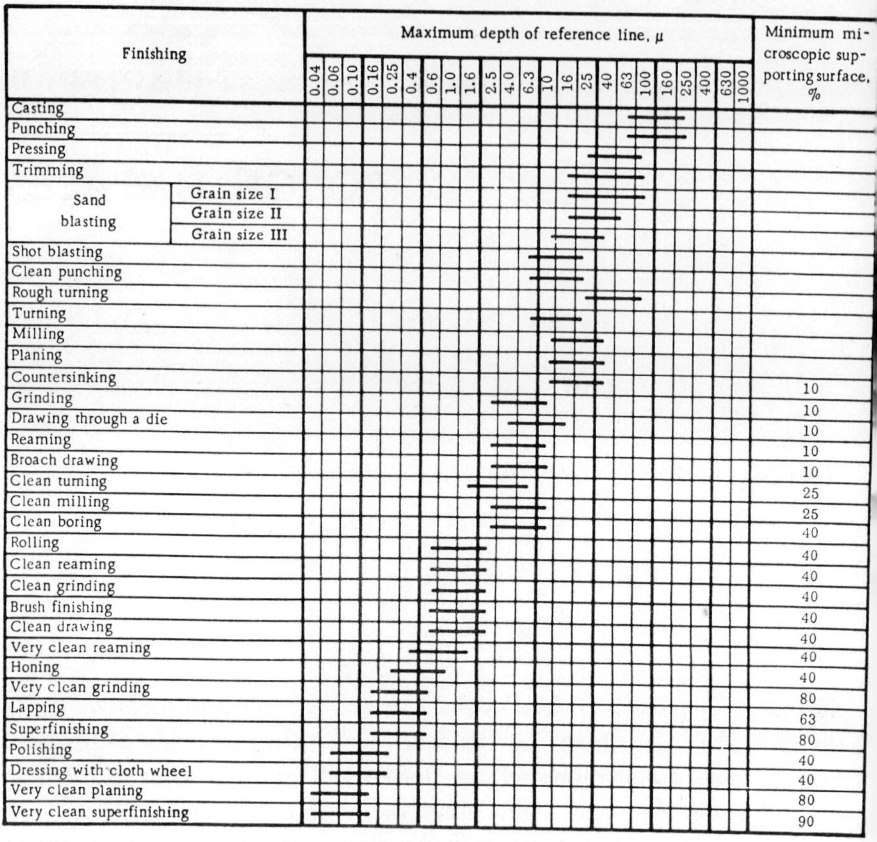

Finishing		Maximum depth of reference line, μ	Minimum microscopic supporting surface, %
Casting		63 – 250	
Punching		40 – 160	
Pressing		40 – 160	
Trimming		40 – 160	
Sand blasting	Grain size I	40 – 100	
Sand blasting	Grain size II	40 – 100	
Sand blasting	Grain size III	25 – 63	
Shot blasting		16 – 63	
Clean punching		16 – 63	
Rough turning		25 – 100	
Turning		16 – 63	
Milling		10 – 40	
Planing		10 – 40	
Countersinking		16 – 63	
Grinding		2.5 – 10	10
Drawing through a die		6.3 – 25	10
Reaming		6.3 – 16	10
Broach drawing		6.3 – 16	10
Clean turning		4.0 – 16	10
Clean milling		6.3 – 25	25
Clean boring		6.3 – 25	25
Rolling		1.6 – 4.0	40
Clean reaming		1.6 – 4.0	40
Clean grinding		1.6 – 6.3	40
Brush finishing		1.6 – 6.3	40
Clean drawing		1.6 – 4.0	40
Very clean reaming		1.0 – 2.5	40
Honing		0.6 – 2.5	40
Very clean grinding		0.4 – 1.6	80
Lapping		0.4 – 1.0	63
Superfinishing		0.4 – 1.6	80
Polishing		0.1 – 0.4	40
Dressing with cloth wheel		0.1 – 0.4	40
Very clean planing		0.06 – 0.25	80
Very clean superfinishing		0.04 – 0.16	90

The depth scale along the top of the chart reads: 0.04, 0.06, 0.10, 0.16, 0.25, 0.4, 0.6, 1.0, 1.6, 2.5, 4.0, 6.3, 10, 16, 25, 40, 63, 100, 160, 250, 400, 630, 1000.

Figure 8.1

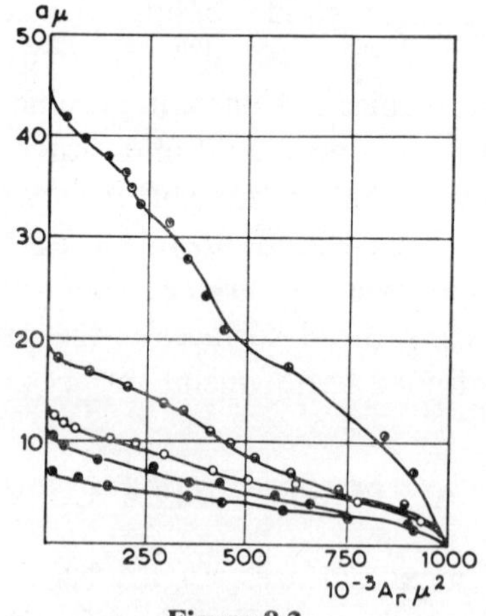

Figure 8.2

of a surface was estimated by the maximum depth of the reference line on the finished surface. The supporting surface was expressed as the percentage of actual contact area in a measured length of the specimen under consideration. The earliest record of specifying surface[106] dates back to 1933 when the notion of bearing area curves was used. Figure 8.2 shows a set of these curves[107] for different classes of surface finish after milling and planing. These are plots of the distance between surfaces in contact, a, versus real area of contact, A_r.

B.   PROFILOMETRIC REPRESENTATION OF SURFACES

In recent years, with the aid of high-speed computers, topographical information has been accrued by stylus profilometers. A typical profile trace is given at the top of Figure 8.5.   This method is most suitable for metal surfaces or replicas of surfaces which may be too soft. A recent method is to take the digital topographical information, whether from profilometric or other forms of measurements, and remove regular waviness leaving the random roughness. This is done by assigning a

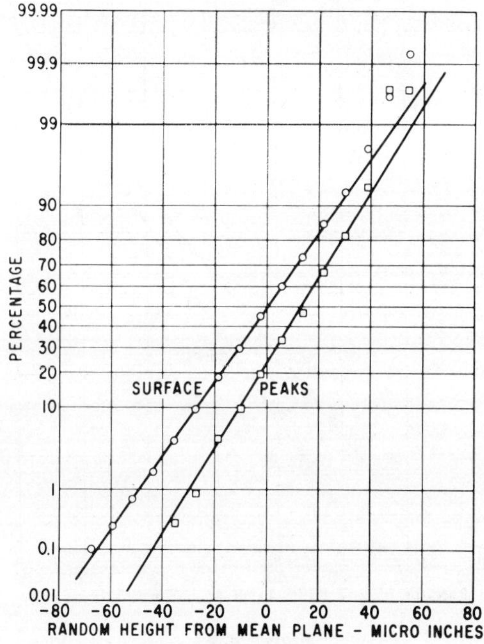

Figure 8.3

double Fourier series to the regular waviness part,[108] and the remainder may then be analyzed for their actual distribution. Such random data have been obtained for two samples that are electrochemically machined. These are shown in the form of cumulative probability distribution data in Figures 8.3 and 8.4 for a smoother and a rougher surface, respectively.

More recently, surface characterization has been further delineated[101] according to the method of preparation: (1) Cumulative processes; (2) extreme value processes; and (3) transitional processes. The first involves a large number of repeated events which occur randomly over the surface. Figure 8.5 shows a plot of cumulative height distribution versus height as an example of the near Gaussian height distribution for this class. The dots, crosses, and triangles represent data for heights, profile peaks, and summits, respectively. The second involves preparation techniques which lead to stratified textures. Figure 8.6 shows an example of the cumulative height distribution versus height plot. The third, an example of which is shown in Figure 8.7, involves starting

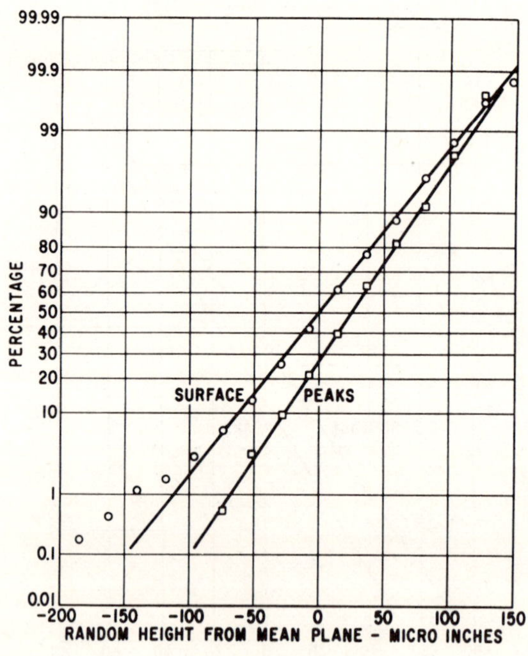

Figure 8.4

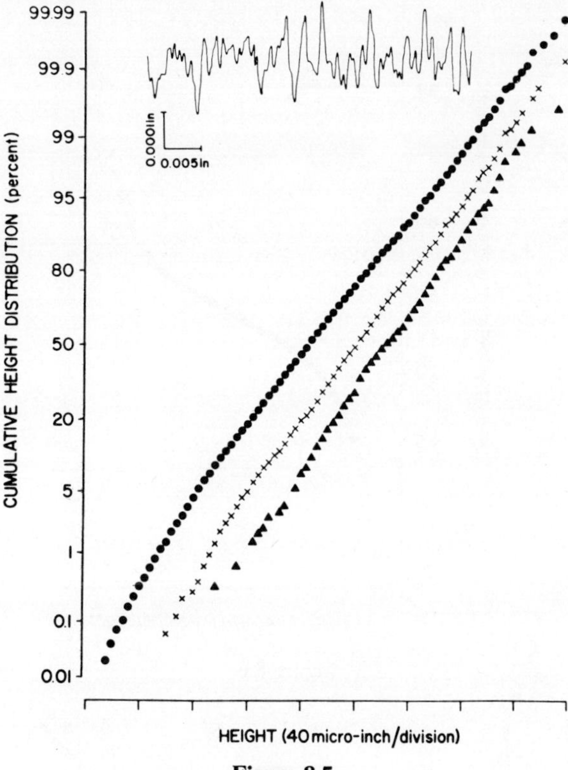

Figure 8.5

out with a Gaussian distribution and having the surface undergo transitions through processes such as wear.

C. CHARACTERIZATION OF SURFACES BY AUTOCORRELATION FUNCTIONS

The use of autocorrelation functions to characterize surface profiles[109] was introduced in 1959. Correlation techniques were applied to all kinds of surfaces[104] and autocorrelation functions and dispersion spectra were used to establish the required reference lengths for surface roughness measurements.[110] Later, additional numbers were proposed to characterize surfaces. These are the first and second derivatives of surface profiles and their ratio.[111] These were intended to give the direc-

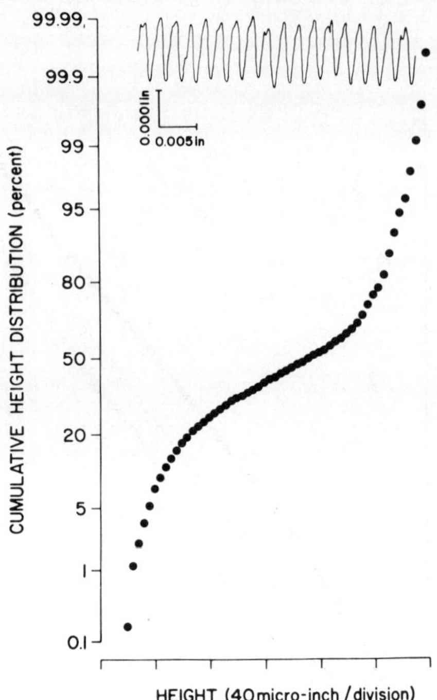

Figure 8.6

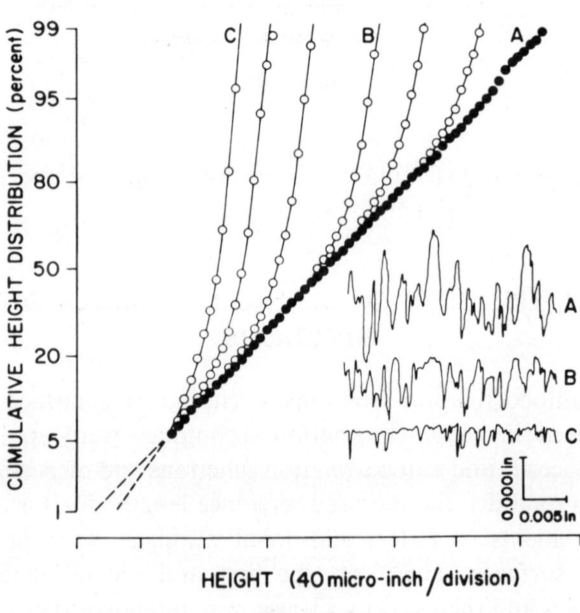

Figure 8.7

[178]

tional properties of surfaces as they are measured. It should be noted, however, it has been shown that for a random surface, a profilometric representation along a straight line is sufficient to characterize the surface.[112]

If $y(x)$ is the height representation of a profile trace with x the distance along the line with respect to which the height is measured, then

$$f(y) = (2\pi)^{-1/2} \exp(-\frac{1}{2} y^2) \qquad [8.1]$$

is the normal distribution. Here y has been normalized by the standard deviation, σ of the height distribution. The autocorrelation function

$$C_y(\beta) = \lim_{L \to \infty} \frac{1}{L} \int_{-L/2}^{L/2} y(x)y(x + \beta)dx \quad . \qquad [8.2]$$

If $P(\omega)$ is the power spectral density function of random waveform,

$$C_y(\beta) = (2\pi)^{-1} \int_{-\infty}^{\infty} P(\omega)\exp(i\omega\beta)d\omega \quad . \qquad [8.3]$$

A three point analysis is used[102] for defining a peak; that is, a peak is said to be occurring where one reading is higher than its neighboring readings in a height sampling given the same sampling interval. It has been shown that if the height distribution is Gaussian, then the peak height distribution is also approximately Gaussian for all sampling intervals. For large sampling intervals, the probability of finding a peak is one-third while, for short sampling intervals, the probability is one-fourth. The peak curvature distribution is a function of peak height. Theory and experiment show that the peak curvature distribution exhibits a general tendency to be skewed towards zero curvature.

While the complete characterization of surface roughness requires the autocorrelation function of surface profiles be known, a simpler measure may be found in a pair of numbers which are theoretically sound[113] and may be computed from surface profiles. Results have been verified by experiment. This is true for stationary random processes. The two numbers are the standard deviation of the slope and the mean thickness of the surface profile. The former contains the behavior of the auto-correlation function of a surface profile.

For a stationary profile, the standard deviation of the slope requires the counting of crossings between the profile $y(x)$ and the level $y = a$ investigated. This gives the average number of crossings per unit length.

D. CHARACTERIZATION OF SURFACES BY ACTUAL AREA OF CONTACT

In most surface contact situations, actual area of contact refers to the collection of areas of contact as they are projected on a plane which passes through the interface zone. It is used because of convenience and it is a good approximation since the slopes of asperities are gentle. In fact, for a large percentage of time, surface roughness is studied because of the fact that actual area of contact is desired. Such information is sought for electrical or mechanical reasons.

During the mid-1950s, several papers appeared addressing to the question of Amonton's second law of friction which says that frictional resistance is directly proportional to the applied normal load. These papers involved models of surfaces engaged in multiple contacts. One paper examined factors influencing the area-load relationship in plastic contact[89] while another examined the elastic contact of surfaces with small protuberances on top of larger protuberances.[114] In the latter, it was found, for the first time, Amonton's second law could hold even for elastic contact though classically it was thought possible through plastic contact only. Moreover, it was shown that the geometric features of the particular model used may not be of the greatest significance in seeking an explanation of Amonton's second law. Any increase of load is shown to be consumed in creating new areas rather than enlarging old ones. Later, based on profilometric data, it was found that the average size of micro contact is nearly constant.[112] Other papers include the elastic and plastic contact of surfaces. One of these gives an elaborate, modelling of the surfaces and method of determining actual area of contact.[105] Another deals with two surfaces that are pressed together under given conditions. The process of change of roughness is made with a transformation law from distribution of the pressed surface profile to that of the initial profile.[115]

A simple analysis which includes elastic and plastic effect is given here in some detail.[107] Letting A_r be the actual area of contact or real area of contact and A_c be the area formed by bulk compression of the waviness of surfaces, a be the distance between surfaces, and $h_{\max}$ be the maximum height of the peak, then it is convenient to introduce the dimensionless variables $\eta_1 = A_r/A_c$ and $\epsilon = a/h_{\max}$. Experimentally, it has been found that

$$\eta_1 = \epsilon^{x_1} \, e^{x_2 \left(1 - \frac{1}{\epsilon}\right)}, \qquad [8.4]$$

where x_1 and x_2 are constants depending on surface treatment and ϵ is shown in Figure 8.8 for five types of finishing processes. It is the same set of bearing area curves shown in Figure 8.2 as they are replotted in the dimensionless variables.

Since ϵ is $\ll 1$, a simpler form is

$$\eta_1 = b \, \epsilon^m, \qquad [8.5]$$

where, for turning, planing, and milling, b is between 1.6 and 3, and $m = 2$; for grinding, $b = 5$ and $m = 3$; for polishing, b is between 10 and 16 and $m = 3$.

Now assuming the asperities have spherical caps of radius r, the critical value of a, a_k, at which transition takes place from elastic to plastic contact, is given by

$$a_k = 5.35r \, \Psi^2, \qquad [8.6]$$

where $\Psi = (1 - \mu^2)p_m/E$, μ is Poisson's ratio, E is Young's modulus, and p_m is the yield pressure.

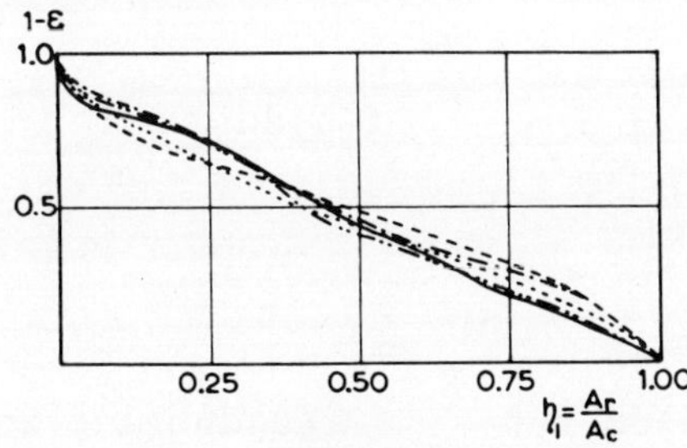

Figure 8.8

Recasting [8.6], the dimensionless parameter

$$(E/p_m) \sqrt{a_k/r} = 2.31 \sqrt{1 - \mu^2} \quad .$$

In a later reference,[112] a dimensionless parameter akin to the one above is referred to as a plasticity index.

E. CHARACTERIZATION OF SURFACES BY COMPLIANCE

Related to the section above but somewhat different in purpose is the compliance of rough surfaces in contact. Here, compliance is defined as the approach of two reference lines, each imbedded in one of the two mating surfaces in their undeformed state. This information may be sought with regard to sealing of micromolecular flow, for example.

Let $f(u)$ be the probability density of asperity heights as given by a stylus traverse along a straight line. Let $-l\sigma \leq u \leq l\sigma$ be the measure of height in terms of the standard deviation σ of the data in a sampling interval, L. In other words, contact begins at $u = l\sigma$. The real area of contact, A, over a rectangular apparent area $L_x \cdot L_y$ is[116] assumed to be represented by

$$A = \frac{\displaystyle\int_0^{L_x} \int_0^{L_y} \int_u^{l\sigma} f(u)\,du}{\displaystyle\int_{-l\sigma}^{l\sigma} f(u)\,du} \quad . \qquad [8.7]$$

Letting $A = W/p_m$, where W is the total load, [8.7] leads to

$$L_x L_y \int_u^{l\sigma} f(u)\,du = \frac{W}{p_m} \int_{-l\sigma}^{l\sigma} f(u)\,du$$

or

$$\int_0^u f(u)\,du = \left[\int_0^{l\sigma} f(u)\,du \right] (1 - 2W/L_x L_y p_m) \quad . \quad [8.8]$$

Letting $u = t\sigma$, and for a normal distribution, [8.8] becomes

$$\int_0^t \frac{1}{\sqrt{2\pi}} \exp\left(-\frac{t^2}{2}\right) dt = \Phi(\ell)\left(1 - \frac{2W}{L_x L_y P_m}\right) \quad,$$

where

$$\Phi(\ell) = \int_0^\ell \frac{1}{\sqrt{2\pi}} \exp\left(-\frac{t^2}{2}\right) dt \quad.$$

The $\Phi(l)$ may be evaluated for given values of l, and, given $L_x L_y$ and W, $u = t\sigma$ may be estimated. Letting the depth of penetration be $l\sigma - u$,

$$U = (\ell - t)\sigma \quad. \tag{8.9}$$

As the maximum height of asperities, H, is given by $H = j\sigma$,

$$U = (\ell - t)Hj \quad. \tag{8.10}$$

For the contact between two surfaces H is replaced by $(H_1^2 + H_2^2)^{1/2}$ where H_1 and H_2 are the maximum heights of asperities of the two surfaces, respectively. These results have been shown[116] to be reasonably good for loads that are not very large. This limit may be expressed in terms of depth of penetration equalling a quarter of the maximum height of asperities in the case of contact between a flat and a rough surface. This is also the range in which the assumption of Abbott's bearing curve is valid and of noninteracting adjacent asperities is permissible.

For larger depths of penetration, a more involved theory is required in the face of experimental evidence which shows a steeper load-compliance data for higher load than the theory above would accommodate, see Figure 8.9.

In the past, several analyses[89,114] of bulk surface quantities such as actual area of contact assume that one of the surfaces is flat and rigid while the other is assumed to be rough and deformable. The rough surfaces are assumed to be composed of asperities distributed in various forms.[117] Table 5 shows four forms of distribution which are normalized. The x is the distance measured from the reference plane, the plane which contains the tip of the outermost asperity in its undeformed state, to the tip of a generic asperity in its undeformed state.

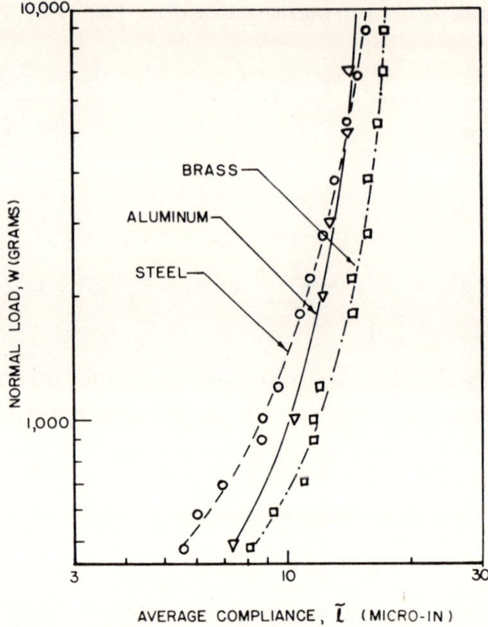

Figure 8.9

Table 5.   Some Distribution Functions, $p(x)$, Used

A	Uniform	$p(x) = 1, \int_0^1 p(x)\,dx = 1$
B	Linear	$p(x) = 2(1 - x), \int_0^1 p(x)\,dx = 1$
C	Normal	$p(x) = (1/\sigma\sqrt{2\pi})\ \exp\ [-(x - h/2)^2/2\sigma^2]$, σ is so selected that $\int_0^1 p(x)\,dx \simeq 1$
D	Poisson	$p(x) = e^{-m}\,m^x/x!, \sum_{x=0}^{m} e^{-m}\,m^x/x! \simeq 1$, where n is the number of discrete divisions; $0 \le x \le n$ is interpreted as the number of integer divisions.

Table 6 shows how the yield load, P, is related to the deformation of these simple shapes assuming constant yield stress; y^* is the distance which the rigid flat has moved toward the rough surface from the reference plane; hence it is the measure of compliance. The $y^* - x$ is the deformation of a single shape originating at x.

Table 6.   Some Simple Shapes Used

---

a	Single rigid, perfectly plastic wedge or hemisphere deformed by a rigid flat; $P \sim (y^* - x)$
b	Single rigid, perfectly plastic cone deformed by a rigid flat; $P \sim (y^* - x)^2$

The bulk load-compliance of the ensemble is

$$W^* = \int_0^{y^*} P(x,y^*)p(x)dx \quad , \qquad [8.11]$$

where W^* is the bulk load. Figure 8.10 shows a log-log plot of the load-compliance characteristics for eight combinations of shapes and distributions, for example, A_a represents the case in which wedges of hemispheres (Table 6) uniformly distributed (Table 5) are compressed.

It should be noted that in all these cases, no consideration is made of interaction of neighboring asperities,[89] nor has the effect of failure of Kicks' law been considered.[118] Note that the slopes of the eight are all below $3\frac{1}{2}$. It is known that interaction of asperities or the fusing of asperities of the same surface would give higher values of slopes to the load-compliance. However, experiments, Figure 8.9, show the possibility of slopes of 6.

The answer for higher slopes must be found in contact of rough surfaces. Take the case in which asperities are in the form of cones. When both surfaces are rough, a cone will carry load only when it bears on another cone on the mating surface. Besides, only those cones which are in direct alignment on opposite surfaces will bear on each other. As the surfaces approach each other, cones which were in the vicinity of being in direct alignment will bear on one another. It is seen that

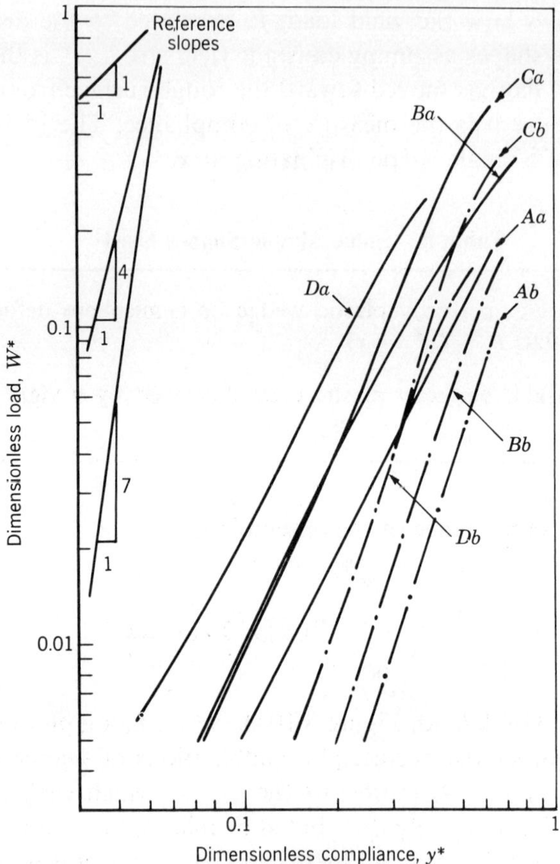

Figure 8.10

this vicinity embraces a larger and larger area as surfaces are drawn closer and closer. The probability of cones coming into contact increases as the surfaces approach each other. This is certainly true for randomly distributed cones.

The progression of the number of contacts with compliance is seen to be geometric. Within limits, the area of contact of a given pair in contact increases as the surfaces approach each other; the deformation load for the given pair increases accordingly. Therefore, it stands to reason that by considering both surfaces rough, the slope of the load-compliance characteristics would be higher than the case in which one of the surfaces is flat and rigid. In the literature, it is often said that contact of two rough surfaces is not different from one rough surface containing a rigid flat.[112] This statement is true in the theoretical contact of surfaces with phantom bodies. In reality, with the general case of two asperities not aligned, the picture is quite different. Not only is the nature of contact different but also the mode of deformation (i.e., elastic or plastic) is different.[119] It has been shown that even under the lightest of loads for surfaces characterized by profilometry and asperities in the form of spheres, about half the load carrying asperities are beyond the range of elastic deformation, that is, the apparent pressure is of the order of 10 psi. By apparent pressure, it is meant load per apparent area. For most pressures encountered in engineering, say 150 psi apparent pressure, 90% of the mating asperities are in the plastic range and therefore results based on elastic analysis have no real meaning.

Figure 8.11 shows schematically two rough surfaces in contact. Reference plane is the plane which contains the tip of the outermost asperity of a given collection of asperities on a surface in its underformed state; thus there are two reference planes. This distance between the reference planes, l, will be designated compliance. Each surface is characterized by a parameter, h, the distance between the outermost

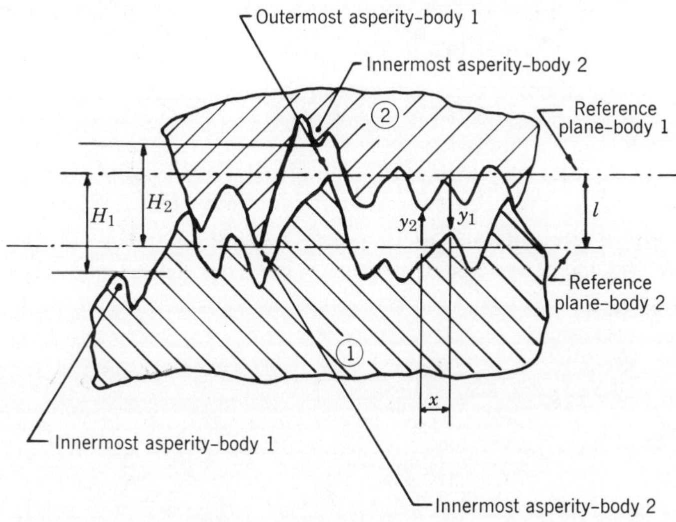

Figure 8.11

and the innermost asperities in their undeformed states. Each asperity is characterized by a parameter, y, the distance from the reference plane to the tip of its undeformed state. Two potential mating asperities are characterized by two parameters, r and θ: the former is the projection of the distance between the tips of the asperities on a reference plane and the latter is the angle between r and some reference direction.

The above is convenient for study of surfaces which do not have waviness underlying asperities. With the advent of the common use of computers, elaborate models of asperities may be used in conjunction with asperity distribution functions to obtain such quantities as compliance, area of contact. This was briefly illustrated recently.[119] However, in view of the result shown in Figure 8.10 and the purpose of illustrating the steepness of load-compliance characteristics, the model advanced in 1958 is shown below.[117]

1. All asperities are in the form of symmetrical cones of semi-opening angle β

2. Two surfaces are similar, that is, $h_1 = h_2$.

3. The probability of a cone on body 1, finding a cone on body 2, in the "immediate vicinity" is S_p. Immediate vicinity means the volume $\pi (h \tan \beta)^2 \cdot h$.

4. Asperities are randomly distributed.

For a large number of asperities, the probability to commence the mating of a pair of asperities in the "immediate vicinity,"

$$S_{r\theta y_1 y_2} = S_p (r \, \Delta r \, \Delta \theta / \pi \, h^2 \tan^2 \beta)(\Delta y_1/h)(\Delta y_2/h), \quad [8.12]$$

where Δr, $\Delta \theta$, Δy_1, and Δy_2 represent, respectively, the average value of r, θ, y_1, and y_2 between two asperities which are the closest. Hence $\Delta y_1/h$ is the probability of one asperity having $y_1 = \Delta y_1$, $2 \Delta y_1$, . . ., $N \Delta y_1$, where $N = h/\Delta y_1$. For the case with cylindrical symmetry, [8.12] leads to

$$S_{ry_1 y_2} = S_p (2r \, \Delta r \, \Delta y_1 \, \Delta y_2 / h^4 \tan^2 \beta) . \quad [8.13]$$

Of course,

$$\sum_r \sum_{y_1} \sum_{y_2} S_{ry_1 y_2} = S_p .$$

Obviously the load developed between a pair of asperities,

$$P = \begin{cases} 0 & (\ell \leq \ell_0) \\ \\ P & (\ell > \ell_0) \end{cases} \qquad [8.14]$$

where $\ell_0 = r + y$ and $y = y_1 + y_2$.

Postulate the following deformation process:

1. Immediately after contact of a pair of asperities has been initiated, the contact area grows in the form of an ellipse having roughly dimensions shown in Figure 8.12.

2. Plastic deformation commences immediately.

3. There exists a constant yield pressure p_m.

4. After this mode of deformation has undergone an amount indicated by an angle Ψ, Figure 8.12, fracture will shear off a piece of one of the tips or the tip of one of the asperities will be pushed off in large plastic flow.

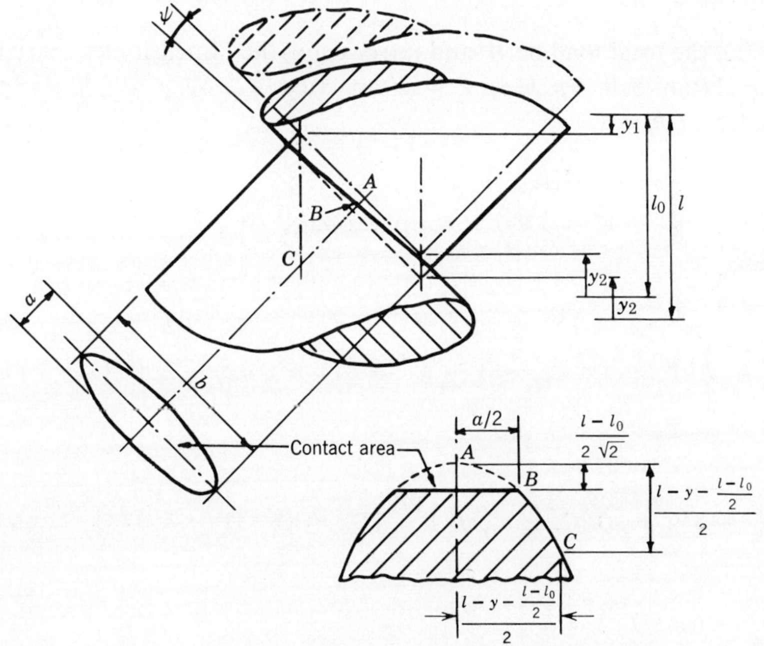

Figure 8.12

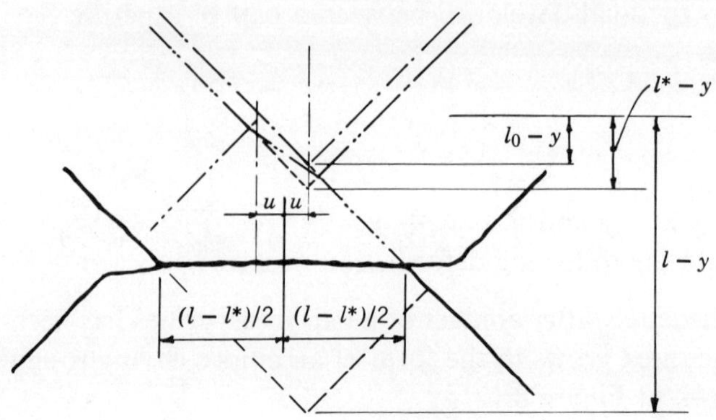

Figure 8.13

5. A second mode of larger deformation which consists of crushing of one cone on another will follow. The new area of contact, which has been enlarged, is shown in Figure 8.13. The l^* is the value of l corresponding to $\Psi = \epsilon$, some value which demarcates the first mode from the second.

Letting the total load be W and introducing the dimensionless variables $\tilde{W} = 2\,(\tan^2\beta)W/\pi p_m h^2 S_p$, $\tilde{P} = 2(\tan^2\beta)P/\pi p_m h^2 S_p$, $\rho = r/h$, $\eta = y/h$, and $\lambda = l/h$,

$$\tilde{W} = \sum^{\rho} \sum^{\eta_1} \sum^{\eta_2} \rho\, \Delta\rho\, \Delta\eta_1\, \Delta\eta_2\, \tilde{P} \quad , \qquad [8.15]$$

where

$$\tilde{P} = \begin{cases} 0 & (\lambda < \lambda_0 \equiv \rho + \eta \equiv \rho + \eta_1 + \eta_2) \\[2mm] \tilde{P}_1 & (\lambda_0 \leq \lambda \leq \lambda^* \equiv \{(1+\epsilon)\lambda_0 - 2\epsilon\}/[1 - \epsilon]) \\[2mm] \tilde{P}_2 & (\lambda > \lambda^*) \end{cases} \quad ,$$

$$\tilde{P}_1 = 8(1 + \mu)[(\lambda - \lambda_0)/2]^{1/2}[(\lambda - \eta) - (\lambda - \lambda_0)/2]^{3/2} \quad ,$$

$$\tilde{P}_2 = (\lambda - \lambda^*)^2 \quad .$$

Figure 8.14 shows a log-log plot of $\tilde{W}$ versus λ for arbitrary values, $\mu = 1$, $\epsilon = 0.25$, and $\Delta\rho = \Delta\eta_1 = \Delta\eta_2 = 0.01$. The circled points repre-

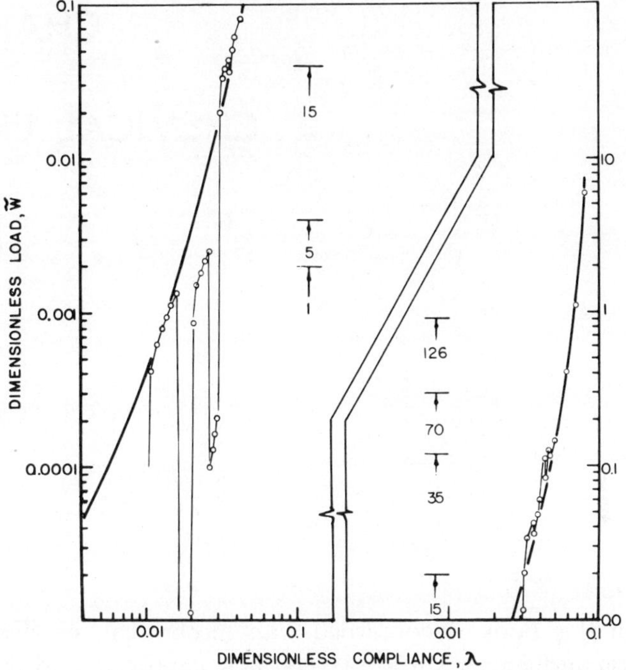

Figure 8.14

sent calculated points from [8.15]. The heavy curve is a curve faring through the calculated points. Each number below a short horizontal line and a vertical arrow indicates the number of pairs of asperities which are engaged in action at that value of load, $\tilde{W}$. At low values of load, because of the nature of the assumed deformation process and only a few pairs of cones are in contact, the $\tilde{W} - \lambda$ characteristics fluctuate a good deal.

As more pairs are engaged in action, the fluctuation diminishes. The significant character of this simple model is the slope of the $\tilde{W} - \lambda$ curve. Starting from small values, the slope increases steadily to beyond what previous theories would allow, Figure 8.10. Moreover, this is so without interaction of neighboring asperities on the same surface. It can be shown that the steepness in slope is due more to the increasing changes of asperities to come into contact as λ increases, and to a lesser extent to the deformation process, hence the justification of the simple deformation process. Also, it may be seen that a more elaborate asperity density distribution would not alter the qualitative character of [8.15].

CHAPTER 9

CHEMICAL EFFECTS

Although this book is concerned with mechanical manifestations of actions on surfaces or surface layers, it is appropriate to offer a brief account of some chemical effects. For example, one of the important reasons for the study of surface temperature is chemical actions that are dependent on temperature; also mechanical properties may, under certain circumstances, be dependent on chemical actions on the surface.

A.   THERMODYNAMICS OF SURFACE PHASES

Equation 1.8 gives the energy relationship in thermodynamics where it was indicated that the sum, $\sum\limits^{n} \tau_\alpha \dot{v}_\alpha$, includes chemical effects. In other words, where v_α relates to masses of constituents, the corresponding τ_α would relate to chemical potential. It would be appropriate to describe the thermodynamics of surface phases here.[120]

Let two bulk phases α and β have volumes v^α and v^β and let these be separated by an imaginary and arbitrary surface s between them. Let c_i be the moles per unit volume of substance i; $c_i{}^\alpha$ and $c_i{}^\beta$ refer to the interior of the bulk phases. The surface phase, $n_i{}^s$, is defined below:

$$n_i^{\,s} = n_i - (c_i^{\alpha} v^{\alpha} + c_i^{\beta} v^{\beta}) = n_i - \sum^{\alpha} c_i^{\alpha} v^{\alpha} \quad , \quad [9.1]$$

where n_i is the total number of moles and the summation extends over all bulk phases.

Similarly, the energy ϵ and entropy η for surface phase in a system are defined, respectively, as

$$\sum^{s} \epsilon^{s} = \epsilon - \sum^{\alpha} \epsilon^{\alpha} \ , \qquad [9.2]$$

$$\sum^{s} \eta^{s} = \eta - \sum^{\alpha} \eta^{\alpha} \ . \qquad [9.3]$$

Let H be the enthalpy, G be Gibbs' free energy, and σ be Helmholtz's free energy. Then

$$\sum^{s} H^{s} = H - \sum^{\alpha} H^{\alpha} = \epsilon + \sum^{\alpha} p^{\alpha} v^{\alpha} - \sum^{\alpha} H^{\alpha} \ , \qquad [9.4]$$

where $P^{\alpha}v^{\alpha}$ represents work. Equation 9.4 may also be written as

$$\sum^{s} H^{s} = \epsilon - \sum^{\alpha} \epsilon^{\alpha}$$

or

$$\sum^{s} H^{s} = \sum^{s} \epsilon^{s} \ . \qquad [9.5]$$

Similarly,

$$\sum^{s} G^{s} = G - \sum^{\alpha} G^{\alpha} = H - T\eta - \sum^{\alpha} H^{\alpha} + T \sum^{\alpha} \eta^{\alpha} = \sum^{s} H^{s} - T \sum^{s} \eta^{s} \ ,$$

and, in view of [9.5],

$$\sum^{s} G^{s} = \sum^{s} \epsilon^{s} - T \sum^{s} \eta^{s} \ , \qquad [9.6]$$

which, by definition, leads to

$$\sum^{s} G^{s} = \sum^{s} \phi^{s} \ . \qquad [9.7]$$

It may be shown in an analogous fashion that equations of the type [9.5] and [9.6] hold for each surface phase separately.

B. SURFACE TENSION

Define a surface tension

$$\gamma = - \left(\frac{\partial W}{\partial A} \right)_{T, \text{ stress composition}} \qquad [9.8]$$

which says that work W, must be done on the system in the creation of unit area of surface, A, without changing any of its properties. Surface tension γ differs from stresses discussed in Chapter 1.

Letting

$$\mu_i = \left(\frac{\partial G}{\partial n_i} \right)_{T, \, n_j \neq i, \text{ stress}} = \left(\frac{\partial F}{\partial n_i} \right)_{T, \, v, \, n_j \neq i}$$

$$= \left(\frac{\partial \epsilon}{\partial n_i} \right)_{W, \, \eta, \, n_j \neq i} \quad ,$$

[1.8] may be written as

$$\dot{\epsilon} = T \dot{\eta} - \dot{W} + \sum^i \mu_i \dot{\eta}_i$$

$$\dot{\epsilon}^S = T \dot{\eta}^S + \gamma^S \dot{A}^S + \sum^i \mu_i \dot{\eta}_i^S \quad . \qquad [9.9]$$

From [9.7] and [9.9]

$$\dot{G}^S = \dot{F}^S = -S^S \dot{T} + \gamma^S \dot{A}^S + \sum^i \mu_i \dot{n}_i^S \quad . \qquad [9.10]$$

For a given phase, then,

$$\gamma^S = \left(\frac{\partial G^S}{\partial A^S} \right)_{T, \, n_i, \text{ stress}} = \left(\frac{\partial F^S}{\partial A^S} \right)_{T, \, n_i, \, v} \quad .$$

Since

$$G^S = \left(\frac{\partial G^S}{\partial A^S} \right)_{n_i, \, T} A^S + \sum^i \left(\frac{\partial G^S}{\partial n_i} \right)_{A, \, T} n_i^S \quad ,$$

it may be expressed as

$$G^S = \gamma^S A^S + \sum_i \mu_i n_i^S \qquad [9.11]$$

or

$$\dot{G}^S = \gamma^S \dot{A}^S + A^S \dot{\gamma}^S + \sum_i \mu_i \dot{n}_i^S + \sum_i n_i^S \dot{\mu}_i \ . \ [9.12]$$

Comparing with [9.10]

$$S^S \dot{T} + A^S \dot{\gamma}^S + \sum_i n_i^S \dot{\mu}_i = 0 \ . \qquad [9.13]$$

For $\dot{T} = 0$, define surface excess concentration

$$\Gamma_i^S = n_i^S / A^S \ . \qquad [9.14]$$

Equation 9.13 becomes the Gibbs adsorption equation

$$\dot{\gamma}^S = - \sum_i \Gamma_i^S \dot{\mu}_i^S \ , \qquad [9.15]$$

which relates the change in surface tension and composition for a given surface phase.

For any process in which the volumes and compositions of the bulk phases remain constant and changes involve shapes, $\sum^\alpha F^\alpha$ remain unchanged. The equilibrium requirement is that $\sum^s F^s$ be a minimum or any virtual change

$$\delta \int_s \gamma \ dA \geq 0 \ . \qquad [9.16]$$

From [9.10],

$$\sum_{system} \mu_i n_i = constant \ , \qquad [9.17]$$

which is the Gibbs-Curie criterion for equilibrium of shapes.

Equation 9.7 for wetting angles of liquids on solid surfaces leads to

$$\gamma_{solid-liquid} + \gamma_{liquid} \cos \theta = \gamma_{solid} \ , \quad [9.18]$$

where θ is the angle between the solid surface and the tangent to the liquid surface. Wetting will occur for $\theta < 90°$ and total spreading occurs when $\theta = 0$.

C. ADSORPTION

There are two types of adsorption physical and chemical. The first is nonspecific and occurs between almost all solids and gases; the forces involved are of the van der Waals type. The second, chemisorption, is a very specific phenomenon and involves the formation of a monolayer of adsorbed gas on the surface. The forces involved are similar to those in chemical bonding. The formation of a chemisorbed layer may be regarded as an attempt of the surface to lower its free energy.

The BET theory predicts the adsorption isotherm in physical adsorption:

$$\frac{1}{v} \left(\frac{x}{1-x} \right) = \frac{1}{v_1 c} + \frac{c-1}{cv_1} x \quad , \qquad [9.19]$$

where v is the amount of gas adsorbed, v_1 is the amount of gas required to form the first layer, c is a constant, and x is the pressure.

A striking set of phenomena regarding chemisorption is that the heats of adsorption on films or wires have very high values at low coverage and decrease by 50 to 100% at full coverage. This decrease could be due to the high mobility for adsorbed atoms so that these are free to seek the lowest energy sites which will become occupied first, resulting in a decrease in heat of adsorption with increase of coverage. Alternatively, the mechanism may be that the heat of adsorption, sticking coefficient and abundance of sites. As such, the region of lowest energy accept molecules most readily. Very little is known about the energies of adsorption, however.

A related phenomenon is catalysis. Mechanistically, there are two types of catalysis. In the first, the catalyst acts as a convenient reservoir of donors and acceptors of H atoms, electrons, and such. The second one involves the breaking or weakening of bonds. Both will go through the following steps: (a) diffusion of reactants to the surface; (b) adsorption of reactants; (c) reaction; (d) desorption of products; and (e) diffusion of products from the surface.

D. CORROSION AND OXIDATION

Although corrosion is most common, the present state of understanding is still very general. In the case of electrochemical nature of liquid corrosion the change in heat content, ΔH, is

$$\Delta H = n(I - \chi) + (L_v - L_s) \quad , \qquad [9.20]$$

where I and χ are ionization potentials and work functions of the atom and bulk metal, respectively. The L_v is the heat of vaporization of the atom and L_s is the heat of hydration of the ion when the latter is immersed in solution.

The gas phase corrosion of metals consists of the formation of oxide, halide or sulfide layers. The Cabrera-Mott theory of oxidation provides a good basis for fitting observations. In terms of layer growth the following facts are important:

1. For very thin films of 20 to 100 Å thickness at low temperatures, oxidation is initially rapid and follows the law:

$$\frac{1}{X} = A - B \ln t \quad ,$$

where X is the film thickness, t is the time, and A and B are constants.

2. After several hundred angstroms, at intermediate temperature range,

$$X = C \, t^{1/2} \quad ,$$

where C is a constant.

3. At higher temperature,

$$X = D \, t^{1/2} \quad ,$$

where D is very different from C.

There are examples of exceptions, it should be noted.

E. PLASTIFICATION OF METALS THROUGH ADSORPTION

Metals may be weakened through adsorption.[121] For example, strength of tin crystals with approximately identical initial orientation for various concentrations of oleic acid in vaseline oil are shown in Figure 9.1.

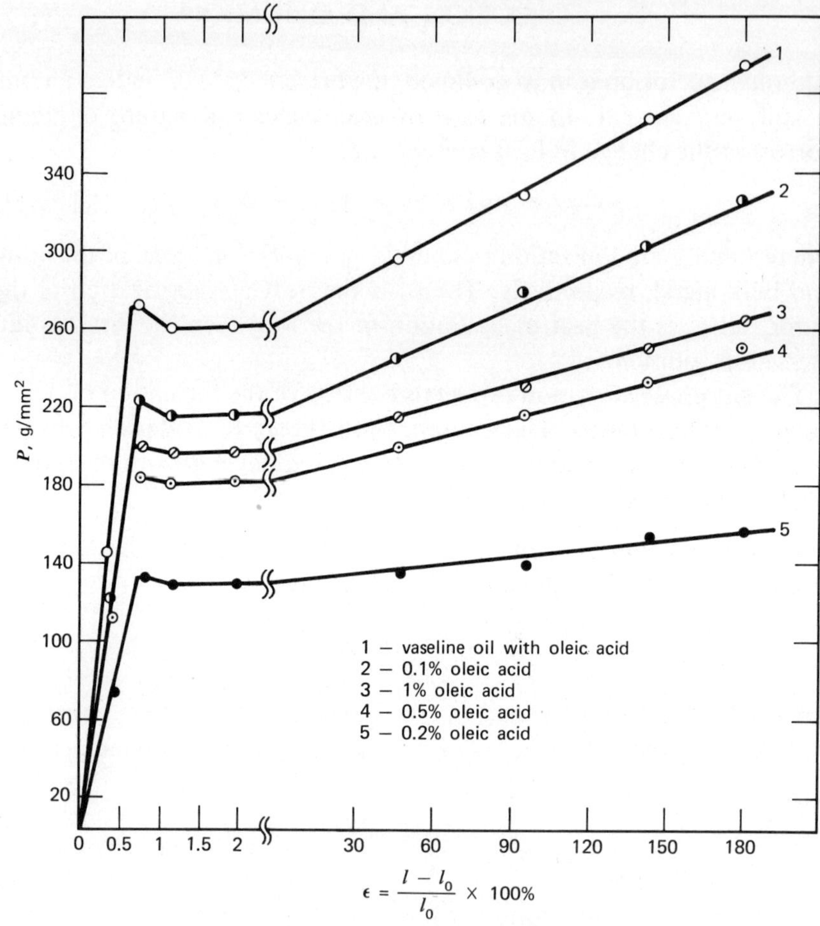

$$\epsilon = \frac{l - l_0}{l_0} \times 100\%$$

Figure 9.1

Adsorption effects may be measured in terms of change in strength

$$\Delta p_m = (p_m)_0 - (p_m)_A , \qquad [9.21]$$

where $(p_m)_0$ is the strength originally and $(p_m)_A$ is the strength upon adsorption.

First of all, temperature, deformation rate, and state of stress play a strong role on the magnitude of the adsorption effect.   The optimum temperature of the plasticizing effect for single crystals is approximately

$$T \doteq \frac{b^2 \sigma}{k \, \ln (\nu \, b / \dot{\epsilon} \, h \, u)} \quad , \qquad [9.21]$$

where b is the Burgers vector, σ is the shear stress, k is the universal gas constant, ν is the Debye frequency of lattice oscillations, $\dot{\epsilon}$ is the relative rate of tensile deformation, h is the average distance between slip lines, and u is the ratio of the total number of slip lines to the number of active lines at any given moment.

Although [9.21] applies to a special model, the result should give some indication as to the magnitude of the temperature. The model is the following: first consider a point on the contour of the active slip plane where the Burgers vector of the dislocation is perpendicular to the contour and the step attains its maximum width $\sim b$. During plastic flow of the crystal, so many dislocations pass every second through this point, and each dislocation forms near the point on a new step of area $\sim b^2$ possessing free energy $\sim b^2 \sigma$.

Second, metal surfaces are covered with oxide films. Surface active substances loosen the oxide film, thus enhancing the plastic flow of metals. An example is the adsorption plasticizing of surface layer of metal by lubricants.

Third, the adsorption effect in the initial range of single crystal deformation follows the equation shown below:

$$P_f = \lambda (\epsilon_0 - \epsilon_1) \quad , \qquad [9.22]$$

where ϵ_0 is the elastic strain related to initial load P_0 through $P_0 = k\epsilon_0$, ϵ_1 is the residual elastic strain related to P_f through $P_f = k\epsilon_1$, and λ is the constant of proportionality.

Under the influence of surface active substances, the load will be denoted by P_f'; this leads to

$$\lambda' = k \, \frac{P_f'}{P_0 - P_f'}$$

therefore,

$$\frac{\lambda}{\lambda'} = \frac{P_f (P_0 - P_f')}{P_f' (P_0 - P_f)} \quad . \qquad [9.23]$$

The ratio of the λ's is, of course, a measure of plastification.

Fourth, the composition and molecular nature of the surface-active medium and the rheological properties of the metals all contribute to the plastification effect.

F.   OTHER EFFECTS

The opposite of plastification through adsorption is embrittlement of metal in the presence of liquid metallic media. Another type of effect is corrosion fatigue. Upon repeated microplastic deformation, the weakest grains in a metallic body are strengthened. When the possibility of strengthening is exhausted, slip will occur with microcracks. Table 7 gives examples of corrosion fatigue of 20Kh steel (pearlitic-ferritic structure) in various media.

Table 7.   Fatigue Limits (σ_{-1}) and Fatigue Limit Coefficients

Group	Medium	σ_{-1} (kg/mm²)	β (%)
I	Air	28.9	100
	Vaseline oil	28.9	100
II	Vaseline oil + 0.1% cetyl alcohol ($C_{16}H_{33}OH$)	—	97
	Vaseline oil + 2% p-butyric acid (p-C_3H_7COOH)	—	97.3
	Vaseline oil + 2% oleic acid ($C_{17}H_{33}COOH$)	27.4	95
	MS oil	27.9	96.5
	MS oil + 2% oleic acid ($C_{17}H_{33}COOH$)	26.9	93
	Water + 1% isoamyl alcohol ($C_5H_{11}OH$)	20	70
III	Water	19.0	66
	Water + 1% saponine	16.4	57

Note.   The fatigue limit was determined: in air at $N = 5 \times 10^6$ loading cycles; in hydrocarbons at $N = 10^7$ cycles; in media of group III and also in solutions of isoamyl alcohol at $N = 2 \times 10^7$ cycles. The first two in Group II were determined at $n = 1590$ cycles/min.

CHAPTER 10

APPLICATIONS

In this chapter selected examples in application are presented. It should be noted, while it can be demonstrated that there is at least one example for each fundamental solution shown prior to this chapter, that only a finite number may be selected by reason of space and time available. To an extent, examples are implicit in Chapters 3 and 8. It remains for the reader to judge the true potential of the fundamental solutions in solving problems involving surfaces, surface layers, and interfaces. Indeed, he may readily arrive at the conclusion that there ought to have been more fundamental solutions presented. This eventuality is considered not only a probability but also a certainty. When this happens in the minds of many, then the author of the first book of this kind will have been exonerated for contributing still another element to the information traffic jam.

A. ON BLOK'S CONJECTURE

In Section 2.A, Figure 2.1 shows a half-space moving with a constant velocity, V, passing an arbitrarily distributed heat source. In the absence of surface radiation, the fundamental solution for surface temperature is given by [2.14]. The complete solution is given by [2.13] with the

[201]

limits of integration from -1 to 1 instead of $-\infty$ to ∞. The dimensionless heat input, $P(\xi)$, is arbitrary.

Now, supposing this body is in contact with another stationary body as shown in Figure 3.4 in Section 3.B; a method was given to estimate the interface temperature and is shown by the dotted curve in Figure 3.4. This is Blok's conjecture.[21] To make this calculation exactly, [2.13] modified as mentioned above should be used with a similar one for the stationary body together with a continuity condition discussed in Section 3.B.

For the stationary body, there is the difficulty with the two-dimensional case as mentioned in Section 2.A. To circumvent this problem, [2.18] is used as the fundamental solution to be used with [2.13] with the appropriate change in the limits of integration and interpretation of $P(\xi)$ as that distribution of heat going into the stationary body. Equation 2.18 amounts to the solution taken along the direction of motion of the mating surface but at the center for a rectangular heat source, $\zeta = 0$, with the width much greater than the length.

Such a formulation for temperature continuity at the interface leads to

$$\int_{-1}^{1} P_s(\xi')K_s(\xi,\xi')d\xi' = \int_{-1}^{1} P_m(\xi')K_m(\xi,\xi')d\xi' \quad , \qquad [10.1]$$

where P_s and P_m are the heat source distribution for the stationary and moving body, respectively. The K_s and K_m represent [2.14] and [2.18], respectively.

In addition, the total heat input which is the sum of $P_s(\xi)$ and $P_m(\xi)$ must be given. For the case of the sum being a constant [10.1] may be written as

$$\int_{-1}^{1} K_s(\xi,\xi')d\xi' = \int_{-1}^{1} P_m(\xi')[K_s(\xi,\xi') + K_m(\xi,\xi')]d\xi' \quad , \qquad [10.2]$$

where P_s and P_m have been normalized by the constant heat input at the interface.

Equation 10.2 is a singular integral equation of the first kind which is discussed in the Appendix. An iterative method[10] has been developed to invert [10.2]. Figure 10.1 shows the first plots of dimensionless interface temperature, subject to the conditions imposed, versus dimensionless distance within the contact length. The dashed, circled, and solid curves represent the initial guess, result of the first and final iterations, respectively. This is for the Peclet number, $Vl/\kappa = 1$ and aspect ratio of the rectangular contact area equalling 4. The κ is the thermal diffusivity. The dashed and dot curve and the dashed curve are Blok's bounds, which were discussed in Section 3.B. These are shown together with the interface temperature curve in Figure 10.2.

B. INTERFACE TEMPERATURE BETWEEN A ROTATING DISC AND A RING SECTOR

The fundamental solution of a rotating disc shown in Figure 2.2 in Section 2.D is given by [2.51]. This is to be used in conjunction with [2.50] for the surface temperature. For large Peclet numbers, [2.52] may be used to replace [2.51]. The surface temperature of a ring sector shown in Figure 2.3 is given by [2.55].

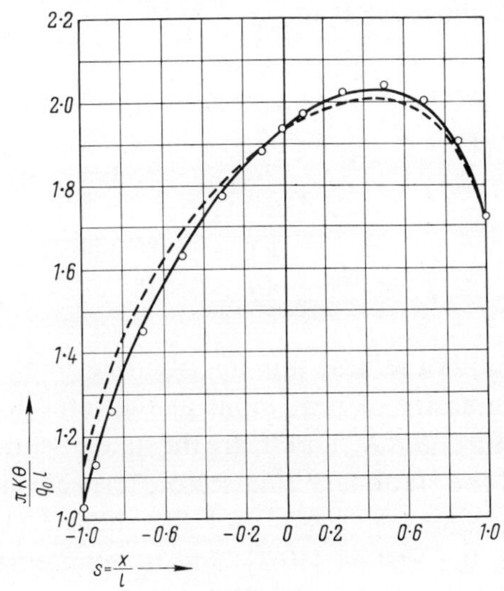

Figure 10.1

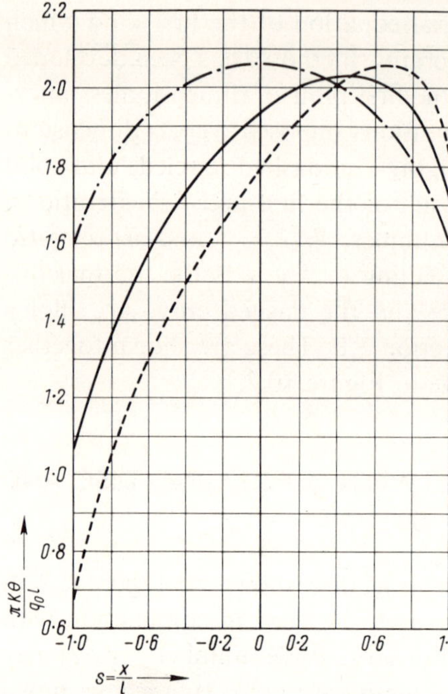

Figure 10.2

For temperature continuity and a uniform heat input at the interface, the above leads to

$$u(1,\phi) = \int_{-\alpha}^{\alpha} [f(\phi') + (2\pi)^{-1} u(1,\phi')] K_m(\phi,\phi') d\phi'$$

$$= \int_{-\alpha}^{\alpha} [1 - f(\phi')] K_s(\phi,\phi') d\phi' \quad , \tag{10.3}$$

where $f(\phi)$ is the heat flux distribution going into the disc normalized by the intensity of heat input at the interface, u is the dimensionless temperature, and K_m and K_s are the fundamental solutions for the moving disc and the stationary ring sector, respectively.

The surface temperature of the slider $u_1(1, \phi)$ can be obtained by inverting the first of [10.3], and is represented by the Fourier series:

$$u_1(1,\phi) = A_0 + \sum_{n=1}^{\infty} (A_n \cos n\phi + B_n \sin n\phi) , \quad (-\pi \leq \phi \leq \pi) . \tag{10.4}$$

With the given data for which $A_0 \gg A_n, B_n,$

$$A_0 = \frac{1}{1 - \alpha/\pi} \int_{-\alpha}^{\alpha} f(\phi)d\phi = 1.058 \int_{-\alpha}^{\alpha} f(\phi)d\phi \quad ,$$

$$A_n = C \cos \pi/4 \int_{-\alpha}^{\alpha} f(\phi)(nT_n)^{-1/2}(\cos n\phi + \sin n\phi)d\phi +$$

$$C \pi^{-1} \cos \pi/4 \, A_0(nT_n)^{-1/2} \frac{\sin n\alpha}{n} \quad ,$$

$$[10.5]$$

and

$$B_n = C \cos \pi/4 \int_{-\alpha}^{\alpha} f(\phi)(nT_n)^{-1/2}(-\cos n\phi + \sin n\phi)d\phi -$$

$$C \pi^{-1} \cos \pi/4 \, A_0(nT_n)^{-1/2} \frac{\sin n\alpha}{n} \quad ,$$

where $C = B(2/N)^{1/2}$.

For the matching, it is convenient to represent the heat flux $g(\phi)$ by a complete Fourier series in the range $-\pi/18 \leq \phi \leq \pi/18$

$$g(\phi) = A_0^* + \sum_{n=1}^{\infty} (A_{18n}^* \cos 18n\phi + B_{18n}^* \, 18n\phi) \quad , \quad [10.6]$$

and of course $f(\phi) = 1 - g(\phi)$.

With the Fourier representation of $f(\phi)$ substituting into [10.5], A_0, A_n, B_n, will be denoted by A_0, A_m, B_m respectively.

Further expand $u_1(1, \phi)$ in the form

$$u_1(1,\phi) = \tilde{A}_0 + \sum_{\rho=1}^{\infty} (\tilde{A}_{18\rho} \cos 18\rho\phi + \tilde{B}_{18\rho} \sin 18\rho\phi) \quad , \quad (-\tfrac{\pi}{18} \leq \phi \leq \tfrac{\pi}{18})$$

$$[10.7]$$

where

$$\tilde{A}_0 = A_0 + \sum_{m=1}^{\infty} A_m \frac{\sin m\alpha}{m\alpha} \, ,$$

$$\tilde{A}_{18\rho} = \sum_{\substack{m=1 \\ m \neq 18\rho}}^{\infty} A_m \left[\frac{\sin(18\rho - m)\alpha}{(18\rho - m)\alpha} + \frac{\sin(18\rho + m)\alpha}{(18\rho + m)\alpha} \right] + A_{18\rho} \, ,$$

and

$$\tilde{B}_{18\rho} = \sum_{\substack{m=1 \\ m \neq 18\rho}}^{\infty} B_m \left[\frac{\sin(18\rho - m)\alpha}{(18\rho - m)\alpha} + \frac{\sin(18\rho + m)\alpha}{(18\rho + m)\alpha} \right] + B_{18\rho} \, .$$

The surface temperature in the rider using the second half of [10.3] for the given data is

$$u_2(1,\phi) = 1.556 \int_{-\alpha}^{\alpha} g(\phi')d\phi' +$$

$$0.398 \sum_{j=1}^{\infty} \cos[9j(\phi + \alpha)] \frac{1}{j} \int_{-\alpha}^{\alpha} \cos[9j(\phi' + \alpha)]g(\phi')d\phi' \, .$$

$$[10.8]$$

Equations 10.7 and 10.8 yield

$$u_2 = 1.556(2\alpha A_0^*) + 0.398 \sum_{j=1}^{\infty} \frac{1}{j} \cos[9j(\phi + \alpha)] \cdot$$

$$\left[\cos 9j\alpha \left\{ \frac{2 \sin 9j\alpha}{9j} A_0^* + \right. \right.$$

$$\left. \sum_{\substack{n=1 \\ 9j \neq 18n}}^{\infty} A_{18n}^* \left[\frac{\sin(9j - 18n)\alpha}{(9j - 18n)} + \frac{\sin(9j + 18n)\alpha}{(9j + 18n)} \right] + A_{9j\alpha}^* \right\} -$$

$$\sin 9j\alpha \left\{ \sum_{\substack{n=1 \\ 9j \neq 18n}}^{\infty} B_{18n}^* \left[\frac{\sin(9j - 18n)\alpha}{(9j - 18n)} - \frac{\sin(9j + 18n)\alpha}{(9j + 18n)} \right] + B_{9j\alpha}^* \right\} \right] \, .$$

Then using the condition of $T_1\,(a,\,\phi) = T_2\,(a,\,\phi)$, the coefficients of $g(\phi)$ are

$$g(\phi) = 0.8718 + 0.00070 \cos 18\phi - 0.00615 \sin 18\phi$$
$$- 0.00392 \cos 36\phi + 0.00548 \sin 36\phi$$
$$+ 0.00560 \cos 54\phi - 0.00742 \sin 54\phi + \ldots$$

The final surface temperature has the form

$$u\,(1,\,\phi) = 0.0471 - 0.000605 \sin\ 9\phi + 0.000024 \cos 18\phi$$
$$+ 0.000082 \sin 27\phi - 0.000068 \cos 36\phi$$
$$- 0.000018 \sin 45\phi + 0.000065 \cos 54\phi + \ldots \ [10.9]$$

Figure 10.3 shows the result: $u - u_0$ versus ϕ $-\alpha \le \phi \le \alpha$ where u_0 is the average value of u. Note that the interface temperature is rather constant for the case of uniform heat generation which is good for average normal loads.

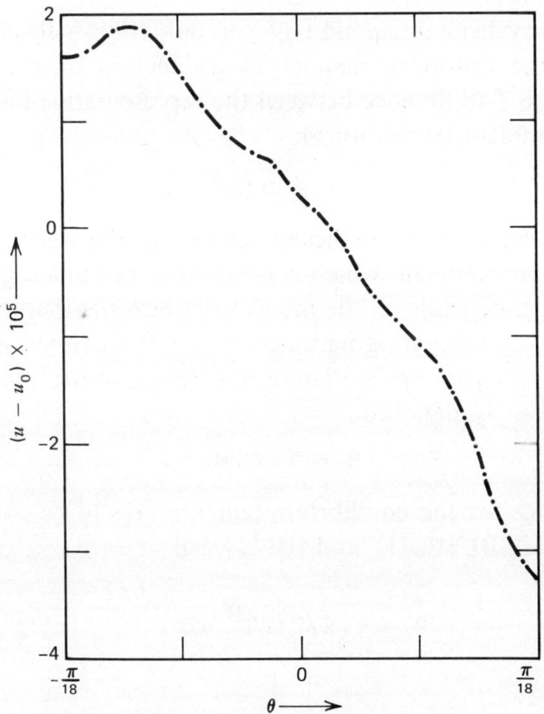

Figure 10.3

C.   DISSIPATION OF FRICTIONAL HEAT

In the text as well as Section 10.A heat dissipation was treated; this mode is referred to sometimes as primary dissipation. Secondary heat dissipation, on the other hand, refers to dissipation of frictional heat from the source toward some cooling surface.

In solid friction, secondary heat- dissipation comprises all states of heat dissipation which occur at positions not in close proximity to the contact surface; in the case of liquid friction it refers to all stages of heat dissipation outside the flowing liquid in which the frictional heat is generated.

Consider the case of journal bearings;[122] with a shaft diameter D and a bearing length L, the frictional heat H developed per unit of time is

$$H = \pi LD^2 \, f \, PN \quad , \qquad [10.10]$$

where f represents the coefficient of shaft friction. N is the speed and P is the load per unit area.

The secondary heat dissipation Q per unit time will, as usual, be expressed in the following manner as a function of the secondary temperature rise T (difference between the representative film temperature and the ambient temperature):

$$Q = \pi LD \, ET^n \quad , \qquad [10.11]$$

where E is a heat transfer coefficient relating to the wetted shaft area πLD, and n is an exponent, which is greater but not much greater than unity, and which depends on the intensity of heat dissipation.

If, for the known "bearing parameter" ZN/P, in which Z is the dynamic viscosity at the representative film temperature, the symbol G is adopted, then by definition

$$Z \equiv (P/N)G \quad . \qquad [10.12]$$

Since $H = Q$ for the equilibrium temperatures to be considered, it follows from [10.10], [10.11], and [10.12] that

$$P = \left\{ (E/D)ZT^n/fG \right\}^{1/2} \qquad [10.13]$$

and

$$N = \left\{ (E/D)(T^n/Z)G/f \right\}^{1/2} \quad . \qquad [10.14]$$

For a selected value of T, that is, for a specified bearing isotherm, the viscosity Z is known for the given oil and for the given ambient temperature; for such an isotherm, which will naturally be a curve on the equilibrium surface sought, P and N are known by [10.13] and [10.14] to be a function of the coefficient of friction f and the bearing parameter G.

Assume that f is exclusively a function of G. In that case, however, for each isotherm, P and N are known by [10.13] and [10.14] to be a function of G only; this means that for each isotherm, that is, for each constant value of T (hence of Z), [10.13] and [10.14] together may be conceived as a parametric formulation (with G as a parameter) of the relation between P and N for that isotherm. As soon as the functional relation between f and G is known, it is therefore possible to construct any desired isotherm. As the required equilibrium surface may be entirely made up of isotherms, it may be constructed accordingly, the method being as follows.

The functional relationship between f and G, which need be known only experimentally and is called hereafter the frictional characteristic, is represented schematically in Figure 10.4 by the full curve. The isotherm desired can be constructed point by point as soon as the values of N corresponding to each relevant value of P are known. Now according to [10.13], for the given values of T and Z, the value of $fG = (E/D)$.

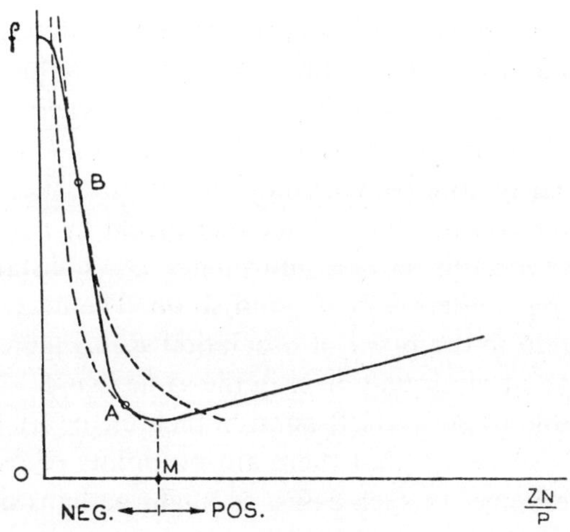

Figure 10.4

(ZT^n/P^2) is established for each value of P selected. The equilateral hyperbola for this value of fG can now be constructed in Figure 10.4 and its points of intersection with the frictional characteristic can be located. For each of these points of intersection one can determine the corresponding value of G/f and substitute this value in [10.14]. In this way the values of N corresponding to each P-value selected are known; hence the isotherm desired can in fact be constructed point by point. This applies to every isotherm desired; hence the required equilibrium surface can be found by combining the frictional characteristic with [10.13] and [10.14] in the manner indicated.

The foregoing method can be simplified by replotting the frictional characteristic in a logarithmic diagram of $\log f$ against $\log G$.

Figure 10.4 shows two equilateral hyperbolae (see broken curves) which osculate the frictional characteristic, the point of osculation being indicated for each one (see points A and B, respectively). At these points of osculation the following condition, which is especially convenient in the method of the logarithmic diagram, holds good

$$\frac{f}{G}\;\frac{dG}{df} = \frac{d\,\log\,G}{d\,\log\,f} = -1 \quad .$$

It will be seen that these two osculating hyperbolae define a certain region in which each equilateral hyperbola between these two has three points of intersection with the frictional characteristic; outside that region each equilateral hyperbola has only one point of intersection with the frictional characteristic. In other words, for each value of fG (hence of P) inside the region in question there are three corresponding values of G/f (hence of N) and outside it only one value of G/f (hence of N). On closer examination it is found that this yields a shape of the equilibrium surface in the three-dimensional P-N-T diagram which may be described qualitatively as indicated in the schematized Figure 10.5, in which the equilibrium surface is depicted by a set of curves of intersection between this surface and planes T = constant, that is, by the corresponding isotherms T_1, T_2, and so on. The maxima of the isotherms correspond to the point of osculation B in Figure 10.4, and the minima to the point of osculation A in the same figure.

It is conceivable where the left-hand branch of the frictional characteristic descends so gently that there are no points of osculation with equilateral hyperbolae. In such cases, as might perhaps occur in gears, the isotherms would not show any maxima or minima, and unstable thermal equilibria could not occur.

The equilibrium surface shows, for the positive values of P, N, and T exclusively considered here, a pronounced valley enclosed between a flank which, for $N \to 0$ and for all values of T, ascends steeply toward $P \to \infty$ (this flank asymptotically approaches the plane $P - T$) and a ridge which, on the side away from the valley, descends for higher values of N towards the plane $N - T$, that is to $P = 0$. By intersecting this surface with a plane $N =$ constant, the shape of $T - P$ diagram for the chosen value of N is arrived at.

Let the right-hand branch of the frictional characteristic ($G > M$) be called the "positive branch" and the left-hand branch ($G < M$) the "negative branch"; "positive" is taken to refer to the fact that for the right-hand branch the value of df/dG is positive. The temperature equilibria which may be deduced from the respective branches will be termed correspondingly "positive" and "negative" equilibria, respectively.

The "positive equilibria" are thermally stable. By thermally stable equilibria are meant equilibria that are automatically restored after the cessation of a temporary thermal disturbance (in the case of thermal instability there is no such restoration of equilibrium, with all consequences resulting therefrom); the load and speed are considered by definition to be undisturbed.

A quantitative view of the limiting transition between thermally stable and thermally unstable temperature equilibria can be obtained with the aid of the general criterion of this transition:[123]

$$\frac{\partial H}{\partial T} = \frac{\partial Q}{\partial T} \, , \qquad [10.15]$$

where T is the secondary temperature rise. If $\partial H/\partial T \leq \partial Q/\partial T$, the equilibrium considered is stable, and if the reverse is the case, it is unstable.

Using the identity

$$\frac{\partial H}{\partial T} = \frac{\partial G}{\partial T} \quad \frac{\partial H}{\partial G} = \frac{N}{P} \quad \frac{dZ}{dT} \quad \frac{\partial H}{\partial G}$$

in [10.10] and substituting the value of N following from [10.14], it is easily seen that criterion [10.15] can be brought into the form:

$$\left\{ \frac{T}{Z} \frac{dZ}{dT} \right\} \left\{ \frac{G}{f} \frac{df}{dG} \right\} = n \, . \qquad [10.16]$$

From condition [10.16] it should be noted that (T/Z) and (dZ/dT) are always negative. Assume that the minimum itself $(G = M)$ would constitute the critical transition concerned.

Figure 10.5 shows schematically the thermally allowable bearing load P as a function of rotational speed N while Figure 10.6 shows the $P - N$ diagram and experimental results.

D. EDGE EFFECT ON THE HERTZ SOLUTION

Consider a sphere loaded upon another body which is either a half-space or a cylinder. As in classical contact theory, the sphere will be

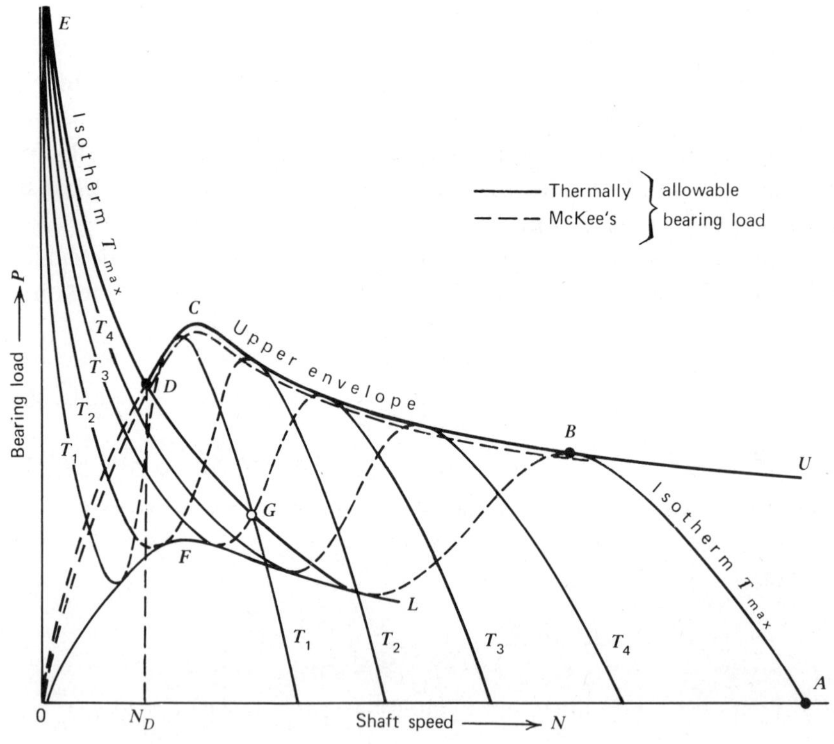

Figure 10.5

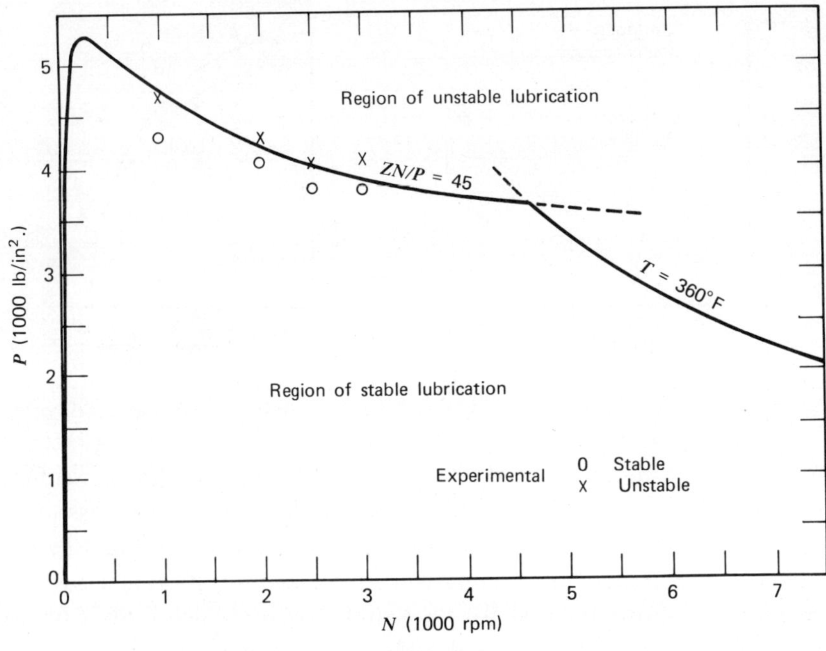

Figure 10.6

approximated by an infinite body with a parabolic surface. Therefore, if the curved surface is given by

$$z = \frac{1}{2}\,\frac{r^2}{R_B}\ ,$$

where R_B is the radius of the sphere, r is the radial distance in a reference plane, and z is the perpendicular distance from the plane to the sphere then the displacement within the contact region is

$$w = \delta - z = \delta - \frac{r^2}{2R_B}\ , \qquad [10.17]$$

where δ is the approach of the center of the sphere.

The Hertz[124] solution on contact pressure amounts to inverting an integral equation formed by equating [10.17] and [4.19] which governs the deformation of a half-space by axi-symmetric loads. Now, if instead of

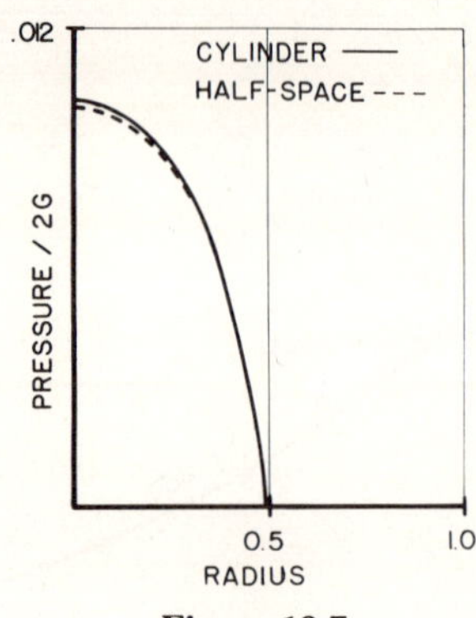

Figure 10.7

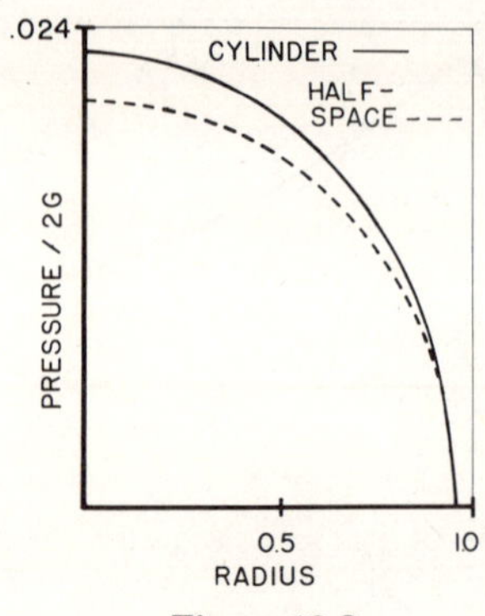

Figure 10.8

the half-space, the end of a finite cylinder is used, results of Section 4.K can again be used.[57] Figure 10.7 shows a plot of dimensionless pressure against dimensionless radius for a contact radius about half of the radius of the cylinder. For comparison, the Hertz solution is shown.

Figure 10.8 shows the case for the contact radius which is 96% of the radius of the cylinder. Again, the Hertz solution is shown for comparison.

E. ON ELASTOHYDRODYNAMIC LUBRICATION

The first successful attempt to examine the combined effect of elastic distortion and viscosity-pressure in a lubricated, concentrated contact was presented in 1949.[125] A new approach was presented in 1959, an inverse method.[126] The subject matter has recently been reviewed.[127]

In hydrodynamic theory of lubrication, the field equation(s) governing the motion of the fluid between a region at the interface of two surfaces, under whatever degree of approximation, and the equation of continuity are solved for the appropriate boundary conditions. In elastohydrodynamic lubrication, the deformation of the surfaces is taken into account as well as the viscosity-pressure effect of the fluid. For the consideration of surface deformation, depending on the geometry and curvature involved, [4.13], [4.15], and [4.21] may be used. Equation 4.13, which is the Flamant solution, is most often used for two-dimensional analyses.

More recently, heat effects have been taken into account.[128] As far as the fluid is concerned, of course, the energy equation must be taken into account. Viscosity-temperature effects may also be accounted for. Moreover, heat conduction through the surfaces may be accounted for by [2.13] or [2.17]. In this case, in addition to the elastic deformation of the surface due to fluid pressure discussed above, deformation of thermal origin may be important. For this purpose [5.75] may be used in conjunction with [4.13]; [5.77] may be used in conjunction with [4.15]. In this regard [5.75] has been applied[65] to a typical set of thermo-elastohydrodynamic data shown in Table 8.

Table 8. Heat Input $q(x)$ Encountered in Elastohydrodynamics

x(in.)	$q(x)(10^{-4} \times$ Btu/in.^{2}-hr)
-0.0125	0.2015
-0.0875	0.3129
-0.005	2.378
-0.0025	6.070
0	12.53
0.0025	10.97
0.005	16.20
0.0075	5.513
0.008125	2.389
0.00875	1.243
0.009375	0.6064
0.01	0.2714
0.01125	0.0565
0.0125	0.1069

Using the above and the appropriate physical constant [5.75] leads to the surface displacement, u_2, shown in Figure 10.9. The difference between the dotted portion and the solid curve represents an approximation where surface radiation is neglected.

F. LUBRICATION OF HEAVILY LOADED SPHERICAL SURFACES

The Reynolds equation, alluded to earlier, is valid as long as the oil is Newtonian. In practice, the oil solidifies as the pressure reaches certain critical value.[129] The solidification is supposed to be instant for the purpose of analysis.

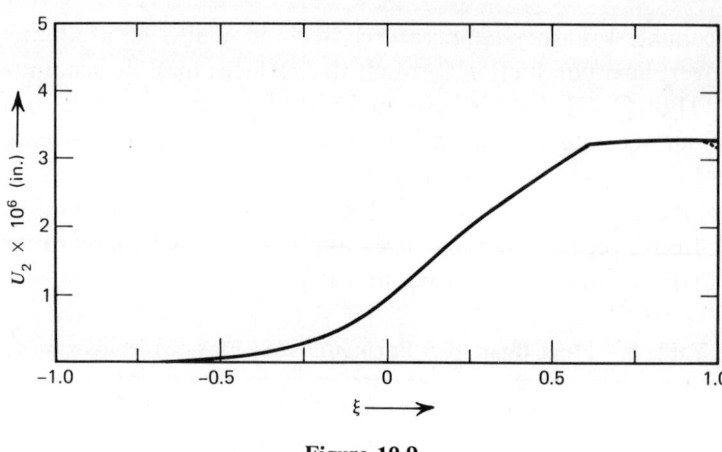

Figure 10.9

Starting out with elastohydrodynamics discussed earlier, for increased pressure at a given temperature, the flow regime may be divided into four parts as shown in Figure 10.10: the liquid oil at the leading edge, the sliding solidified oil, the nonsliding solidified oil, and the cavitation zone.

Again [4.15] and [4.16] governing the deformation of surfaces should be used in calculating the thickness of the lubricant film (see Figure

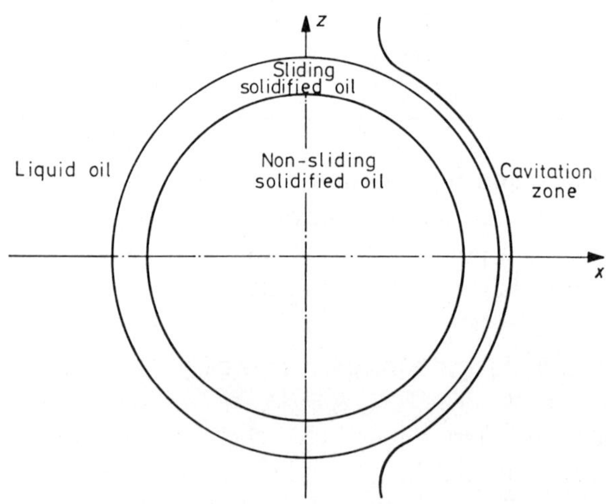

Figure 10.10

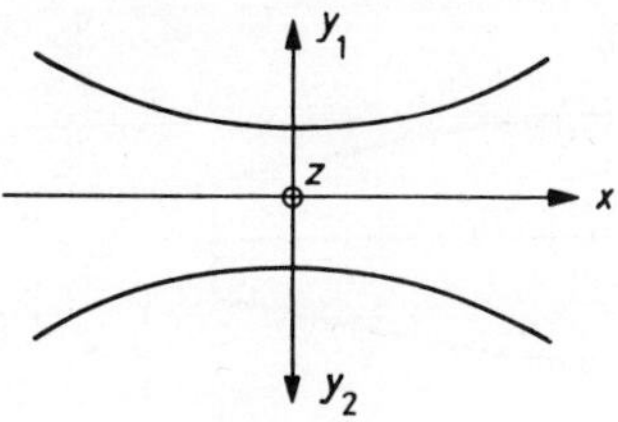

Figure 10.11

10.11) with Reynolds' equation governing the flow in the liquid zone, a quasistatic equilibrium equation for the sliding solidified and non-sliding solidified zone, respectively, and cavitation condition at the trailing zone.[130]

Figure 10.12 shows a typical set of calculated results on a dimensionless film thickness h_0 for a dimensionless deformability of the contact, k_0, and a dimensionless pressure, p_0. The minimum film thickness, h_m, is 0.24 μm.

Figure 10.13 shows the experimental data on film thickness with a minimum value of 0.23 μm.

G. THIN PLASTIC MASS COMPRESSED BETWEEN THE ENDS OF TWO ELASTIC CYLINDERS

High pressure compression of substances[131] between anvils has spanned the first half of the twentieth century by Bridgman. Analyses of elastic

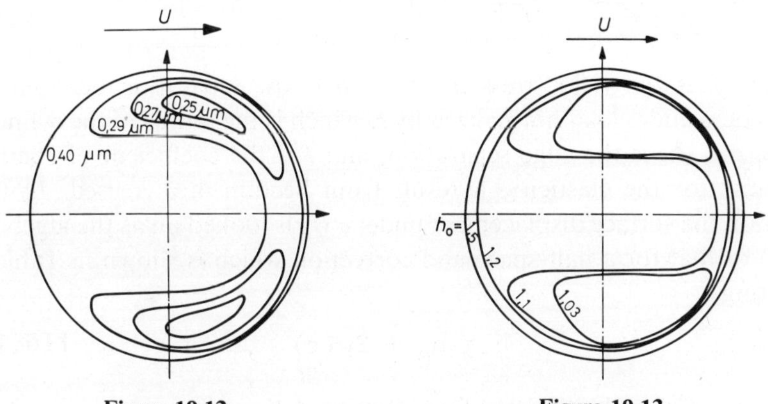

Figure 10.12   **Figure 10.13**

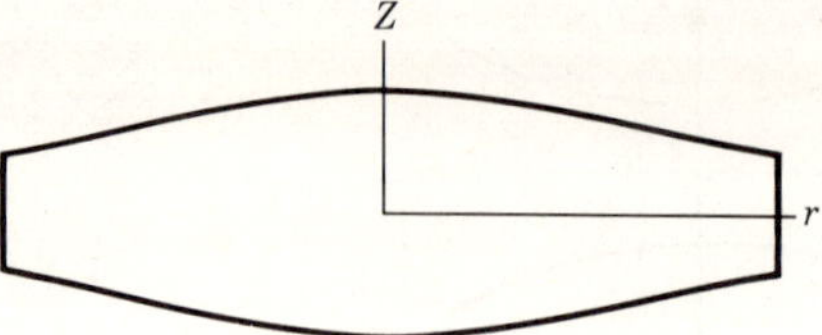

Figure 10.14

or plastic mass being compressed have been made.[132,133] Also, a recent method for determining coefficient of friction has been adopted by using the bridgman anvil.[134] In all cases, the die or anvil have been assumed rigid. In what follows, the effect of elasticity of the anvil will be taken into account. Results bear heavily on the aforementioned method for measuring friction.

Figure 10.14 is a schematic diagram of the deformed shape of a plastic mass being compressed between the ends of two elastic, circular cylinders. The coordinates (z, r) are the cylindrical ones for the case with axial symmetry.

Using rigid, perfect plasticity and neglecting any z-dependence for a thin film, the normal stress, σ_z, for equilibrium and continuity has been shown to be[135]

$$\sigma_z = \begin{cases} -k \left\{ \ell n \left(\dfrac{h}{h_0} \right) + \exp\left[2f \displaystyle\int_\rho^1 \dfrac{d\rho}{h} \right] \right\} & \left(|\sigma_z| < \dfrac{k}{f\sqrt{3}} \right) \\[4ex] -k \left\{ \dfrac{1}{f\sqrt{3}} + \dfrac{2}{\sqrt{3}} \displaystyle\int_\rho^1 \dfrac{d\rho}{h} + \displaystyle\int_h^{h_1} \dfrac{\sigma_z \, dh}{h} \right\} & \left(|\sigma_z| > \dfrac{k}{f\sqrt{3}} \right) \end{cases}$$

$$[10.18]$$

where k is the yield stress in shear, h is the separation between the cylinders under load normalized by r_0 which is the radius of the cylinder, $\rho \equiv r/r_0$, h_0 is the edge separation, and f is the coefficient of friction.

Now for the elasticity, a result from Section 4.K is used. In that section the surface displacement under $\sigma_z(r)$ is looked on as the algebraic sum of that for a half-space and correction which is shown in Table 2. Letting

$$h = h_0 + 2w(r) \quad , \tag{10.19}$$

where $w(r)$ is to be adopted from Section 4.K.

Thus an iterative method is easily formulated. Using a starting value of $h = h_0$, $\sigma_z(r)$ may be computed from [10.18], say $\sigma_z^{(0)}$. With $\sigma_z^{(0)}$, the next estimate of h, say $h^{(1)}$, may be found using [10.19] and the associated table from Section 4.K. Some results, obtained by this process until a convergence was achieved, are shown below.

First of all, Table 9 shows the effect of elasticity of the anvils for various k/G ratios and edge thicknesses for $f = 0.2$. The G is the shear modulus in elasticity. The last column indicates that as k/G ratio increases the maximum pressure obtained by assuming the anvil rigid becomes poorer. In fact, for $k/G = 0.002$, the pressure calculated is only a little above 13% that obtained by assuming the anvil rigid.

Table 9. Some Results for a Plastic Mass Between Elastic Cylinders

k/G	h_o	h_{max}	P_{ave}	P_{max}	$\dfrac{P_{max}}{P_{max,rigid}}$
0.0001	0.02	0.022	18.	51.5	0.90
0.0001	0.01	0.014	31.7	84.1	0.73
0.0001	0.004	0.009	54.8	125.	0.44
0.0001	0.002	0.008	70.6	146.	0.25
0.0005	0.02	0.028	15.4	40.2	0.70
0.0005	0.01	0.02	23.2	54.1	0.47
0.0005	0.004	0.017	32.8	66.7	0.23
0.0005	0.002	0.016	38.	72.5	0.13
0.001	0.04	0.049	8.4	23.	0.82
0.001	0.02	0.034	13.7	33.7	0.59
0.001	0.01	0.027	19.1	42.1	0.37
0.001	0.004	0.023	25.2	49.5	0.18
0.001	0.002	0.022	27.	51.7	0.09
0.002	0.04	0.056	7.8	20.1	0.72
0.002	0.02	0.042	11.7	27.	0.47
0.002	0.01	0.035	15.3	32.	0.28
0.002	0.004	0.031	19.1	36.	0.13
0.0	0.02	—	—	57.	—
0.0	0.01	—	—	115.	—
0.0	0.004	—	—	287.	—
0.0	0.002	—	—	580.	—

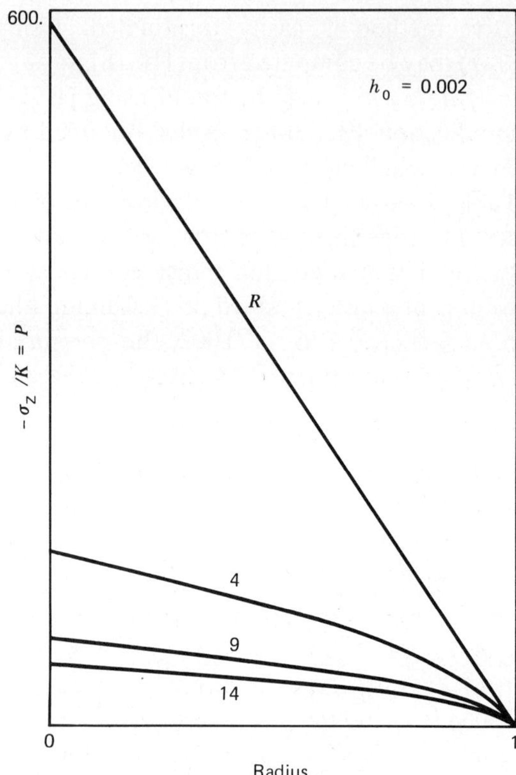

Figure 10.15

Figure 10.15 shows a dimensionless pressure distribution versus radius curve for $h_0 = 0.002$. The one designated R is for rigid anvil.

For large separation, the effect of f becomes more pronounced. Near $r = 1$, the pressure is highly dependent on f. As loads increase, everything else being equal, the importance of f decreases.

H.   TRANSIENT PHENOMENA IN ROLLING CONTACT WITH DRY FRICTION

Two elastic cylinders, which are pressed together, are shifted with respect to each other over a small distance without rolling and are then allowed to roll over each other. The results give rise to transient phenomena in traction[136] quite different from the steady state case.[137]

Figure 10.16 shows schematically two cylinders in contact. Let $K = A^{-1}[(1 + \nu_1)(1 - 2\nu_1)/E_1 - (1 + \nu_2)(1 - 2\nu_2)/E_2]$, $A = (1 - \nu_1^2)/E_1 + (1 - \nu_2^2)/E_2$, where E is the Young's modulus and ν is the Poisson's ratio with subscripts 1 and 2 referring to the top and the bottom cylinders, respectively.

Figure 10.17 shows the case of similar material (i.e., $K = 0$) so that the compression generates no frictional traction and the traction[138] distribution, x, is shown in Figure 10.17a. The contact length is represented by a dimensionless measure $-1 \leq x \leq 1$. When rolling starts, the singularity at the edge moves toward the trailing edge with rolling velocity which means that it remains frozen on the surface of the cylinder. The singularity is replaced by a small traction force, slowly building up, see Figures 10.18b and c. After $t = 2$, the traction distribution becomes steady,[139] and t is the time taken for rolling contact.

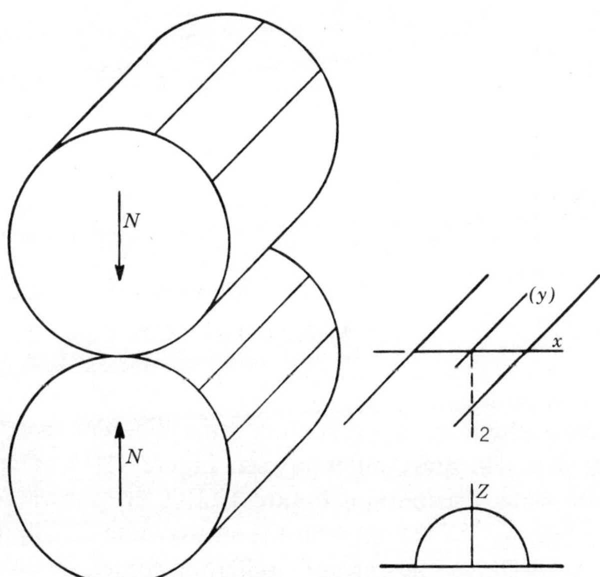

Figure 10.16

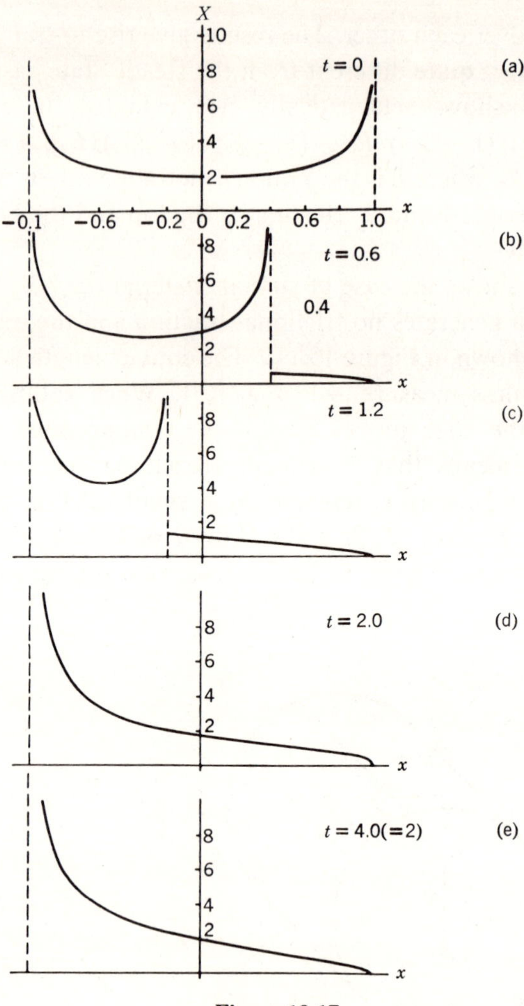

Figure 10.17

In the case where the materials are quite different (e.g., $K = -1$), the results are also quite different, see Figure 10.18. The deviations from steady state distribution, Figure 10.18d, are shown for different times, and they are compared with the steady state distribution.[140]

Figure 10.19 shows the case of oscillating forces, $T = 0.2\pi \cos \pi t$ and $K = 0$. Figure 10.20 shows again the case $K = 0$ and $T = 0.2\pi \cos q_1 t$, where q_1 is as shown.

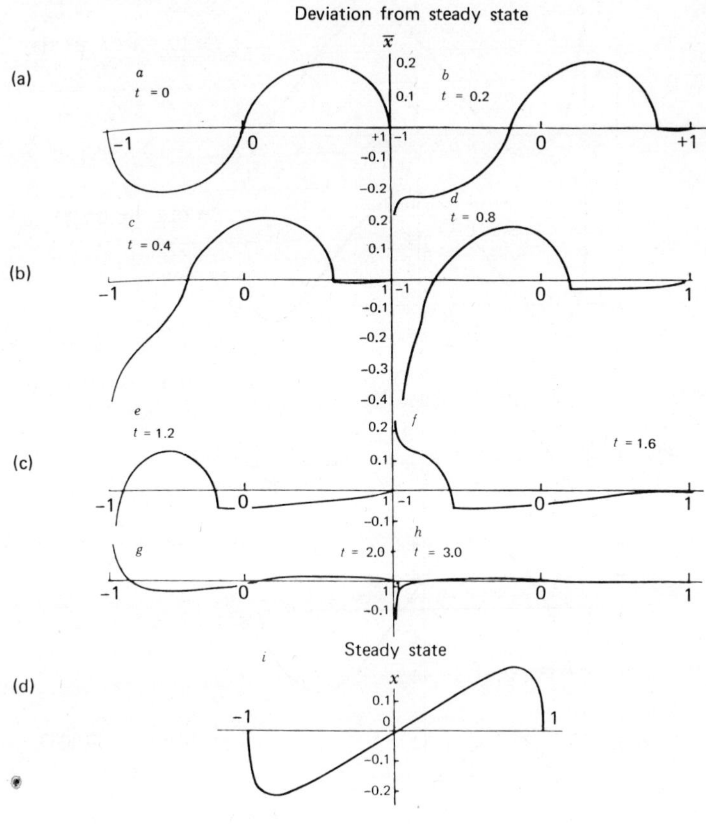

Figure 10.18

I. EFFECT OF LAYER-SUBSTRATE INTERFACE CONDITION ON CONTACT STRESSES OF AN INDENTED ELASTIC LAYER

Similar to the elastic case of Section 10.U, Figure 10.21 shows a rigid cylindrical indenter is in contact with a layer-substrate composite. For the two-dimensional case, [4.15] should be used instead of [4.19].

If p_0 is the maximum contact pressure, R the radius of curvature of the indenter, a the half width of contact, h the thickness of layer, E_l the Young's modulus of the layer material, and ν_l the Poisson's ratio of the layer material, Figure 10.22[141] gives a dimensionless maximum pressure versus a dimensionless a for two cases: (1) when the layer is

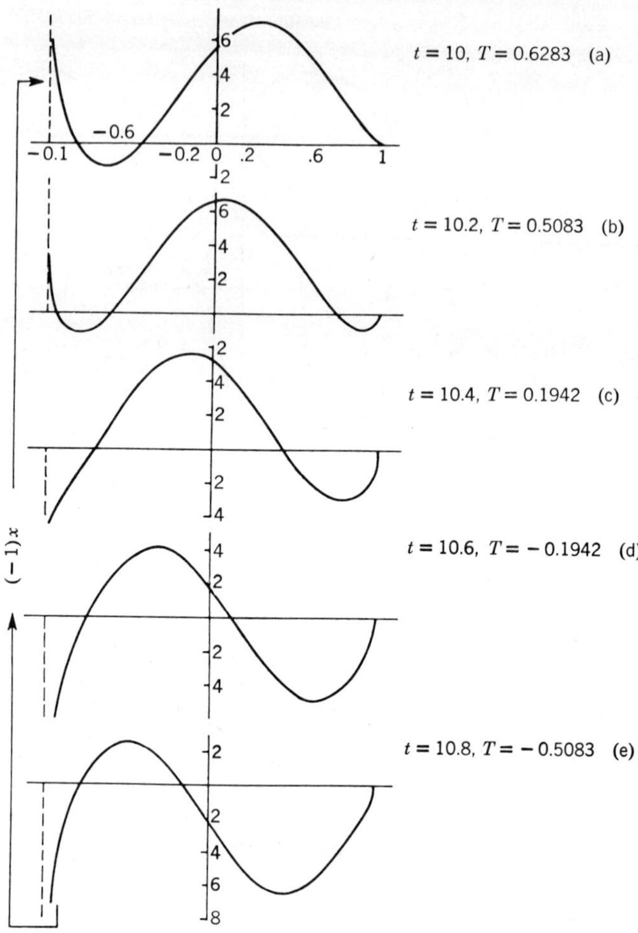

Figure 10.19

rigidly adhered to a rigid foundation; and (2) when the layer is in frictionless contact with rigid foundation.

With a given thickness, the higher the loads and consequently the higher the a's, the wider are the differences between maxima of the two cases. In referring to the figure, it should be noted that the same a/h in the two cases means two different loads in general.

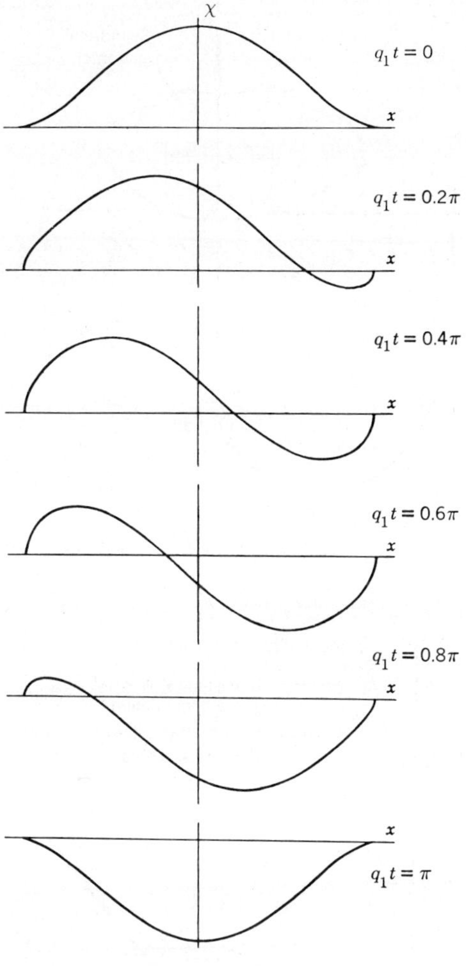

Figure 10.20

J.   RESPONSE OF AN ELASTIC SOLID TO EXPANDING SURFACE LOADS

The problem is shown schematically in Figure 10.23 in which a disc load is expanding at a constant rate c. The load over a circular area is the integral of the ring load. At a given instant, the diagram corresponds to the model shown in Section 10.U. However, for a transient problem,

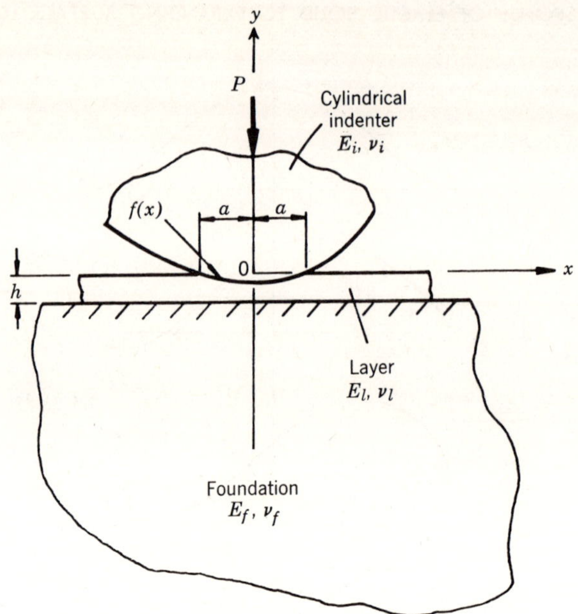

Figure 10.21

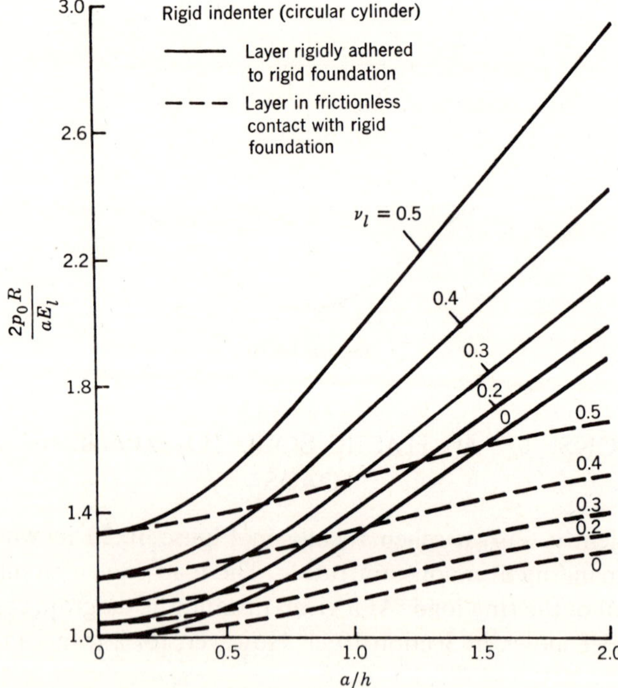

Figure 10.22

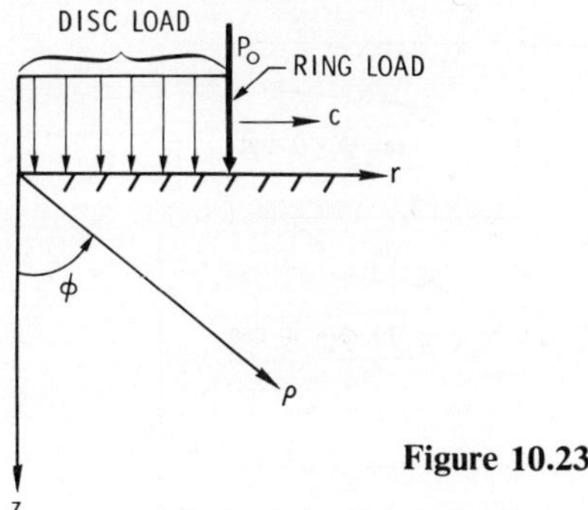

Figure 10.23

the equation of motion must be used. Often in cases like these the Laplace transform may be used; in the transformed state, of course, time is dormant. The solution of the transformed equation is very similar to that of Section 10.U.

Upon inverse transform, quantities such as displacement may be obtained. In this case, the inverse transform may be aided by numerical method.[142,143]

In the dynamic case, as in Section 4.H, depending on the material properties, the speed of expansion of load may be said to be superseismic, transeismic, or subseismic. The first refers to the case where c is larger than the speed of propagation of dilatational waves; the third refers to the case where c is smaller than the speed of propagation of shear waves; and the second refers to the case where c is between the two.

Figures 10.24, 10.25, and 10.26 show dimensionless vertical displacement as a function of dimensionless time for various angles ϕ for a disc expanding at rates that are superseismic, transeismic, and subseismic cases, respectively. The notations (dc, sc, and sd) relate to conical wave fronts while (d, s) relates to hemispherical wave fronts. The R refers to Rayleigh wave, and the L to the load.

K.   CONTACT OF A PLATE PRESSED BETWEEN TWO SPHERES

Figure 10.27 shows a plate pressed between two spheres. In the case of two equal spheres, it could be considered a special case of a sphere on a plate which, in turn, rests on a rigid foundation. In general the elastic properties of the spheres are different. For this reason and the fact that

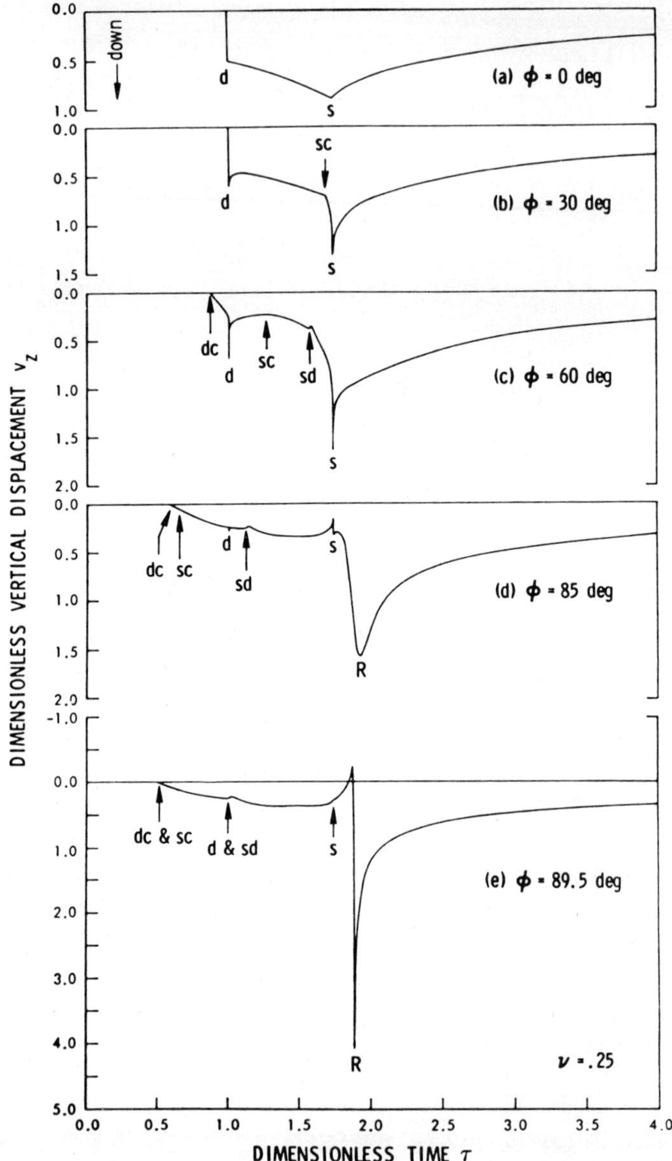

Figure 10.24

[228]

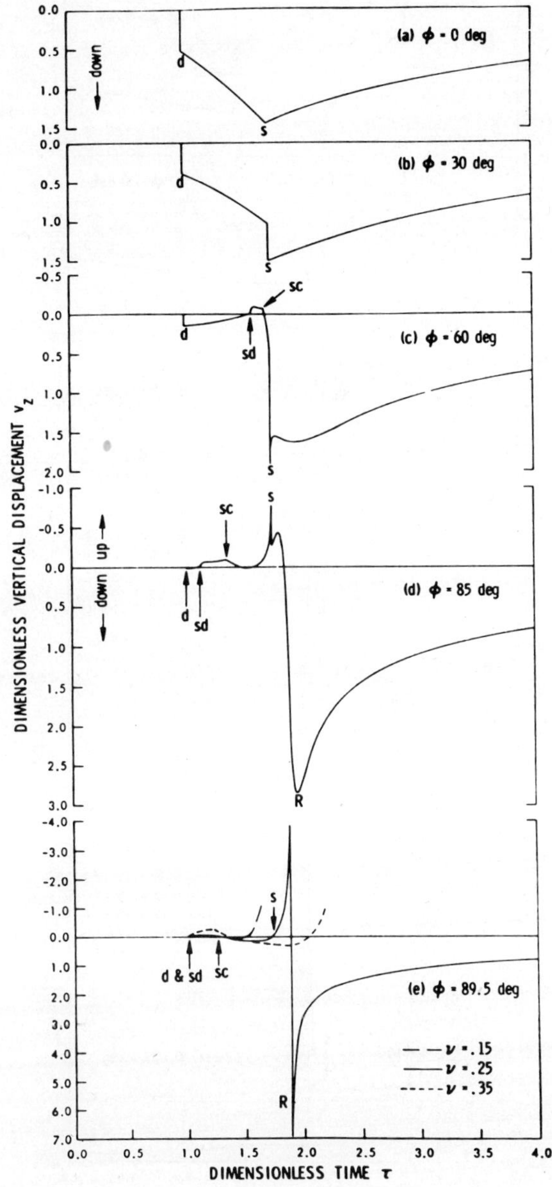

Figure 10.25

[229]

DIMENSIONLESS VERTICAL DISPLACEMENT v_z
down
(a) φ = 0 deg
d
s
(b) φ = 30 deg
d
s
(c) φ = 60 deg
s
d
sd
(d) φ = 85 deg
R
s
up
down
d
sd
L
(e) φ = 89.5 deg
R
ν = .15
ν = .25
ν = .35
d & sd
s
L
DIMENSIONLESS TIME τ

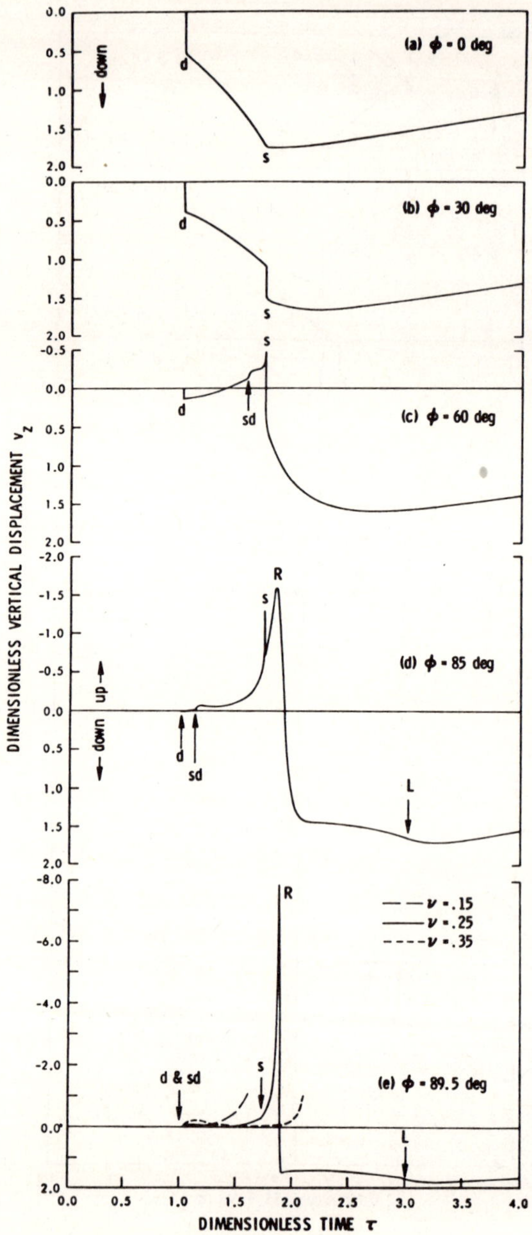

Figure 10.26

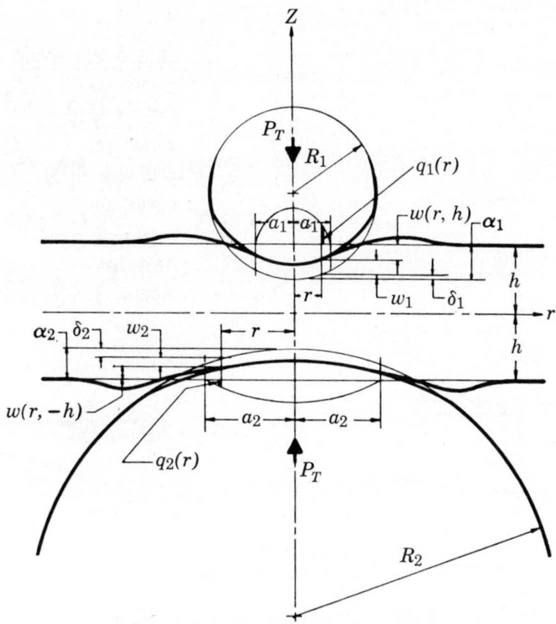

Figure 10.27

the spheres are different in size, the subscripts 1 and 2 are used. Most of the symbols in the figure are self explanatory. The α refers to the approach of the body relative to midsurface of the plate and w is the displacement in the z-direction.

Again the contact problem is formulated by using [10.24] for the fundamental solution for the spheres which are taken to be semi-infinite solids with the contact being changed from a spherical cap. An analogous equation to [10.24] is used for the plate.[144] The continuity at the interfaces lead to two integral equations. These have been solved with the aid of high-speed computers for the case of equal spheres.[145] The results are shown in Figure 10.28 and 10.29: (A) maximum pressure, (B) contact radius, (C) relative approach, (D) dimensionless radial displacement U/a, and (E) dimensionless radial displacement U/R. The symbol ASYM refers to asymptotic solutions, for example, (A)ASYM is the asymptotic solution of (A). The abscissa in both cases is h/R. The k is $2(1 - v^2)/E$, where v and E are the Poisson's ratio and Young's modulus, respectively. The subscript 1 applies to the sphere.

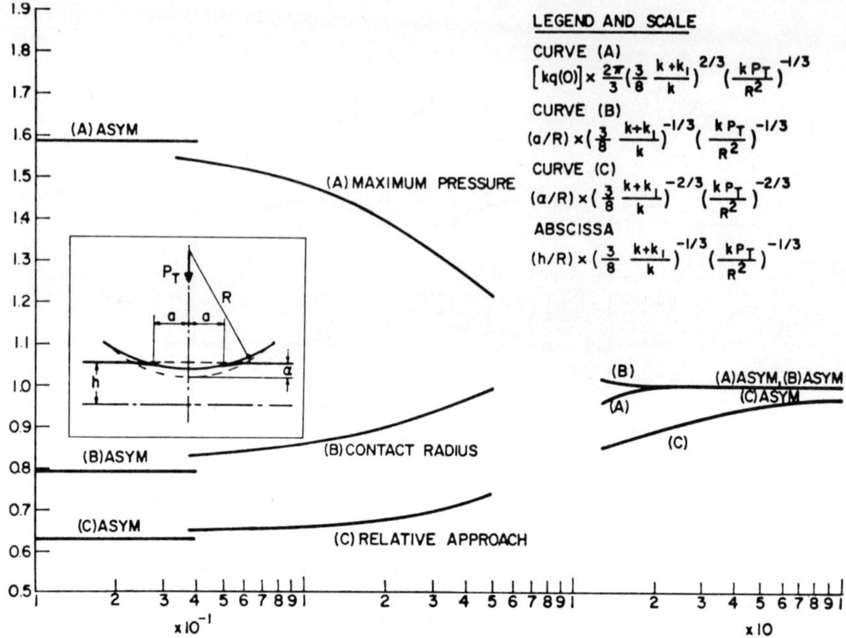

$$\text{CURVE (A)} \quad [kq(0)] \times \frac{2\pi}{3}\left(\frac{3}{8}\frac{k+k_1}{k}\right)^{2/3}\left(\frac{kP_T}{R^2}\right)^{-1/3}$$

$$\text{CURVE (B)} \quad (a/R) \times \left(\frac{3}{8}\frac{k+k_1}{k}\right)^{-1/3}\left(\frac{kP_T}{R^2}\right)^{-1/3}$$

$$\text{CURVE (C)} \quad (a/R) \times \left(\frac{3}{8}\frac{k+k_1}{k}\right)^{-2/3}\left(\frac{kP_T}{R^2}\right)^{-2/3}$$

$$\text{ABSCISSA} \quad (h/R) \times \left(\frac{3}{8}\frac{k+k_1}{k}\right)^{-1/3}\left(\frac{kP_T}{R^2}\right)^{-1/3}$$

Figure 10.28

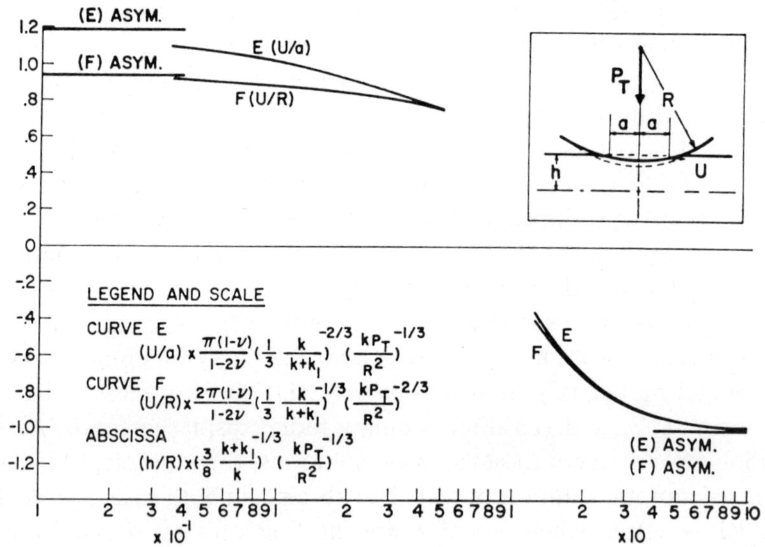

$$\text{CURVE E} \quad (U/a) \times \frac{\pi(1-\nu)}{1-2\nu}\left(\frac{1}{3}\frac{k}{k+k_1}\right)^{-2/3}\left(\frac{kP_T}{R^2}\right)^{-1/3}$$

$$\text{CURVE F} \quad (U/R) \times \frac{2\pi(1-\nu)}{1-2\nu}\left(\frac{1}{3}\frac{k}{k+k_1}\right)^{-1/3}\left(\frac{kP_T}{R^2}\right)^{-2/3}$$

$$\text{ABSCISSA} \quad (h/R) \times \left(\frac{3}{8}\frac{k+k_1}{k}\right)^{-1/3}\left(\frac{kP_T}{R^2}\right)^{-1/3}$$

Figure 10.29

L.   CONTACT BETWEEN PLATES AND AN ELASTIC MASS

The contact between an elastic plate and a large elastic mass under
load may be viewed in the manner of Section 10.K. For the large
elastic mass [4.19] and [4.20] give the displacement due to a generic
ring load, that is, for problems with axisymmetry. A similar formula
may be used for the plate.[146] At the interface of the plate and mass,
there is continuity of displacement; the condition as seen from [4.19] and
its counterpart for the plate results in an integral equation with the
pressure distribution at the interface as unknown.

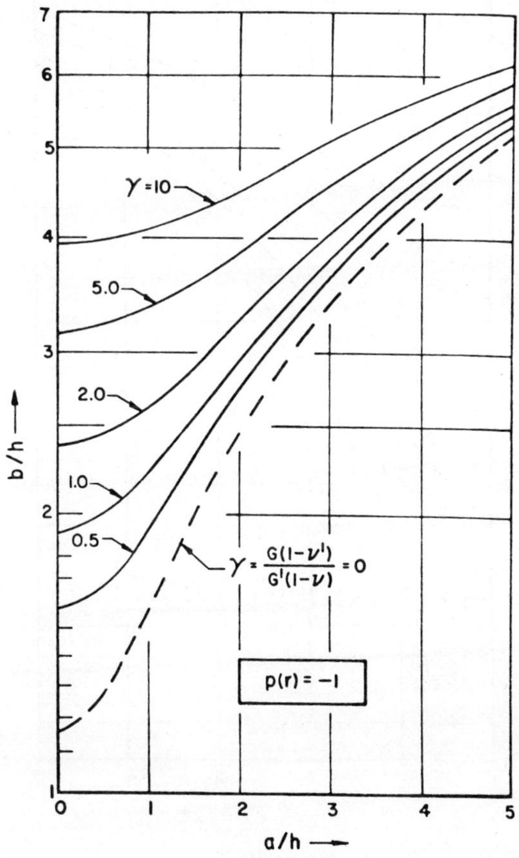

Figure 10.30

This integral equation has been inverted approximately for the case of various values of $\gamma \equiv G(1 - \nu')/G'(1 - \nu)$, where G and G' are the shear modulus of the plate and large mass, respectively, while ν and ν' refer to the Poisson's ratio of the plate and the large mass, respectively. The thickness of the plate is h and a is the radius of the load zone on the upper face of the plate. The radius of the contact between the plate and the large mass at the interface is b.

Figure 10.30 gives b/h versus a/h for a uniform compression[147] and Figure 10.31 gives b/h versus a/h for a parabolic distribution.

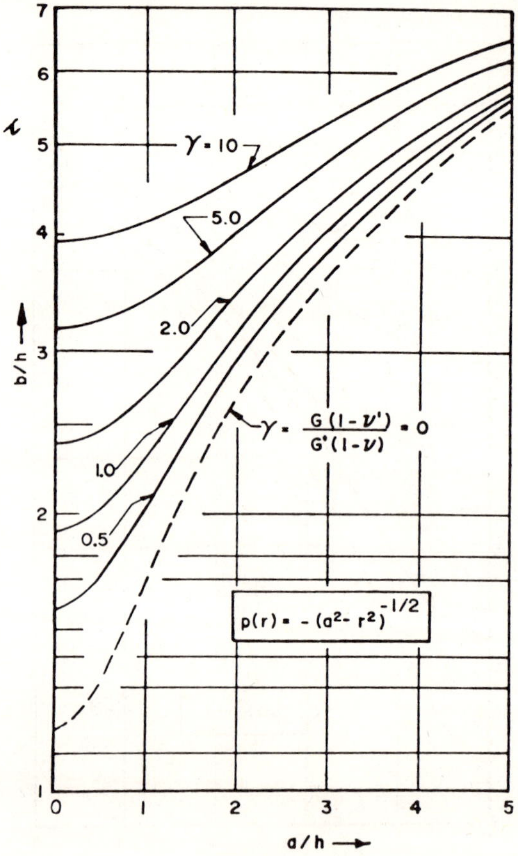

Figure 10.31

M.   FORCED VIBRATIONS OF TWO MASSES ON AN ELASTIC SOLID

Consider the case of a large elastic mass with two small masses, each taking the form of a circular cylinder, placed on top at some distance apart. One of the questions of technical interest is if mass number one is excited, how will mass number two be affected?

If one looks at the problem as a lumped parameter problem, the two masses are coupled with a spring which is the elastic base. In this regard, the spring coupling, in the lumped parameter sense, can be obtained from [4.15] and [4.16]. In short, the average displacement under a rigid circular cylinder is related to the force causing the dis-displacement by integrating [4.19] over a circular area of radius R in the variables (ξ, η) and dividing by πR^2, while the integration in the

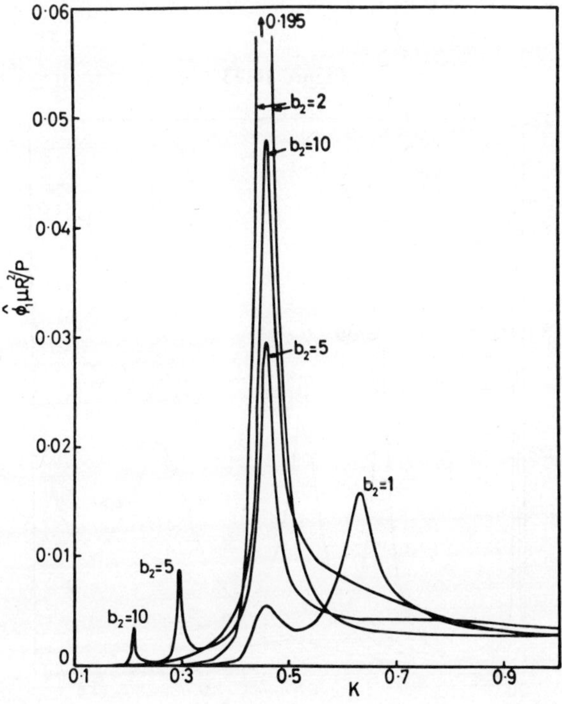

Figure 10.32

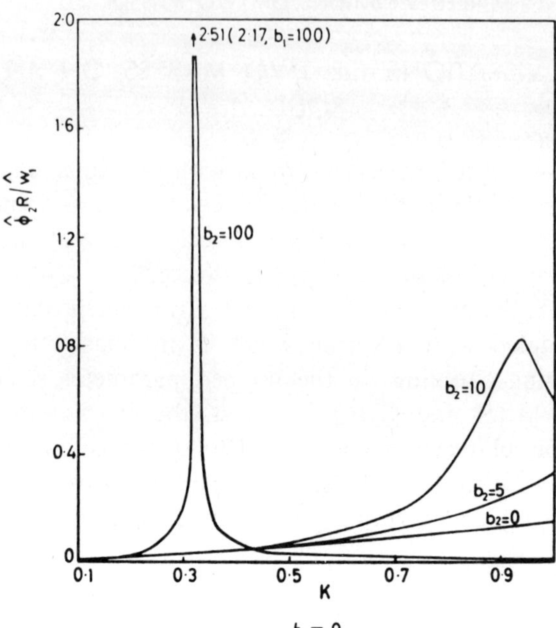

Figure 10.33

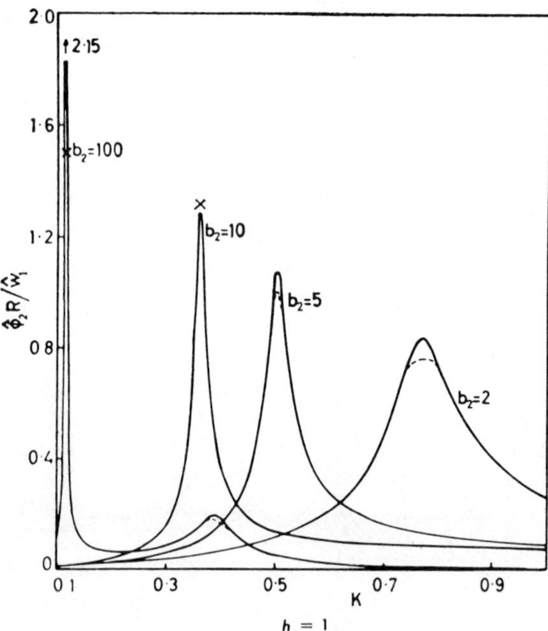

Figure 10.34

[236]

variables (ξ', η') would also be with respect to a circular area. The load will be the integral of q. This establishes the relationship of a load at one circular base to the displacement at another circular base at any location including the same location.

Similar calculations may be made for contacts which are not parallel to the surface of the solid, thus establishing angular motion of the base of the circular masses.

When these relationships are found, the equation of motion of the two masses can be written[148] and their response to excitations studied.

Letting h be the height of the cylinder, b be the mass ratio ($\equiv m/\rho R^3$), m be the mass, ρ be the mass density, w be the amplitude, ϕ be the amplitude of angular motion, K be the frequency of excitation, d be the distance between centers, $\eta (\equiv d/R)$, P be the amplitude of vertical applied force, μ be the shear modulus, and ν be the Poisson's ratio, Figure 10.32 shows the dimensionless amplitude $\hat{\phi}_1 \mu R^2/P$ for values of b_2 with $b_1 = 2$, $h = 2$, $\eta = 10$, and $\nu = 0$. Subscripts 1 and 2 refer to masses 1 and 2, respectively. Figures 10.33, 10.34, and 10.35 show the transmission ratio $\hat{\phi}_2 R/\hat{w}_1$ for values of b_2 with $\eta = 10$ and $\nu = 0$.

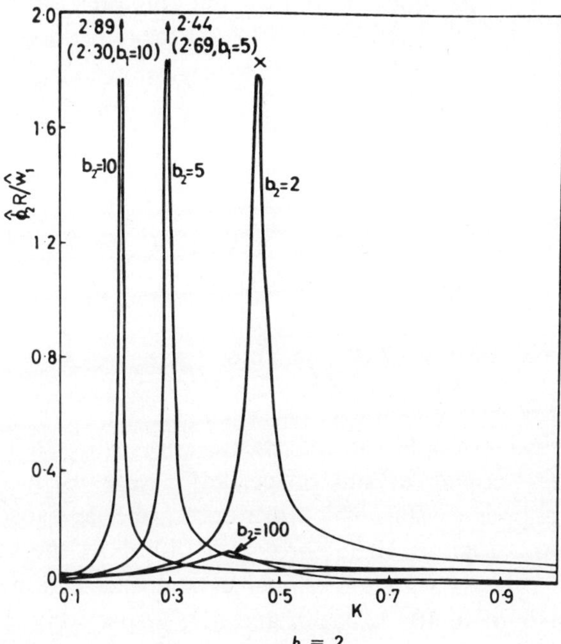

Figure 10.35

N. INDENTATION OF ANISOTROPIC MATERIALS

For an anisotropic elastic solid in plane strain or stress, [1.15] takes the form[149,150]

$$e_{\alpha\beta} = c_{\alpha\beta k\ell}\, \sigma_{k\ell} \qquad (\alpha,\beta,k,\ell = 1,2) \quad .$$

In the case of isotropic material in plane strain,

$$c_{1111} = c_{2222} = (1 - v^2)/E \quad ,$$

$$c_{1112} = c_{2212} = 0 \quad ,$$

$$c_{1122} = -(v + v^2)/E \quad \text{and}$$

$$c_{1212} = 2(1 + v)/E \quad ,$$

where E is the Young's modulus and v is the Poisson's ratio. For plane stress,

$$c_{1111} = c_{2222} = 1/E \quad ,$$

$$c_{1112} = c_{2212} = 0 \quad ,$$

$$c_{1122} = -v/E \quad \text{and}$$

$$c_{1212} = 2(1 + v)/E \quad .$$

In a numerical example for the elastic body with an indenter of circular shape,[149] copper crystal properties[151] were used. The elastic properties of copper crystal, which are cubic, are described by three elastic constants. Figures 10.36 through 10.40 show the contours of maximum shear stress with the copper crystal axis making an angle with the $0\text{-}y$ axis of 0, 10, 22.5, 30, and 45°, respectively. Note, symmetry is restored at 45° again. Both x and y are normalized such that

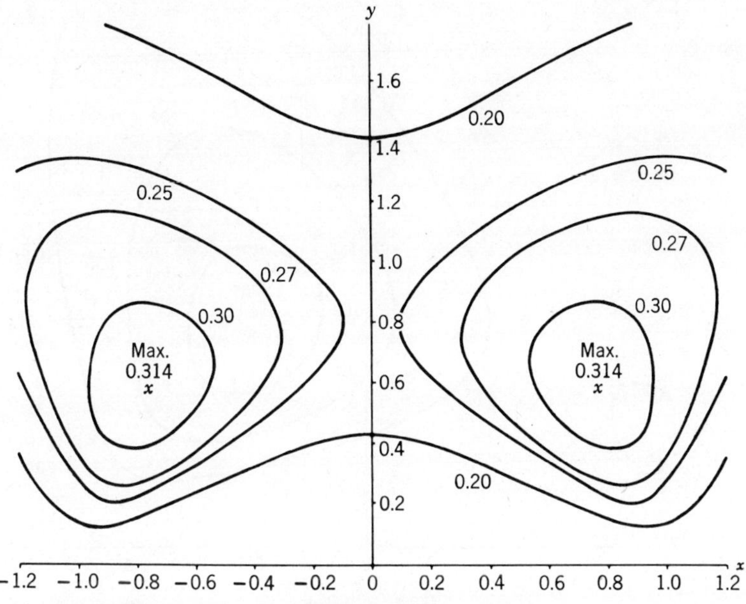

Figure 10.36

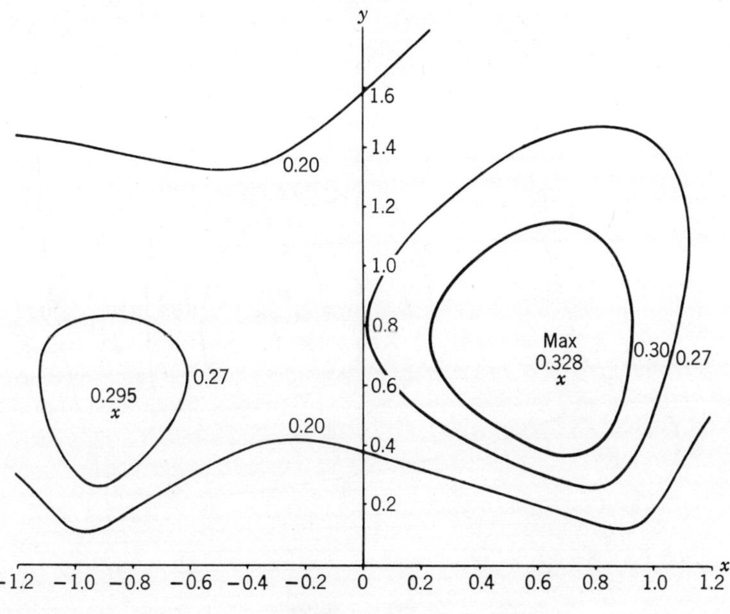

Figure 10.37

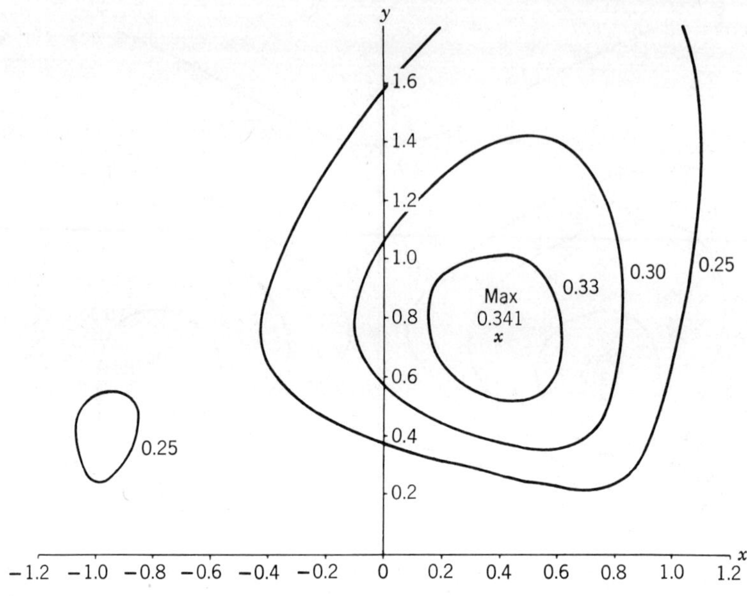

Figure 10.38

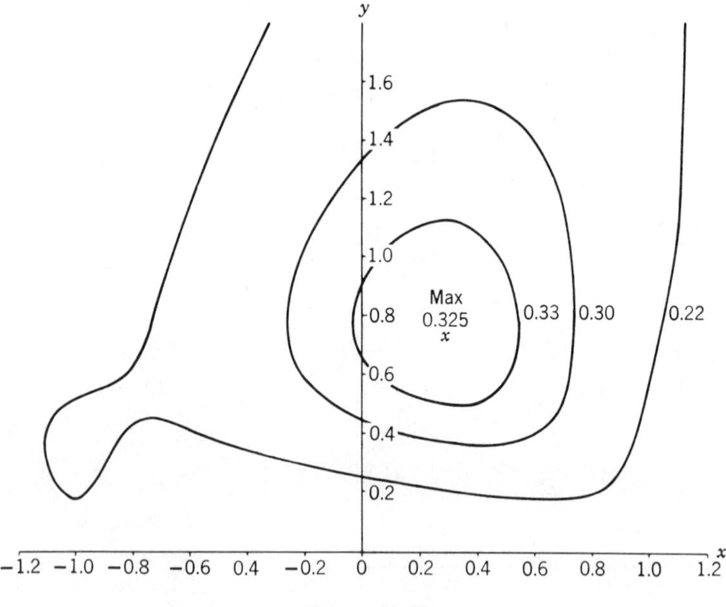

Figure 10.39

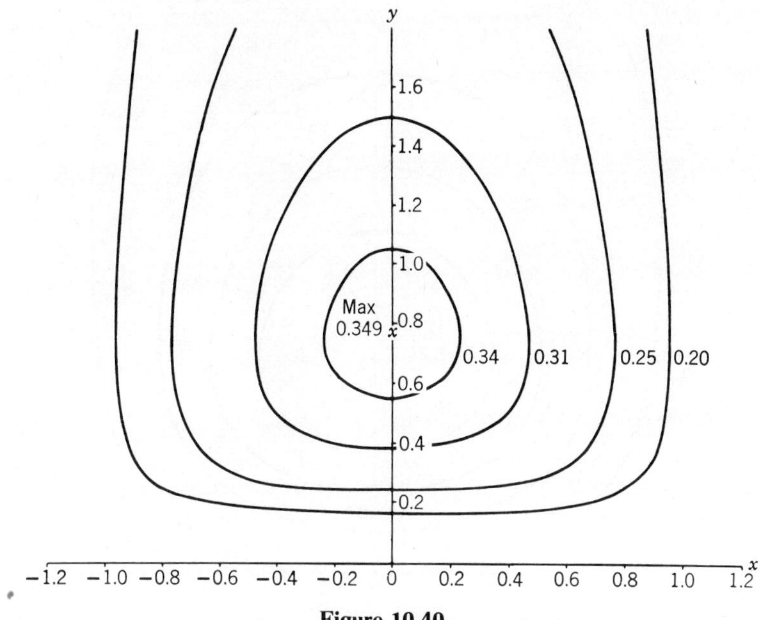

Figure 10.40

the contact length is 2 and the maximum pressure within the contact zone is unity.

Next, zinc, which is a hexagonal crystal and whose elastic properties are embodied in five constants, is studied numerically. Figures 10.41 through 10.44 are for the only axis of symmetry of the crystal making an angle with the 0-y axis of 0, 45, 60, and 90°, respectively.

The formulae used in the above and those for a three-dimensional transversely isotropic material have been worked out.[149]

O.   IMPACT AND CONTACT STRESSES IN MULTILAYER MEDIA

In a method similar to the one described in Section 10.U, an approximate method suitable for calculating contact stresses in multilayer problems has been developed.[152]* Results are shown for static contact and impact contact stress together with experimental results.

* The author acknowledges the kind permission to use these results which are as yet unpublished in the open literature.

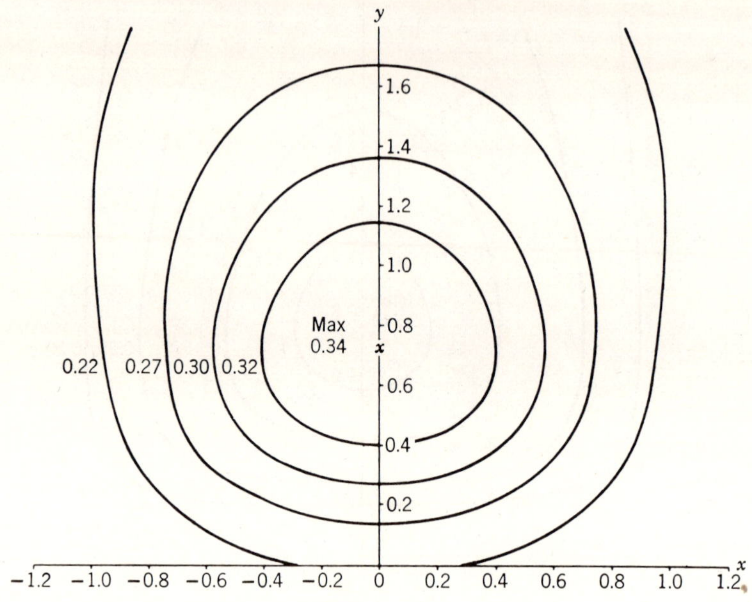

Figure 10.41

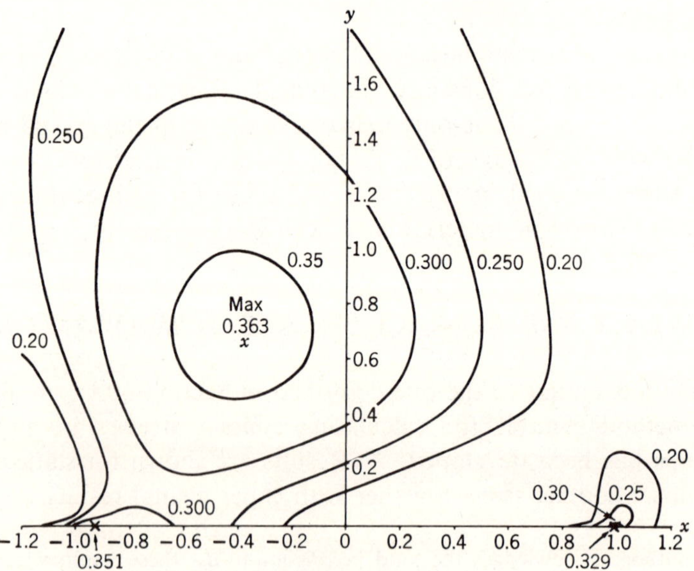

Figure 10.42

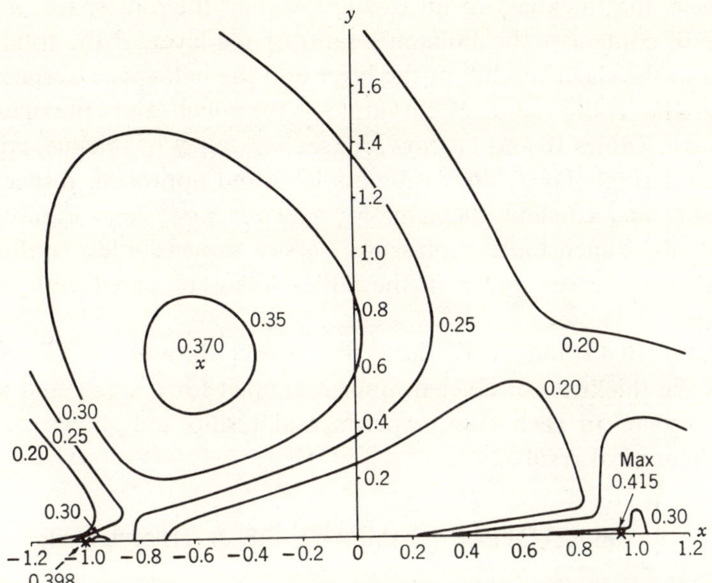

Figure 10.43

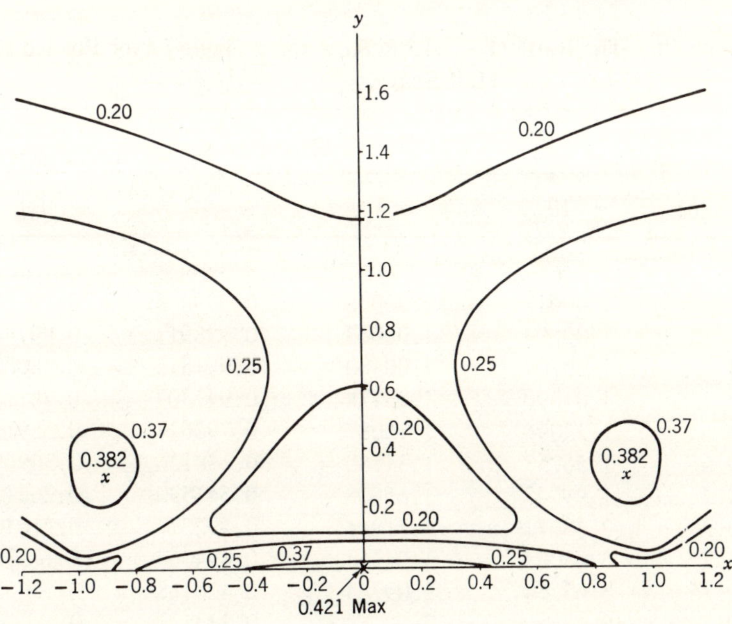

Figure 10.44

If h_1 is the thickness of an elastic layer on the half space, a is the radius of contact, ν the Poisson's ratio of the layer, P the total load, μ_1 and μ_2 the shear moduli of the layer and the half space, respectively, $\beta = a/2R\delta$, R the radius of curvature of the punch, and δ the rigid body approach, Tables 10 and 11 show, for several values of thickness parameter for a single layer, dimensionless load and approach, respectively. Tables 12 and 13 show the same for a two-layer system. Figure 10.45 shows the dimensionless pressure versus dimensionless radius for several of the cases shown in the tables when compared with Hertz's solution.

Figures 10.46 and 10.47 show the impact duration of a steel ball versus the thickness of layer in impact contact for a single and a two-layer system. In each case, experimental results are shown together with theoretical results.

P.   ELASTOPLASTIC IMPACT OF A THIN LAYER

If, in addition to the analysis used in Section 10.U, the material is allowed to yield plastically (i.e., an elastic, perfectly plastic material

Table 10.   The Ratio $(1 - \nu)\, PR/8\mu_1 a^3$ for a Single Layer Bonded to a Half-Space, $\eta_1 = \eta_2 = \frac{1}{3}$

$(2h_1/a)$	(μ_2/μ_1)			
	10	3	$\frac{1}{3}$	0.1
∞	1	1	1	1
16	1.00012	1.00008	0.999863	0.999678
8	1.00093	1.00063	0.998917	0.997457
4	1.00704	1.00480	0.991812	0.980804
2	1.04690	1.03178	0.947707	0.881113
1	1.23096	1.15087	0.793627	0.589116
0.6	1.56753	1.34560	0.638126	0.369493
0.4	1.99743	1.55573	0.541520	0.263115
0.3	2.39671	1.71941	0.492277	0.216407
0.2	3.08094	1.94764	0.444064	9.175325
0.15	3.63554	2.09764	0.420168	0.156621
0	10	3	0.333333	0.1

Table 11. The Ratio $\delta R/a$ for a Single Layer Bonded to a Half-Space, $\eta_1 = \eta_2 = \eta_3 = \frac{1}{3}$

$(2h_1/a)$	(μ_2/μ_1)			
	10	3	$\frac{1}{3}$	0.1
16	0.966880	0.976270	1.05714	1.20001
8	0.934543	0.953085	1.11319	1.39685
4	0.874623	0.810100	1.21740	1.76316
1	0.686853	0.783394	1.43415	2.32244
0.6	0.647928	0.771666	1.35677	1.95716
0.4	0.640572	0.784485	1.27201	1.67294
0.3	0.646590	0.801586	1.21908	1.51763
0.2	0.667250	0.830434	1.16087	1.35967
0.15	0.688213	0.851286	1.12954	1.27905

Table 12. The Ratio $3(1 - \nu) PR/8\mu a^3$ for Two Layers Bonded to a Half-Space, $h_1 = h_2$, $\mu_1 = \mu_3$, $\eta_1 = \eta_2 = \eta_3 = \frac{1}{3}$

$(2h_1/a)$	(μ_2/μ_1)			
	10	3	$\frac{1}{3}$	0.1
∞	1	1	1	1
16	1.00011	1.00007	0.999883	0.999733
8	1.00086	1.00057	0.999078	0.997895
4	1.00651	1.00428	0.993071	0.984180
2	1.04255	1.02766	0.956567	0.902688
1.5	1.08472	1.05429	0.918091	0.821161
1	1.19209	1.11844	1.837486	0.664501
0.8	—	—	—	—
0.6	1.37748	1.21510	0.740170	0.493691
0.4	1.44827	1.24294	0.710060	0.432339
0.3	1.41318	1.22389	0.715235	0.423731
0.2	1.30385	1.17188	0.746916	0.446414
0.15	1.233115	1.13608	0.777619	0.479147
0	1	1	1	1

Normalized Pressure Distribution q(r)/q(0)
1
.8
.6
.4
.2
D
B
Homogeneous
Half-Space
(Hertz Solution)
A
C
Two layers on half-space
2h_1 /a = 2h_2 /a = .3
A   μ_2 /μ_1 = 10, μ_3 /μ_1 = 1
B   μ_2 /μ_1 = .1, μ_3 /μ_1 = 1
One layer on half-space
2h_1 /a = .3
C   μ_2 /μ_1 = 10
D   μ_2 /μ_1 = .1
All Poisson's ratios 1/3
Radius r/a
.1   .2   .3   .4   .5   .6   .7   .8   .9   1

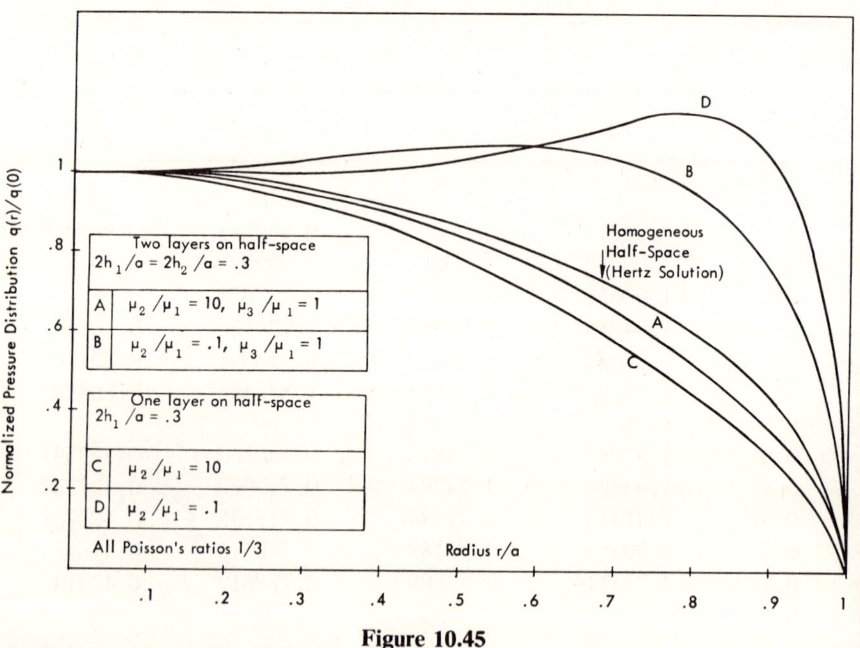

Figure 10.45

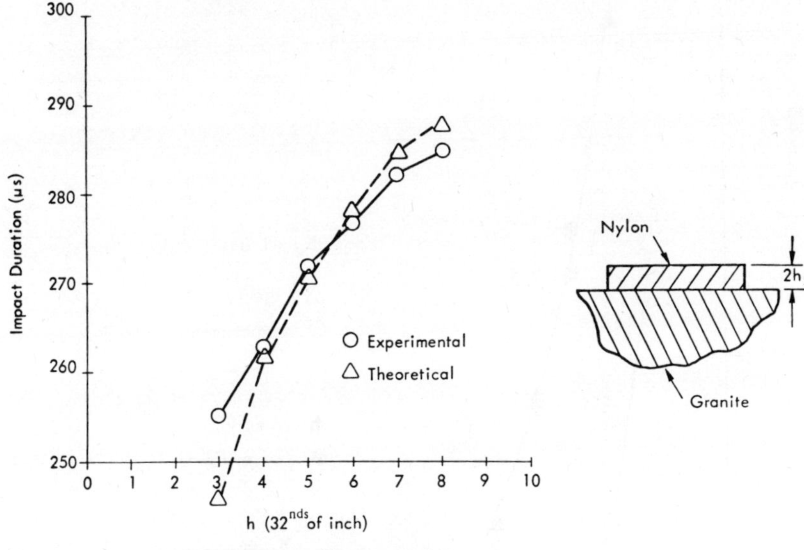

Figure 10.46

with a yield stress Y and Young's modulus E), the analytical results may be represented in Figures 10.48, 10.49, and 10.50 shown with experimental data.[153] These data are for impact of a rigid sphere on layers of aluminum, paper, and mylar layers which are placed under rigid support. The V_0 and V_f are the initial and final velocity, respectively, for the impact period; the other symbols are explained in the figures.

Q. SURFACE CORRUGATIONS GENERATED AT ROLLING CONTACT

Corrugations are frequently observed on surfaces which are subjected to repeated rolling loads. Earth roads and railway tracks are the best known examples; but machine elements, on occasion, may develop corrugations or ripples.

Equation 4.13 in conjunction with [A.9] leads to the two-dimensional Hertzian pressure distribution [A.14] which applies to circular, cylindrical surfaces in contact. From these a relationship between total load, P, at the contact and the approach, δ, are obtained: $\delta \sim P^{2/3}$ with the

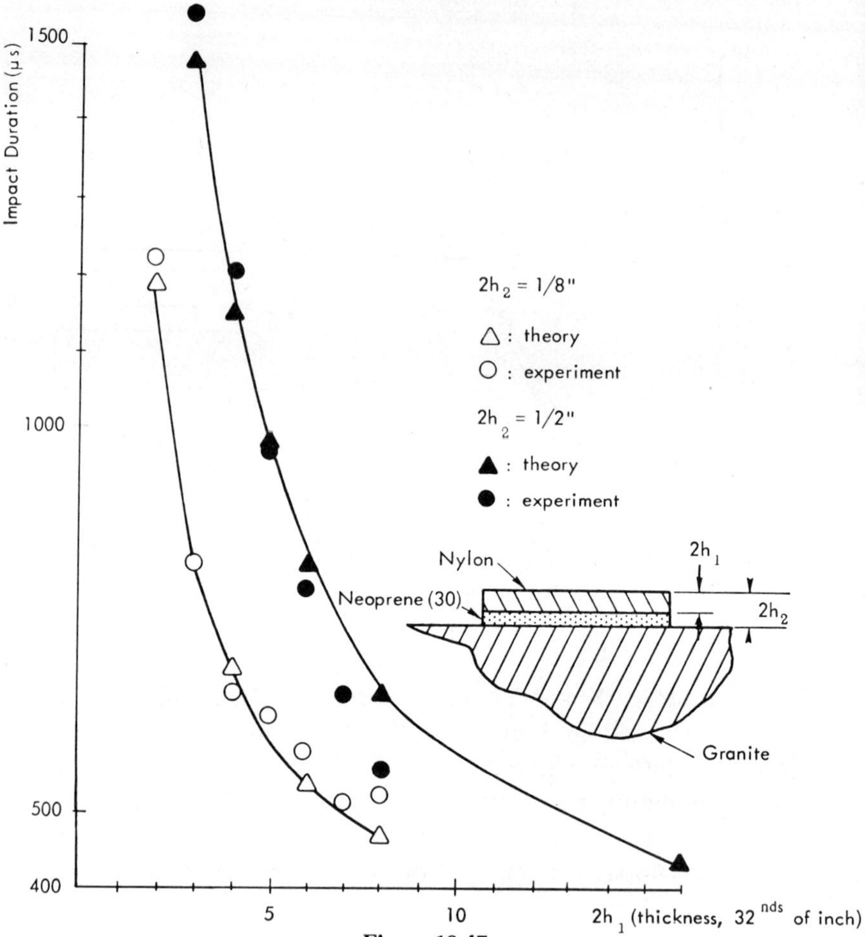

Figure 10.47

constant of proportionality a function of the geometry and mechanical properties. In fact, $dP/d\delta$ may be used as the spring constant of the contact, S_0. Supposing a simple spring mass system shown in Figure 10.51 is used to simulate the contact of two rolling discs, then the natural frequency of oscillation, f_0, is

$$f_0 = (2\pi)^{-1}[2S_0/(M_1 + M_2)]^{1/2} \quad . \quad [10.20]$$

Equation 10.20 has been used[154] to compare with experimental data obtained on a specially built disc machine in which contact resonance

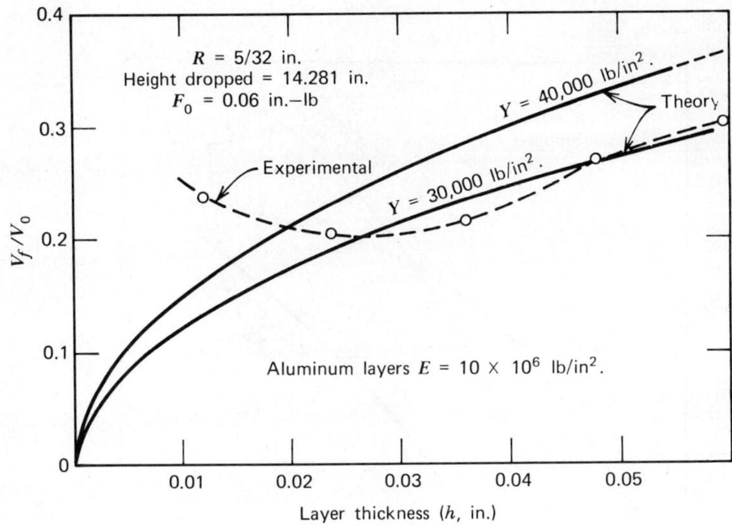

Figure 10.48

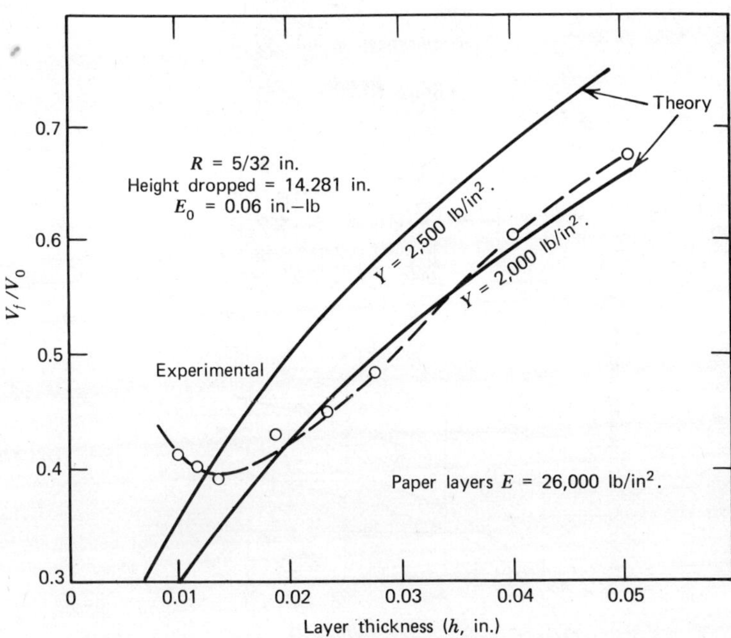

Figure 10.49

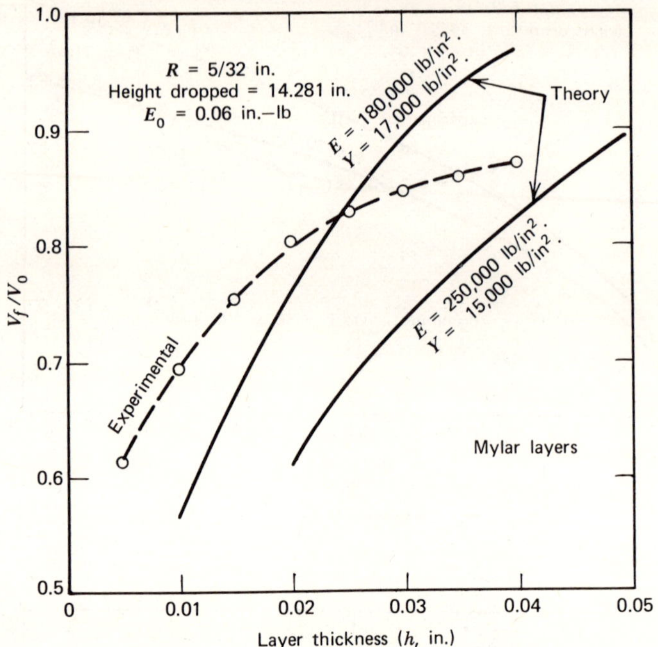

Figure 10.50

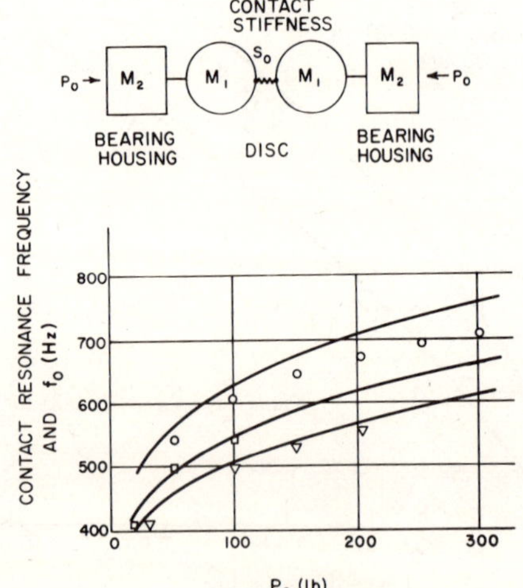

Figure 10.51

is measured. These frequencies are shown with the theoretical estimate f_0 in Figure 10.51. The correlations are even better when the springiness of the bearing housing is taken into account. Under this sort of vibration, corrugation will develop when the surface of the disc is moved by plastic deformation or removed by wear or corrosion. Wave length of all corrugations was consistent with a vibration in the contact mode.

R.   ON TWO-DIMENSIONAL ROLLING IN VISCOELASTIC MATERIAL

Figure 10.52 shows a rigid two-dimensional body rolling on a half-space made of viscoelastic material. Rolling friction is usually defined in the following fashion.

The moment required to turn the roller, M, is

$$M = \int_{x_1}^{x_2} x\, P(x)\, dx \quad ,$$

where $x_1 \leq x \leq x_2$ gives the extent of contact and $P(x)$ is the pressure distribution. Of course, if F_x is the traction component in the x direction and R is the radius of the roller, then

$$M = F_x R \quad .$$

Therefore, from the above two,

$$F_x = R^{-1} \int_{x_1}^{x_2} x\, P(x)\, dx \quad . \qquad [10.21]$$

Given the load, W, the coefficient of friction χ can be calculated.

In [10.21], $P(x)$ is required. To find it, a contact problem similar to the Hertz problem must be solved. In this case, the profile of the rigid roller must be matched by the deformed surface given by [6.53] with the fundamental solution [6.57]. This leads to an integral equation, the inversion of which would lead to $P(x)$.

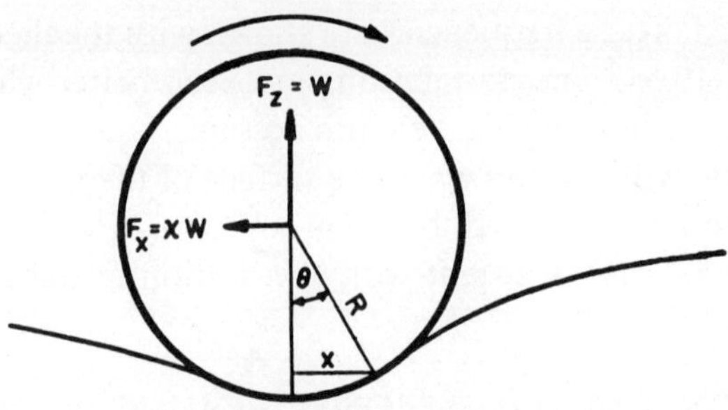

Figure 10.52

Figure 10.53 shows plots of R_X/a_0 versus $V\tau/a_0$ for two values of f, where a_0 is the semicontact length for zero velocity, f the strength of the single-line retardation spectrum, τ the retardation time, and V the linear speed of the rolled cylinder. The result checks qualitatively with experimental data.[73]

Figure 10.54 shows the relationship between dimensionless pressure and distance within the contact zone. The first case corresponds to static or elastic situation. The third case corresponds to material behaving as an elastic solid with shear modulus μ_D while the second case is a typical viscoelastic situation.

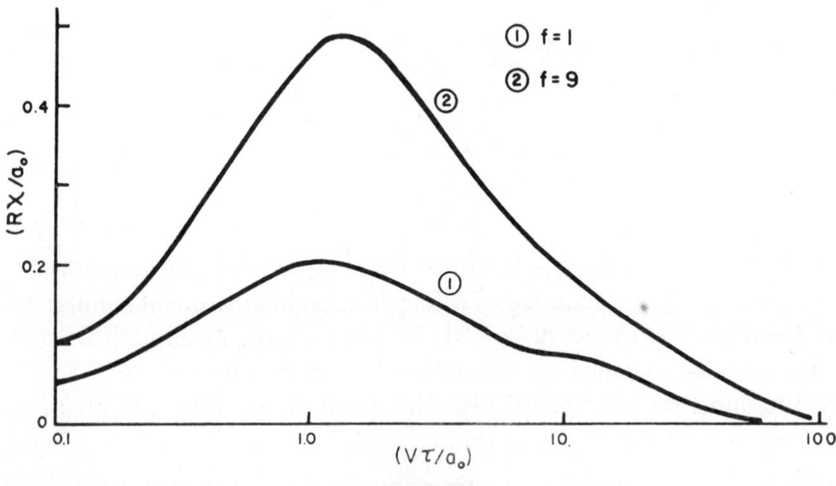

Figure 10.53

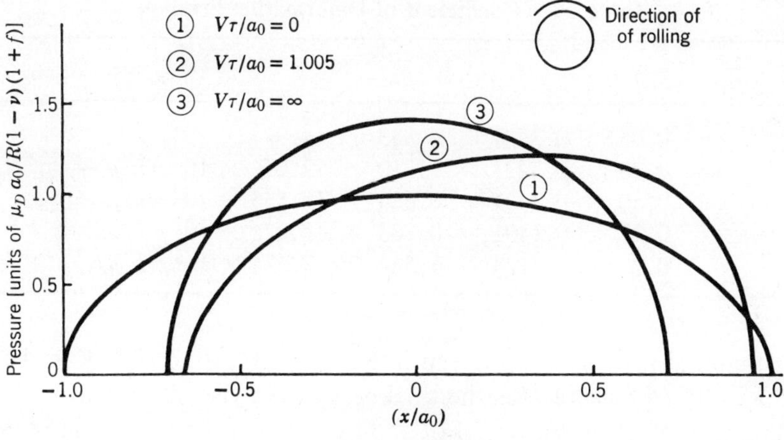

Figure 10.54

S.   ON DEFORMATION FRICTION

Let x be the coordinate reference along the surface of an elastic or viscoelastic body in sliding or rolling contact, $v(x)$ be the normal displacement of the body under load at the surface, and $p(x)$ be the pressure distribution at the interface. The work of friction in the absence of adhesion due to an incremental travel of the force F is

$$F\delta x = \int_{c^+} (\partial v/\partial x)\delta x\, p(x)dx \quad ,$$

where c^+ denotes that portion of the contact surface at which $\partial V/\partial x$ is positive. Of course, here, a simple two-dimensional picture is given. The total normal load is $L = \int_{c} p(x)\,dx$, where c is the whole contact area. Thus the coefficient of friction $f = F/L$ is, according to the above equation,

$$f = L^{-1}\int_{c^+} (\partial v/\partial x)p(x)dx \quad . \qquad [10.22]$$

This definition,[155] which is distinct from any other known definition, may be applied to layered system as well. Using the result from Section 6.H for $v(x)$ and taking the special case of $p(x) = $ constant $= p_0$, [10.22] gives the results[77] in Table 14.

Table 14.   Coefficient of Deformation Friction

h	$\bar{V}$	$\bar{r}$	f
0.05	0.3421	6.843	$4.977 \times 10^3\, p_o h/\mu_e$
0.20	0.3421	1.710	$12.096 \times 10^3\, p_o h/\mu_e$
0.50	0.3421	0.684	$15.535 \times 10^3\, p_o h/\mu_e$
0.2	0.1406	0.703	$18.628 \times 10^3\, p_o h/\mu_e$
0.2	0.7128	3.564	$7.674 \times 10^3\, p_o h/\mu_e$

The material properties are given in Section 6.H; $h \equiv H/l$, $\bar{V} \equiv V/l\zeta$, and $\bar{r} \equiv V/H\zeta$, where H is the thickness of the viscoelastic layer, l the length of contact, V the speed of sliding, μ_e the elastic shear modulus, and ζ one of the material constants.

In the above, for p_0 of the order of 10^3 psi, the maximum value of f shown in Table 14 corresponds to 0.6.

T.   ON THERMAL SOFTENING MECHANISM OF FADING OF LUBRICATED BRAKES AND CLUTCHES

Surface roughness plays a dominant role in many surface or interface phenomena. The plastic collapse of asperities on surfaces in sliding contact, which are separated by a fluid as in the case of lubricated brakes and clutches, causes the nature of the sliding process to change from the state wherein the asperities are deformed only elastically. The outward manifestation may be a relatively high frictional resistance in the elastic case and a decreasing frictional resistance subsequent to such a collapse.

Figure 10.55 shows an experimentally determined load and speed curve[156] which forms the threshold of fading for copper-based sintered material against steel in paraffinic mineral oil. The dots are the actual experimental points, obtained on an annular contact tester.[157]

The mechanical model is based on the identification of the threshold point as being in the microelastohydrodynamics region. As thermal softening causes the area of the asperities to grow in size, other factors being constant, friction coefficient decreases as the Hersey[158] number increases. This number is the ratio of the product of viscosity and speed to pressure. In other words, this portion of the friction coefficient—Hersey number characteristics—is associated with the fading phenomenon.

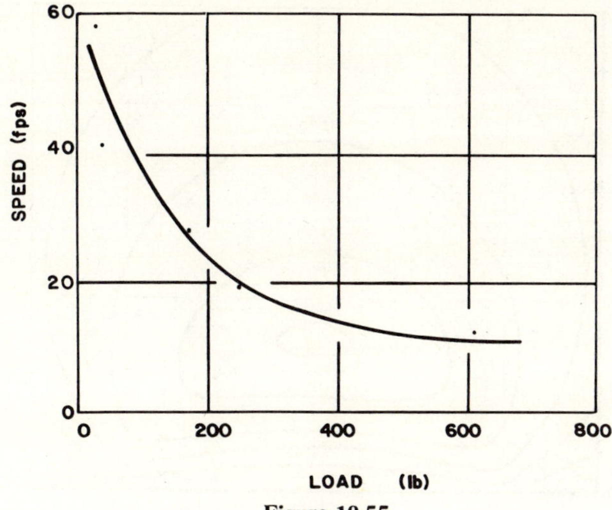

Figure 10.55

Figure 10.56 shows a typical contour map[159] of temperature in a half-space, θ, in the dimensionless (x_1, x_2) plane of Figure 2.1, (ξ, η). This is characteristic of the high Peclet number regime, $R \gg 1$, showing the boundary layer phenomena. The surface temperature would correspond to that provided by [2.17].

The state of stress due to mechanical loading is obtainable by Hertz formula.[124] Figure 10.57 shows a contour map Σ_2 in the (ξ, η) plane for zero shear traction as indicated by friction coefficient $\lambda = 0$. The Σ_2 is

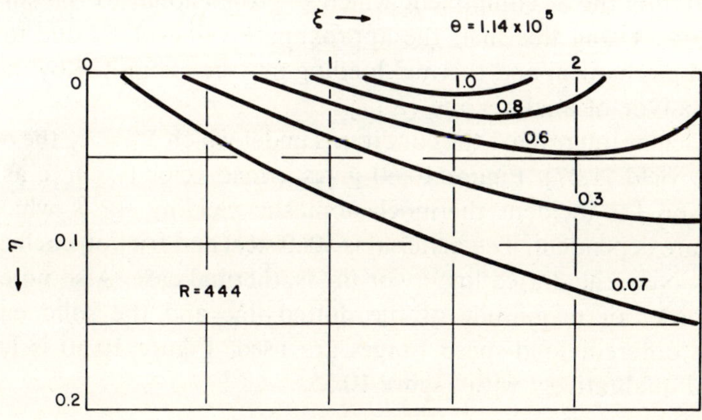

Figure 10.56

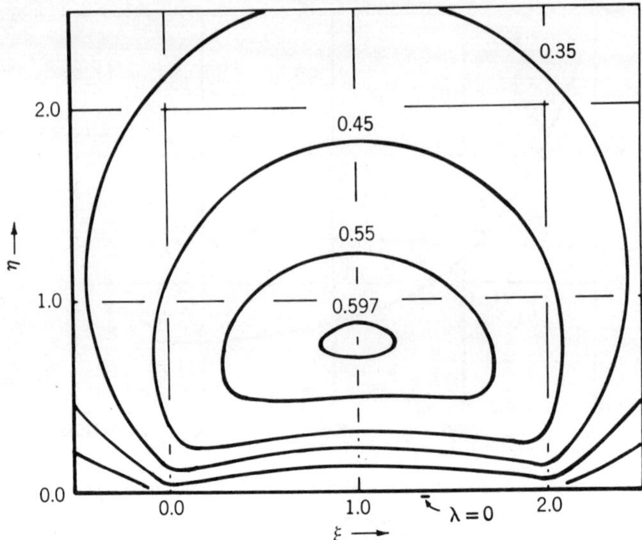

Figure 10.57

the second invariant of the stress tensor, see Section 1.H. Note the maximum is interior to the body. As λ increases, say 0.6, the contour map is as shown in Figure 10.58. When $\lambda = 1$, the contour map is as shown in Figure 10.59. In other words, the maximum is approaching the surface as shear traction is increased.

For the thermoelastic part, a recent analysis[160] has shown that the maximum for $R \gg 1$ is on the surface. Moreover, the most important contribution is the σ_ξ component which is proportional to the surface temperature. Using this fact, the approximate value of Σ_2 due to the combined mechanical and thermal loading may be computed by means of a Hertz type of analysis and [2.17].

Of particular interest are the conditions under which $\Sigma_2 = k_2$, the onset of plastic yield [1.27]. Figure 10.60 gives a load-velocity curve as the demarcation for incident thermoelastoplastic yielding for k which is temperature dependent. The material is 1060 steel and friction coefficient $\lambda = 0.14$. Note the dotted line is for the isothermal case. Also note the relative order of magnitude of the dotted line and the solid curve. Although different load-speed ranges are used, Figure 10.60 is to be compared qualitatively with Figure 10.55.

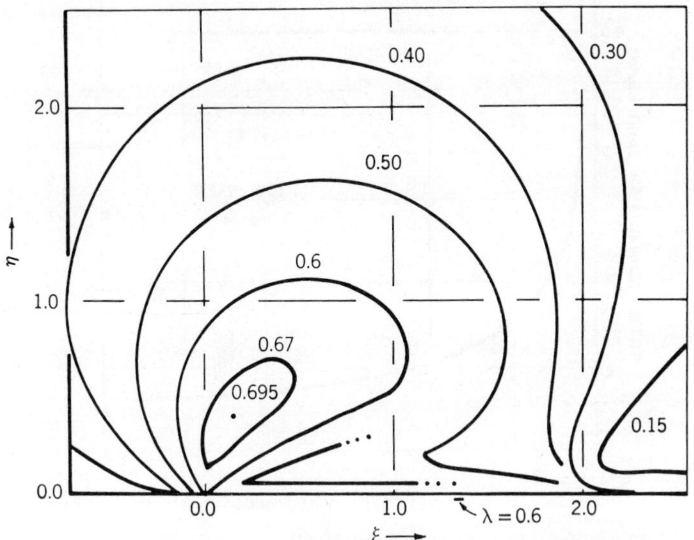

Figure 10.58

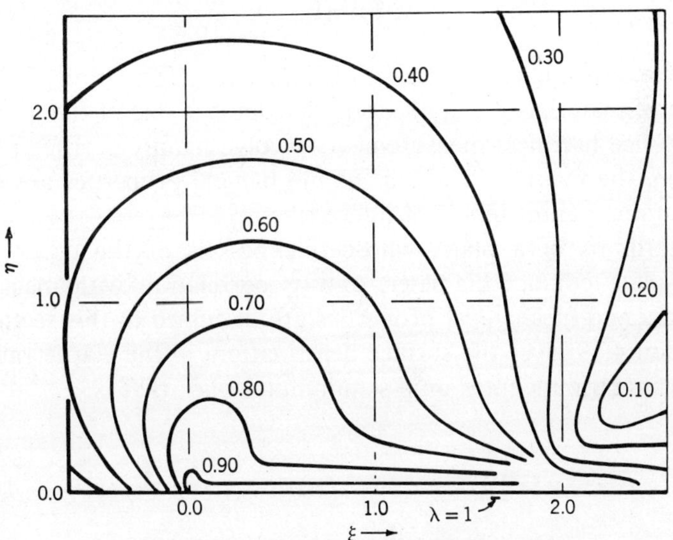

Figure 10.59

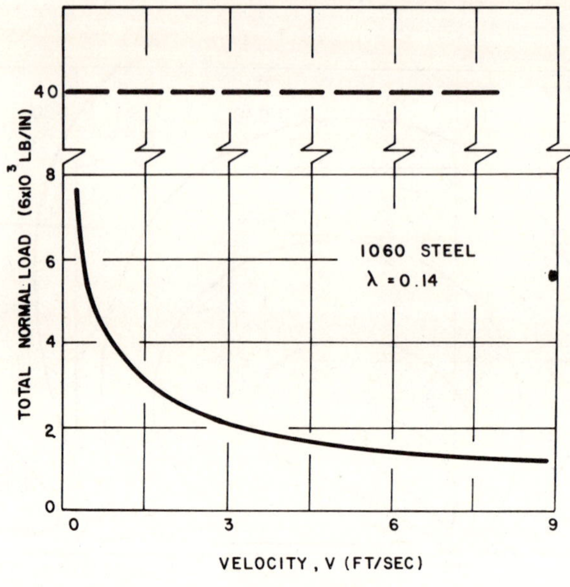

Figure 10.60

U. INDENTATION OF AN ELASTOPLASTIC WORK-HARDENING PLATE

In simplest terms, hardness means resistance to deformation, and in the selection of materials for application, hardness is one of the important indices. Since hardness measurements involve mainly a process of deformation, the manner in which the mechanical properties are related to measurable parameters is of interest.

The hardness of a plate, whose thickness is of the order of the indenter-plate contact diameter, and its correlation with measurable parameters and mechanical properties are discussed in this section.

Equation 4.19 gives the surface deformation, in the elastic range, of an axi-symmetric indenter on a semiinfinite solid; that is,

$$w(r,0) = \int_0^1 q(\rho)K(r,\rho)d\rho \quad , \qquad [10.23]$$

where the physical model is shown in Figure 10.61.

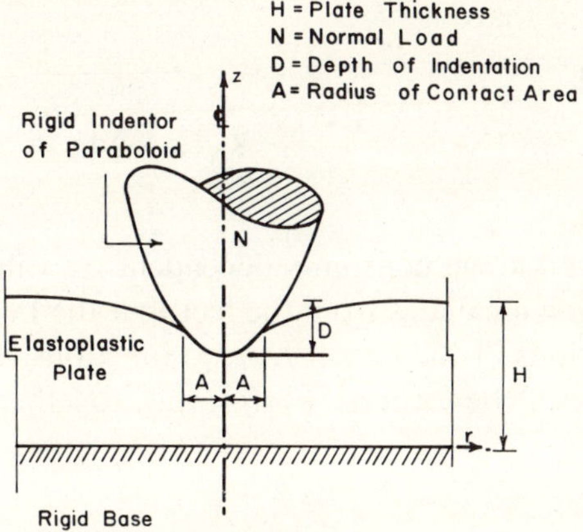

Figure 10.61

Now, again in the elastic range for a plate of thickness H as shown in the figure, the surface displacement can again be expressed in the form [4.15] except now

$$K(r,\rho) = K_0(r,\rho) - K_1(r,\rho) \quad , \qquad [10.24]$$

where K_0 is for the semiinfinite body and

$$K_1(r,\rho) = (1 - \nu)r \int_0^\infty J_0(p,r)J_0(p,\rho)$$

$$\cdot \left[\frac{2e^{-2p\lambda}}{1 + e^{-2p\lambda}} + \frac{2p\lambda}{\sinh 2p\lambda + 2p\lambda} \right.$$

$$\left. - \frac{4p\lambda e^{-2p\lambda}}{(1 + e^{-2p\lambda})(\sinh 2p\lambda + 2p\lambda)} \right] dp \quad ,$$
$$[10.25]$$

where J_0 is the Bessel function of the first kind and order zero, $\lambda \equiv H/A$, and p is the transform parameter.

It is observed that

$$2\pi \int_0^1 \rho(1 - \rho^2)^{i - 1/2} K_0(r,\rho)\,d\rho = \sum_{j=0}^{i} A_j^i\, r^{2j} \quad ,$$

where A_j^i's are known constants dependent on values of i and j only. This observation emanates from the fact that the Legendre polynomials are eigenfunctions of the kernel $K_0(r, \rho)$ functions.[161]

Letting w be of the form $\phi_0 + \phi(r)$; this, together with [10.24], leads to

$$\phi_0 + \phi(r) = \sum_{i=0}^{n} \phi_i\, r^{2i} = \int_0^1 \rho q(\rho) K_0(r,\rho)\,d\rho - \int_0^1 \rho q(\rho) K_1(r,\rho)\,d\rho$$

$$[10.26]$$

Since $K_1(r, \rho)$ is a well-behaved function and, in view of the axisymmetry, it can be approximated by an even-ordered polynomial symmetric with respect to r and ρ. Thus the $(2n)$th ordered polynomial approximation for $K_1(r, \rho)$ can be written as

$$K_1(r,\rho) = \sum_{m=0}^{n} \sum_{i=0}^{m'} c_i^m\, r^{2i}\, \rho^{2i} (r^{2m - 4i} + \rho^{2m - 4i})$$

$$[10.27]$$

where $m' = m/2$ for m to be even, and $m' = (m - 1)/2$ for m to be odd. Assume the form of $q(\rho)$ to be

$$q(\rho) = (1 - \rho^2)^{1/2} \sum_{i=0}^{n} q_i(1 - \rho^2)^i \quad . \quad [10.28]$$

No generality will be lost if the values of n in [10.26], [10.27), and [10.28] are taken to be the same. One may take the largest of them all, for example. Using [10.27] and [10.28] in integration, one obtains

$$2\pi \int_0^1 \rho q(\rho) K_1(r,\rho)\,d\rho = \sum_{i=0}^{n} \sum_{j=0}^{n} b_{ij} q_j\, r^{2i} \quad ,$$

$$[10.29]$$

where b_{ij} are constants depending on i and j, and can be calculated from given $C_k{}^m$'s. Now, using [10.29] and [10.28] in [10.26] and identifying the terms with the same power of r, one obtains, after some rearrangement,

$$\phi_i = \sum_{j=1}^{n} B_{ij} q_j \qquad (i = 1, 2, \ldots, n) \qquad [10.30]$$

and

$$N = \sum_{j=1}^{n} a_j q_j \qquad [10.31]$$

when the q_j's and a have been obtained, the depth of indentation ϕ_0 can be calculated, that is,

$$\phi_0 = \sum_{j=1}^{n} B_{0j} q_j \ . \qquad [10.32]$$

In the region of plasticity near the incipient yield, the stress-strain relation is assumed to be of the incremental work-hardening type. In addition to this nonlinearity, the surface of the stress boundary conditions also changes with the total load applied in case of the indentation hardness analysis. A step-by-step procedure with the assistance of a finite element computer program is adapted for this part of the problem.

The finite element method has become increasingly important to the solution of practical problems of continuum mechanics. Only the following points are to be noted here:

1. The use of the finite element method in contact problems is hindered by the unknown boundary conditions. This is overcome here by a step-by -step iteration procedure.

2. The stress-strain relation in the state of plasticity can be quite general. Various work-hardening rules as well as unloading can be incorporated with some programming considerations.

The procedure for solutions is now described. In the preceding pages it was shown that the elastic solution of a thick plate under a rigid indenter can be reduced to the solution of a set of algebraic simultaneous equations.

In the region of plasticity, the material is assumed to behave according to a linear work-hardening rule of Mises yield condition and Prandtl-Reuss flow equation,

$$f(\sigma_{ij}) = J_2' = \frac{1}{2}\,\sigma_{ij}'\sigma_{ij}' = \frac{1}{3}\,\bar{\sigma}^2 \quad , \qquad [10.33]$$

$$d\epsilon_{ij} = \frac{d\sigma}{3K}\,\delta_{ij} + \frac{1}{2G}\,d\sigma_{ij}' + H\sigma_{ij}'\sigma_{k\ell}'d\sigma_{k\ell}' \quad , \qquad [10.34]$$

where $\sigma_{i'}'$, $\bar{\sigma}$ are deviatoric stress components and uniaxial yield stress in tension; J_2' is the second deviatoric stress invariant; ϵ_{ij} denotes strain components; K, G are the bulk and shear moduli, respectively, and σ is the hydrostatic component of the stress tensor. Also

$$H = \frac{3}{4J_2'}\left(\frac{1}{E_t} - \frac{1}{E}\right) \quad , \qquad [10.35]$$

where E, E_t are Young's modulus and tangent modulus, respectively.

The inverse relation of [10.34] can be shown[162] to be

$$d\sigma_{ij} = 2G\left[d\epsilon_{ij} - \sigma_{ij}'\,\frac{\sigma_{k\ell}'d\sigma_{k\ell}'}{S} + \frac{\nu}{1-2\nu}\,d\epsilon_{kk}\delta_{ij}\right] \quad ,$$

$$[10.36]$$

where

$$S = 2J_2'\left[1 + \frac{1}{3G\left(\dfrac{1}{E_t} - \dfrac{1}{E}\right)}\right] \quad .$$

In matrix notation and for the axisymmetric state of stress, [10.36] becomes

$$\{d\sigma\} = [D]^P\{d\epsilon\} \quad ,$$

where

$$\{d\sigma\} = \begin{bmatrix} d\sigma_{rr} \\ d\sigma_{\theta\theta} \\ d\sigma_{zz} \\ d\sigma_{rz} \end{bmatrix}, \qquad \{d\epsilon\} = \begin{bmatrix} d\epsilon_{rr} \\ d\epsilon_{\theta\theta} \\ d\epsilon_{zz} \\ d\epsilon_{rz} \end{bmatrix},$$

$$[D]^P = [D]^e - [D]^S,$$

with

$$[D]^e = \frac{2G(1-\nu)}{1-2\nu} \begin{bmatrix} 1 & & & \text{SYM.} \\ \dfrac{\nu}{1-\nu} & 1 & & \\ \dfrac{\nu}{1-\nu} & \dfrac{\nu}{1-\nu} & 1 & \\ 0 & 0 & 0 & \dfrac{1-2\nu}{2(1-\nu)} \end{bmatrix}$$

$$[D]^S = \frac{2G}{S} \begin{bmatrix} \sigma'^2_{rr} & & & \text{SYM.} \\ \sigma'_{rr}\sigma'_{\theta\theta} & \sigma'^2_{\theta\theta} & & \\ \sigma'_{rr}\sigma'_{zz} & \sigma'_{\theta\theta}\sigma'_{zz} & \sigma'^2_{zz} & \\ \sigma'_{rr}\sigma'_{rz} & \sigma'_{\theta\theta}\sigma'_{rz} & \sigma'_{zz}\sigma'_{rz} & \sigma'^2_{rz} \end{bmatrix}$$

A perturbed solution for incipient plastic stress state is now sought for this problem. The actual steps of calculation are as follows:

1. Let the elastic solution be known just prior to incipient plasticity.

2. Apply the finite element analysis using the pressure distribution function with a contact circle of radius $A' = A + \Delta A$. This calculation will lead to new values of ϕ's. In the process of calculating, if any element is found to have become plastic, its stiffness matrix is to be adjusted for the next cycle of iteration.

Results of the calculation are shown in Figures 10.62, 10.63, and 10.64 for D, A, and $f_{\max}$ versus $\overline{N}$, respectively, where $f_{\max}$ is the maximum contact pressure and $\overline{N}$ is the normal load. These curves are for $E = 26 \times 10^6$ psi, $\nu = 0.3$. For plastic work-hardening solutions, they are given for $E_t = E/10$, $E_t = 0$, $\bar{\sigma} = 3.0 \times 10^6$ psi, and $\bar{\sigma} = 6.0 \times 10^6$ psi as shown by the short branch curves.

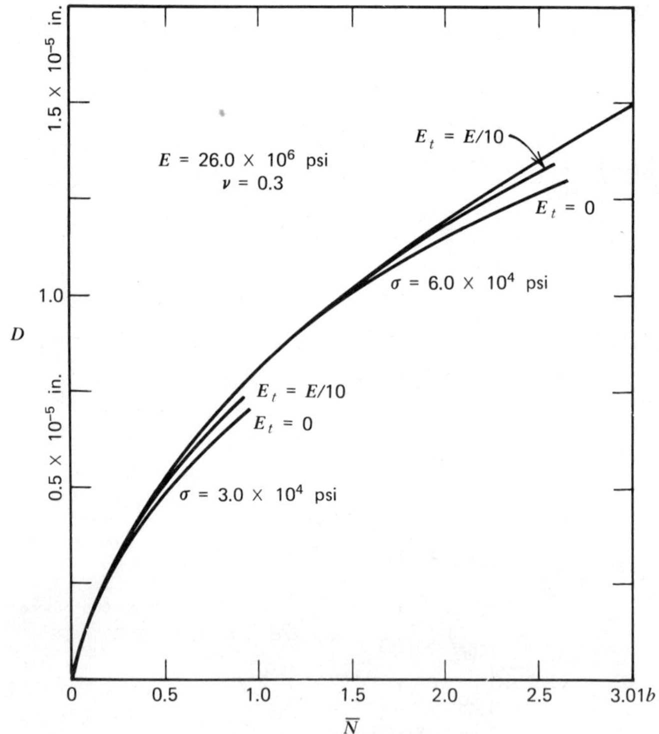

Figure 10.62

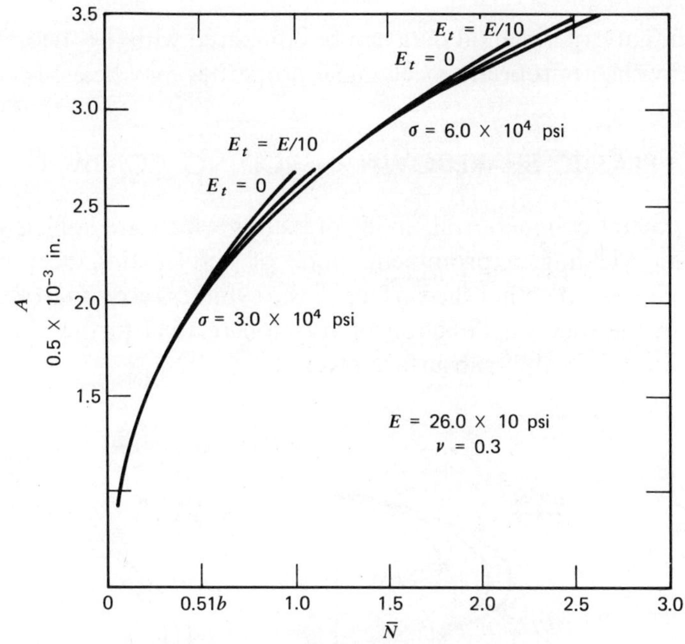

Figure 10.63

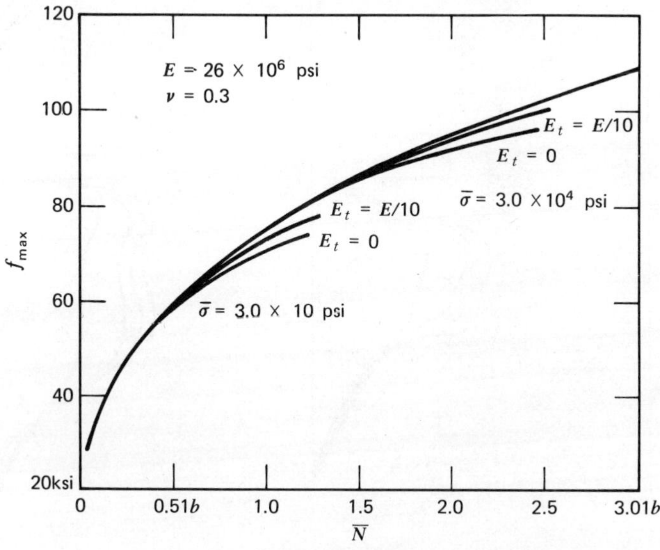

Figure 10.64

It is seen that experimental data can be compared with the theoretical one whereby the appropriate mechanical properties may be assessed.

V. PLASTIC SHAKEDOWN IN ROLLING CONTACT

When two metal cylinders roll under pressures which are sufficient to cause plastic yielding, a prominent mode of deformation occurs. In 1957 it was observed[163] that the surface of the cylinders are progressively displaced in the forward direction of rotation relative to the cone by plastic shearing in a thin subsurface layer.

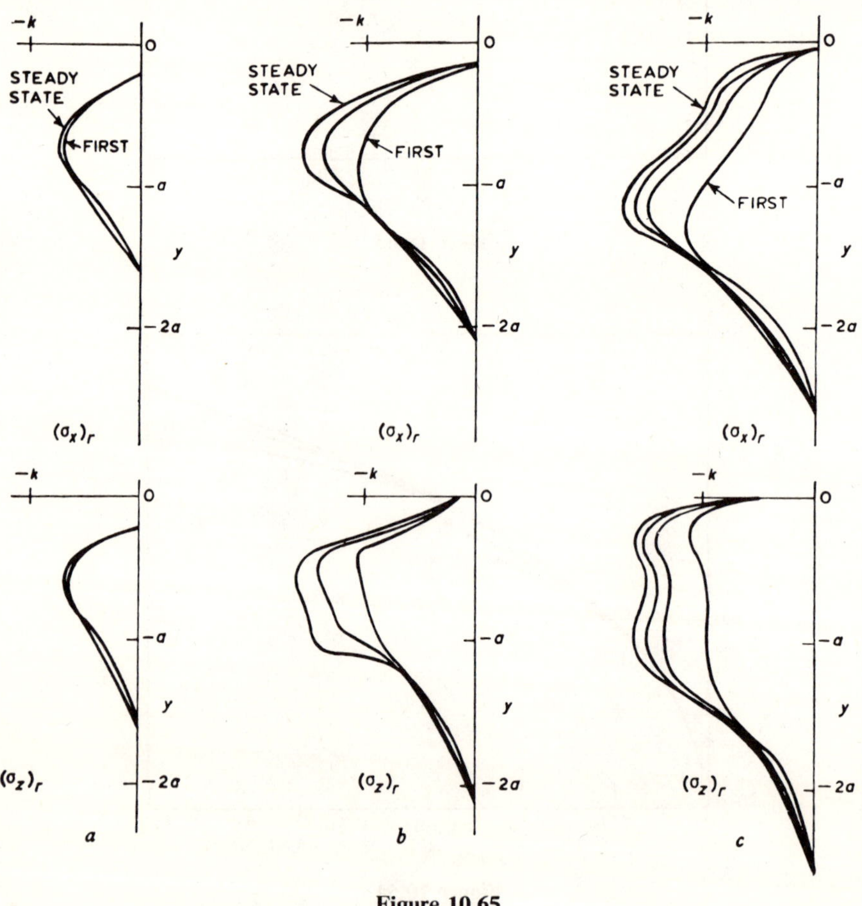

Figure 10.65

In an analysis, similar in nature to the one described in Section 10.U but without the layer and using a numerical analysis[164,165] for an elastic perfectly plastic material in cyclic deformation, the following states of stress have been computed. Figure 10.65 shows the residual stress components $(\sigma_x)_r$ and $(\sigma_z)_r$, where x is parallel to the surface of the body being rolled and in the direction of rolling and y is perpendicular to the surface into the material while z is perpendicular to x and y. The a is the radius of contact, k is the yield stress of the material, and P_0 is the load compressing the rollers together. The shakedown limit referred to in the diagram is illustrated in Figure 10.66; M is the rolling moment, G is the shear modulus, and R is the radius of the roller. The steady state is reached after several passages of load.

W.   SOFT METALS IN STATIC AND DYNAMIC LOADING

The theory behind the loading of a large mass of material which is rigid perfectly plastic represents a special case of Section 10.U.

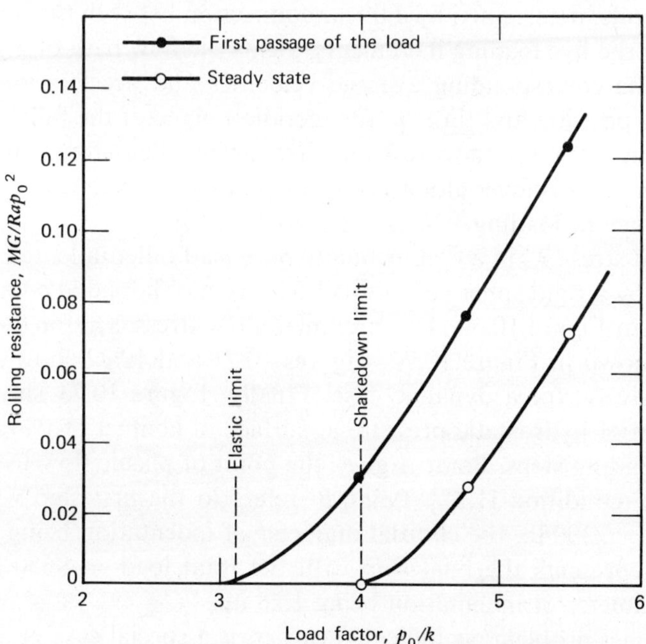

Figure 10.66

In this section, the method of visioplasticity is used to give the experimental information from which the stress patterns are calculated.

The visioplasticity method consists of halving the massive solid and inscribing on the cut face grids as shown in the first frame in Figure 10.67.[166,167] The two halves are then clamped together tightly. For the loading experiment with a rigid ball as the indenter, the point of contact is aligned with the crack where the two halves come together; the load axis is parallel to the severed and inscribed interface. Upon loading and unloading the two halves are separated. The photograph of the resulting grids in a lead billet during five indentation steps under quasistatic loading are shown in Figure 10.67. Under impact loading, photographs of grids in lead billet and oscilloscopic pictures obtained by crystal transducer during five indentation steps are shown in Figure 10.68.

This unit is disassembled after each incremental loading step, and the billet half containing the deformed grid lines photographed. The negatives of the photographs are then projected to the same size on a large sheet of tracing paper, and the intersections of the grid lines for each loading step are then located and plotted. The trajectory of each of these points, determined by a linear connection between the positions for each of the five loading increments, yields the flow path of a generic particle. The corresponding average velocities and accelerations as a function of position and time on the meridian plane of the billet, which serves as the basis for the stress and strain-rate calculations, are evaluated from the distance along each flow path and the time of contact for each stage of loading.

The axial stress σ_{zz} in a commercially pure lead billet indented slowly at 0.1 ipm by a rigid sphere of radius 1.5 in. versus the radial coordinate r is shown in Figure 10.69. The circumferential stress, $\sigma_{\theta\theta}$, for the same model is shown in Figure 10.70. Figures 10.71 and 10.72 show σ_{zz} and $\sigma_{\theta\theta}$, respectively, for a dynamic case. Finally, Figure 10.73 shows the distribution of hydrostatic pressure at surface of contact of two sets of dynamic loading steps. Point A gives the point of plastic flow using the Mises yield condition [1.27]. Point B' refers to the first step with the total load = 2300 lb, the chordal diameter of indentation being 0.7 in. Point C' represents the final step with the total load = 9650 lb, the chordal diameter of indentation being 1.26 in.

The contact problem, as indicated earlier, is a special case of Section 10.U. It may be shown that the method of visio-plasticity gives results which are within 10 to 20% of the analytic results.

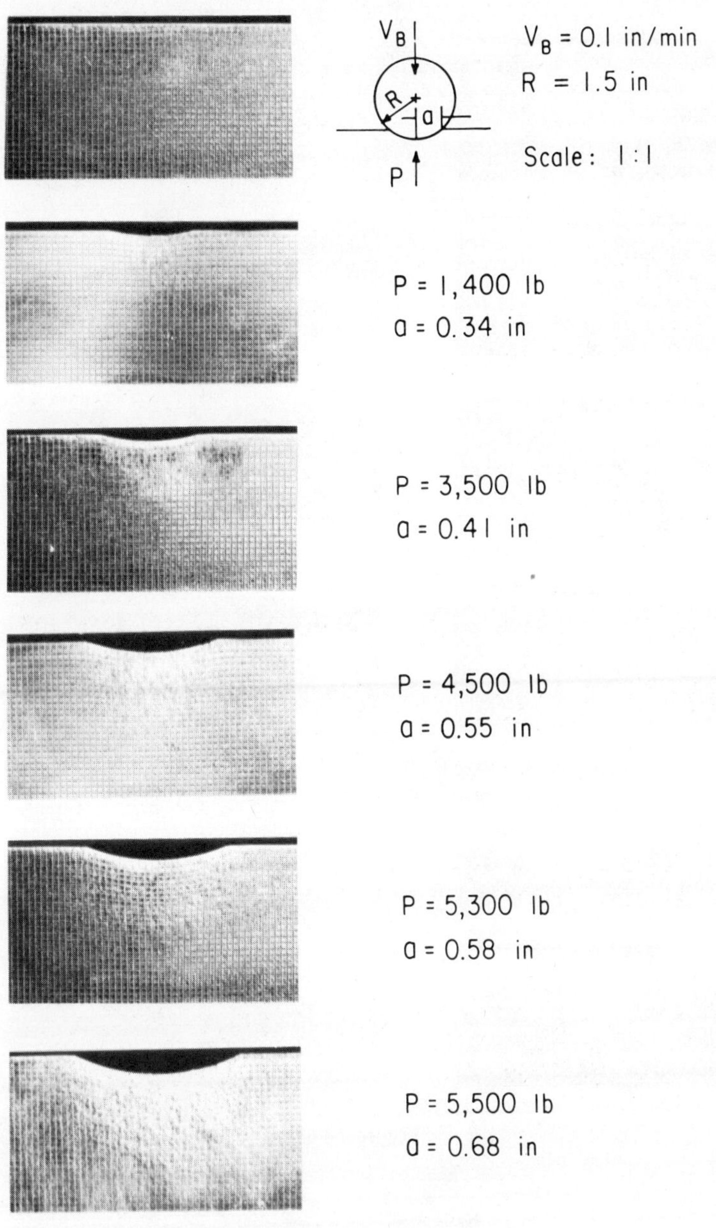

Figure 10.67

[269]

CRYSTAL
TRANSDUCER
V
R
a
R = 1.5 in
Scale 1:1
Pave = 2,250 lb
a = 0.38 in
Pave = 3,800 lb
a = 0.48 in
Pave = 5,400 lb
a = 0.56 in
Pave = 7,000 lb
a = 0.60 in
Pave = 9,600 lb
a = 0.63 in

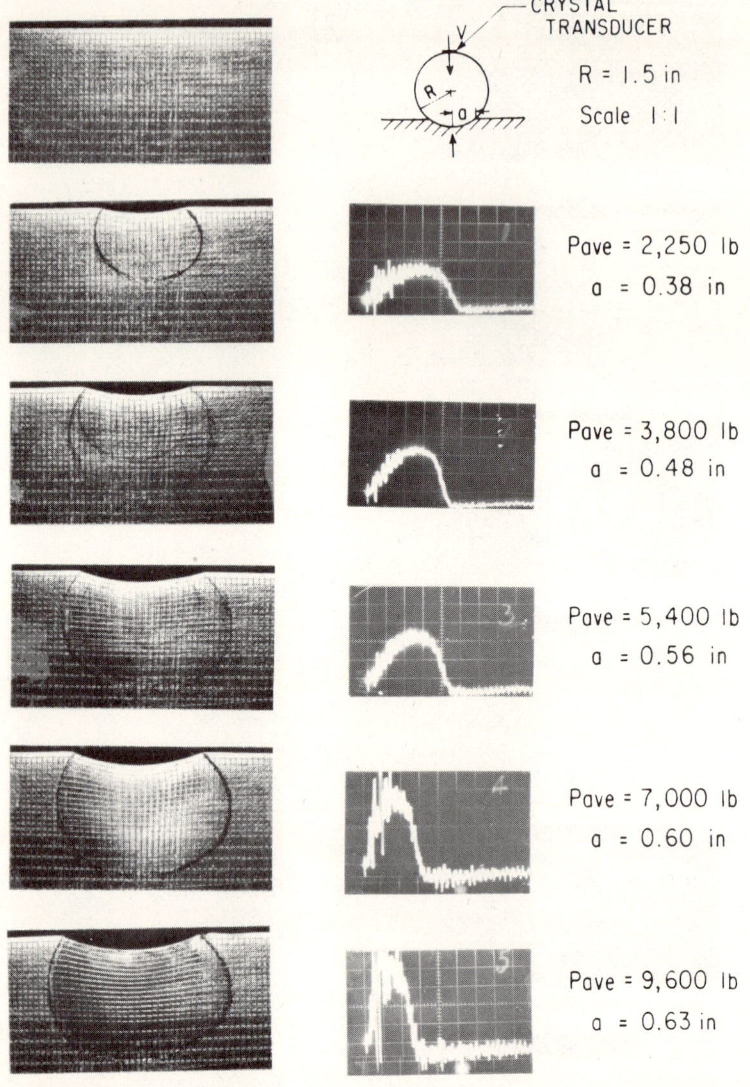

Figure 10.68

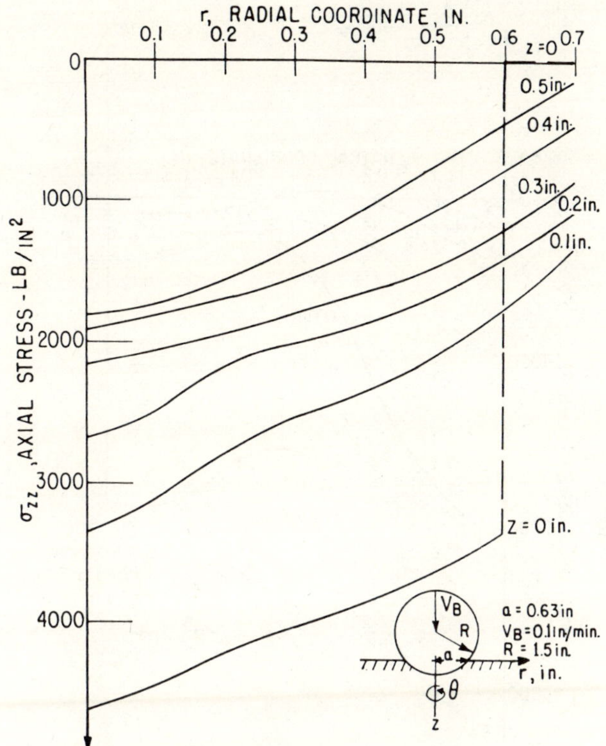

Figure 10.69

X.   HARDNESS MEASUREMENTS

The indentation hardness on yield pressure p of metals is essentially a measure of their plastic properties.[168,169] For a wide range of metals p is approximately three times the uniaxial flow stress of the metal. For pyramidal or conical indenters, when the size is sufficiently large, the indentation hardness is independent of size. This is not so with spherical indenters.

If the material shows an appreciable elasticity as in the polymers and rubber-like material, p would be appreciably lower than three times the uniaxial flow stress.

Much of the above observation can be quantitatively calculated by the methods developed in Chapters 4, 6, and 7. It is of interest to recall here the definition of some of the widely used hardness measurements.

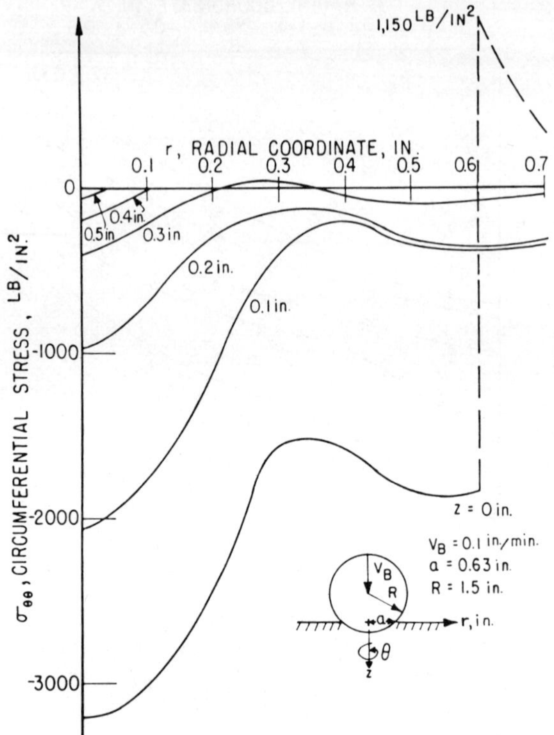

Figure 10.70

Vickers Hardness

Figure 10.74 shows (a) standard Vickers diamond pyramid indenter, (b) the indentation it produces, and (c) the qualitative picture of the material section showing the resulting plastic zone. The yield pressure indicated above is the ratio of load to the projected area of indentation.

Rockwell Hardness

The Rockwell hardness, R_c, uses a conical indenter; it is related to the Vickers hardness by the formula

$$R_c = c - k_2 \, p^{-1/2} , \qquad\qquad [10.37]$$

where c is a constant and k_2 is a constant which is a function of the cone geometry.

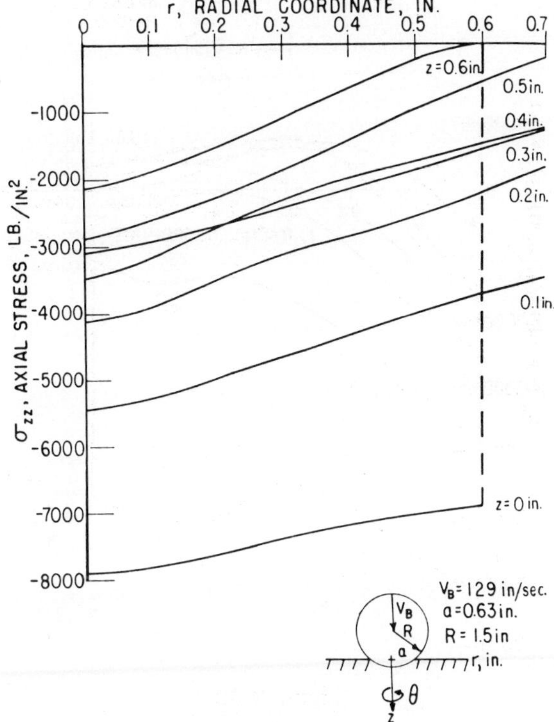

Figure 10.71

Brinell Hardness

The Brinell hardness uses a hard sphere as indenter. Figure 10.75 shows (a) the onset of plastic flow below the surface, and (b) the state when the plastic zone has extended to the surface. It should be noted, only for this case, that the hardness measure is comparable to the Vickers hardness. Associated with this form of indentation, there is the Meyer index, *n*, which is the exponent used in the relation

$$w = \frac{Ad^n}{D^{n-2}} , \qquad [10.38]$$

in which *w* is the load, *D* is the diameter of the indenter, *d* is the chordal diameter of the indentation, and *A* is a constant.

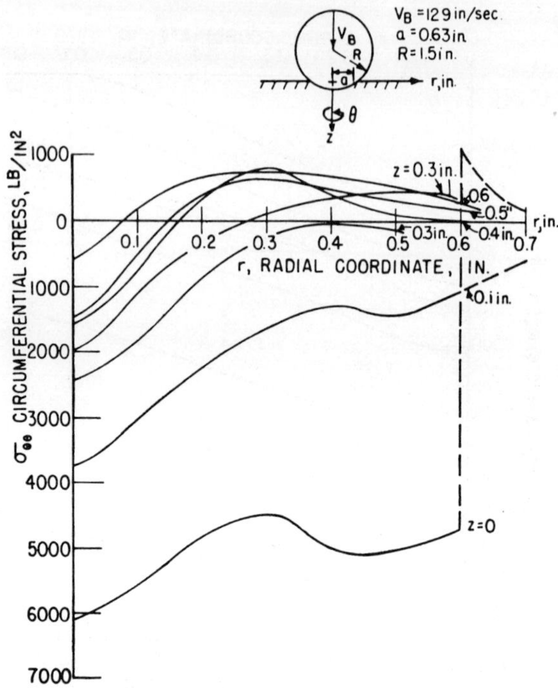

Figure 10.72

Scratch Hardness

For brittle solids such as minerals, the high hydrostatic pressures developed around the deformed region are often sufficient to inhibit brittle fracture. Under these conditions, the deformation is primarily plastic; as such the Mohs scratch-hardness scale is reasonably well correlated with the Vickers and Knoop hardnesses. Figure 10.76 shows the correlations. The Knoop indenter is a modified Vickers indenter. Figure 10.77 shows the same on a log-log plot.

Y. FRICTION UNDER METAL-WORKING PROCESSES

While Amonton's second law of friction[170] states that friction is independent of the normal load, it is certainly invalid in many cases associated with metalworking processes. That is, this law fails to describe

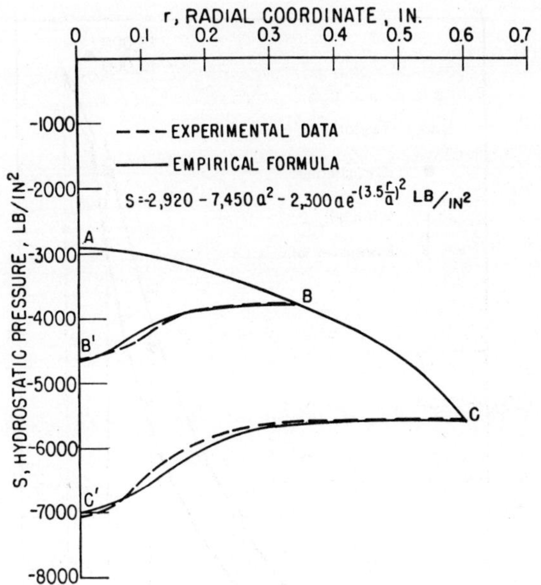

$$S = -2,920 - 7,450\,a^2 - 2,300\,a\,e^{-\left(3.5\frac{r}{a}\right)^2} \; \mathrm{LB/IN^2}$$

S, HYDROSTATIC PRESSURE , LB/IN²

Figure 10.73

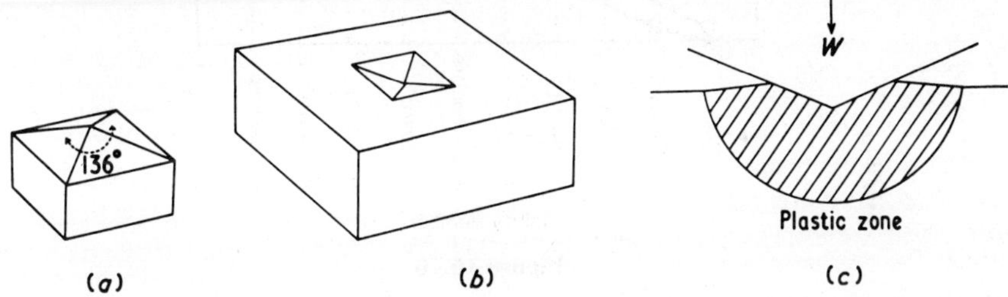

Figure 10.74

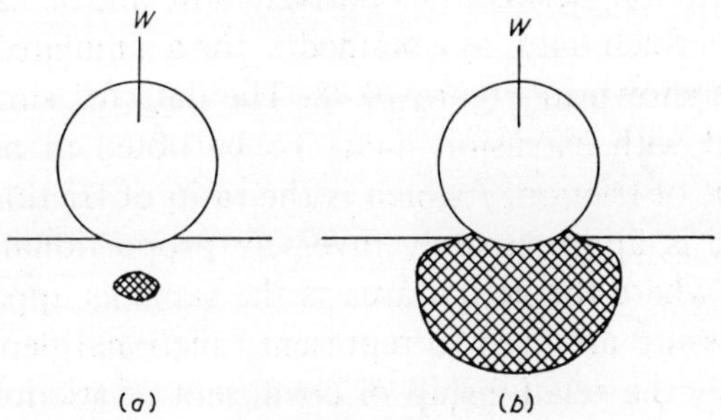

Figure 10.75

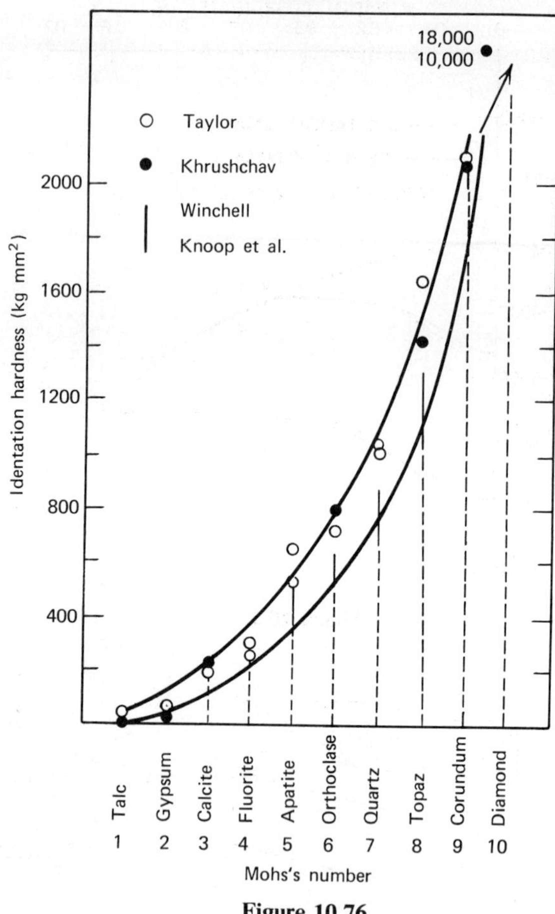

Figure 10.76

the sliding friction between a relatively soft and a hard metal under high pressure. Such data, as obtained[171] for a simulated workpiece-tool interface, are shown in Figure 10.78. The data for simulated test have been checked with extrusion data. To be noted in particular is that the coefficient of friction, f which is the ratio of frictional resistance to normal load, is approximately inversely proportional to pressure, p. In this case, where the actual area is the same as apparent area, both load and pressure are used to represent functional dependence. Figure 10.79 shows[172] the relationship of coefficient of friction versus center-line-average of surface roughness.

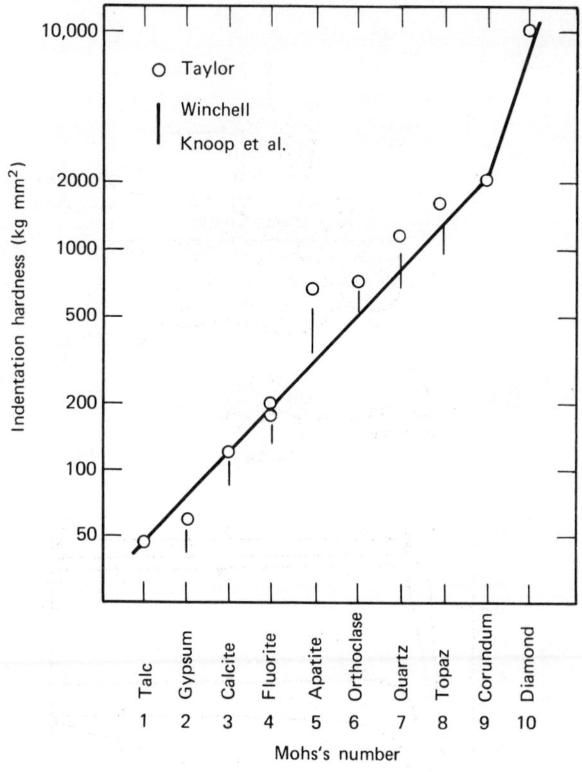

Figure 10.77

Metalworking friction may be divided into four regimes:

1. Classical
2. Transition
3. Deformation
4. Extreme pressure

By and large, Amonton's laws hold in the first category. In the last category, Amonton's second law also holds.

The third is primarily the metalworking regime with the intervening regime designated as transition regime, that is, between the deformation and the classical. Figure 10.78 shows the relationship between f and

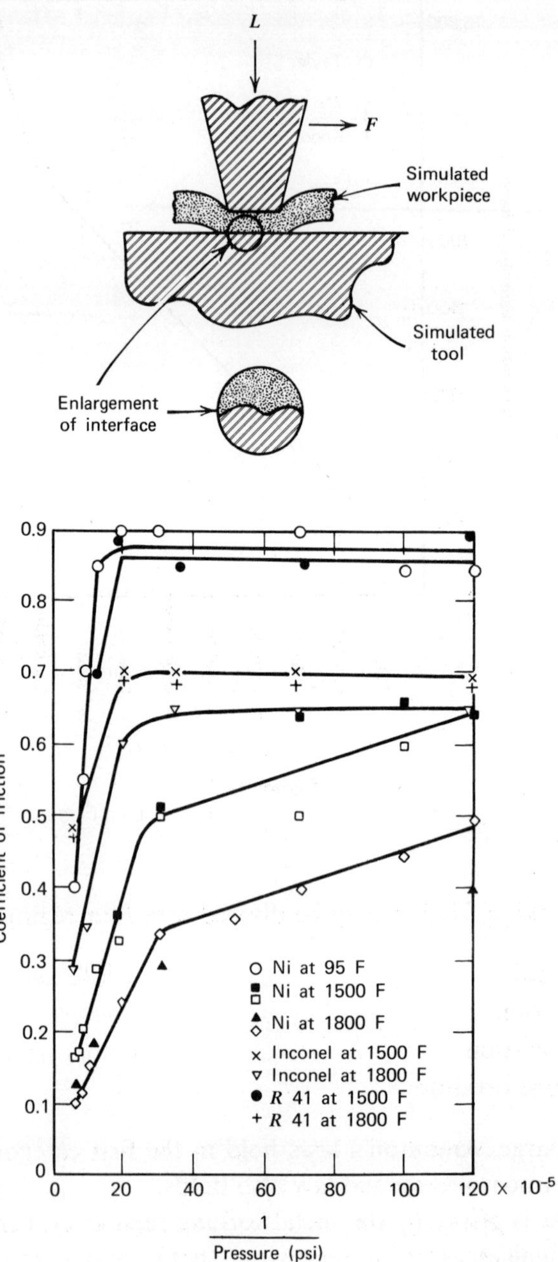

Figure 10.78

[278]

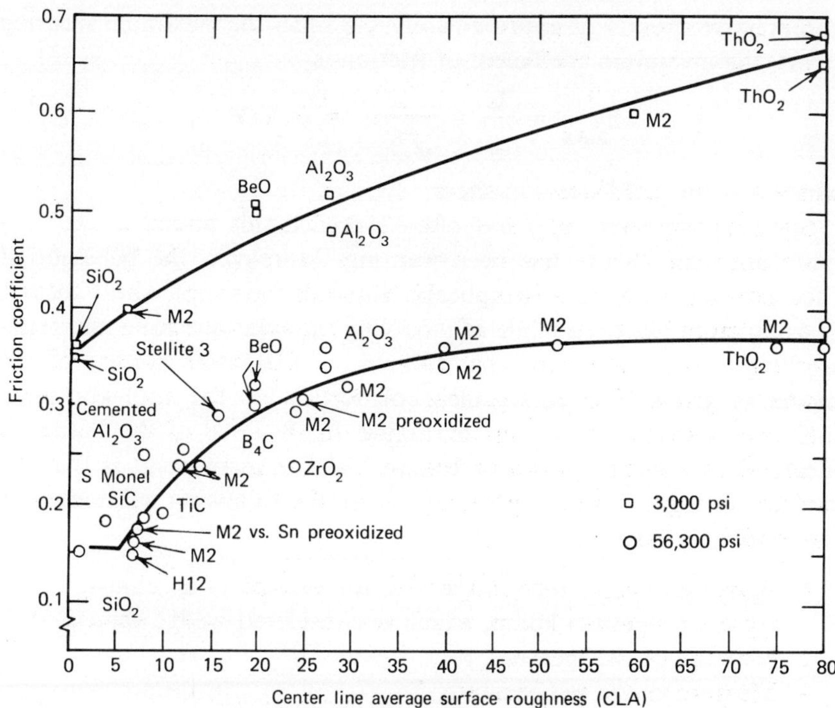

Figure 10.79

the reciprocal of pressure p except for the range covering extreme pressures. Generally speaking:

For the deformation regime, $f \sim p^{-1}$
For the classical regime, $f = $ constant
For the transition regime, f varies in such a way as to connect
the above two.

In the deformation case, f is proportional to p^{-1} or, recalling the definition of f and p, the frictional resistance

$$F = \text{constant.}$$

The conventional argument is that the constant must be a product of area, which does not change with pressure in this regime, and the shear strength of the workpiece material. This is the so-called constant shear-stress theory. Through an ideal plasticity argument that, since the compressive stress, σ, is approximately $\sqrt{3}$ times the maximum shearing stress, the maximum coefficient of friction is

$$f_{max} = \frac{k}{\sigma} = \frac{1}{\sqrt{3}} = 0.577 \quad ,$$

where k is the yield stress in shear.

Since measurement of f has often exceeded this bound of 0.577, a more adequate theory has been wanting. Moreover, the behavior of f for extreme pressure is inexplicable through the simple shear theory.

An attempt has been made at modeling the interface zone for metal-working conditions.[155] In what follows, a qualitative picture of the model is given. The enlargement of section of the tool-workpiece interface is shown at the top of Figure 10.78. Each of the humps is idealized as a wedge shown in Figure 7.4. The main point is that the surface is rough. With the physical model, the following parameters of the model are essential

1. A measure of surface roughness, for example, the semiangle of a typical roughness hump, which is considered wedge shaped, β.
2. Measure of the pressure, p.
3. Measure of the temperature, T.
4. A measure of the strength of the workpiece material, for example, the yield strength in shear $k(p, T)$ which is a function of pressure p and temperature T.
5. A measure of interfacial surface effects, for example, a surface effect index $S_i(p, T)$ which is a function of pressure and temperature and in units of force per unit area. This index gives, at once, a measure of adhesion for dry systems and lubricity of lubricated systems of surfaces.
6. A measure of friction per unit area, F/A.

Dimension analysis shows, for the above, five dimensionless groups, namely,

$$\frac{F}{Ap} \;,\quad \frac{k}{p} \;,\quad \frac{S_i}{p} \;,\quad \beta \;,\quad \frac{E}{RT} \;,$$

where E is an activation energy and R the universal gas constant. For simplicity of discussion here, let the last group be implicit in k and S_i, that is, at each temperature level, no reference to this group is necessary. Of course, F/Ap is just f. Note the first and the second form an alternate group, f_p/k, which is merely the ratio of friction to shear strength of the workpiece material. Thus for an isothermal case, four groups exist:

$$\frac{F}{Ap} \ , \quad \frac{k}{p} \ , \quad \frac{S_i}{p} \ , \quad \beta \ .$$

According to the Buckingham II theorem, the above groups must be related functionally

$$G\left(\frac{fp}{k} \ , \quad \frac{k}{p} \ , \quad \frac{S_i}{p} \ , \quad \beta\right) = 0 \ ,$$

where G is some function whose form is not specified.

Indeed, if one uses a simple slip-line theory of ideal plasticity akin to [7.17], one arrives at a relation of the form shown below.[155] That is, f is proportional to k/p explicitly and a factor of proportionality, B, is in turn a function of k/p, S_i/p, and β:

$$f = B\frac{k}{p} + \frac{S_i}{2p} \ , \qquad\qquad [10.39]$$

where

$$B = \frac{1}{2}\left[1 + \sqrt{1 - (S_i/k)^2} + 4\beta - \frac{\pi}{2} - \cos^{-1}(-S_i/k)\right]\cot\beta \ .$$

Equation 10.39 is the generalization of the constant shear-stress theory of friction with k/p as the usual measure of coefficient of friction. This is, of course, the case if $B = 1$ and $S_i/2p = 0$.

Generally, [10.39] says much more than the simple theory in that roughness through β, surface effects through S_i/p as well as deformation effects through k/p are brought into play. Note S_i/p is a measure of coefficient of friction for interfacial slip.

While the form of [10.39] suggests that a plot of f versus p^{-1} or k/p should be linear, it need not be at all. In fact, the form of [10.39] with $B = B(k/p, S_i/p, \beta)$ can qualitatively explain all the observed f versus k/p data, functionally and empirically speaking. Some conclusions are described below.

First of all, [10.39] gives the theoretical result indicating the effect of semiwedge angle β on friction. Figure 10.80 shows fp/k versus β for two values of surface index $S_i = k$ and $S_i = k/2$. The larger angle corre-

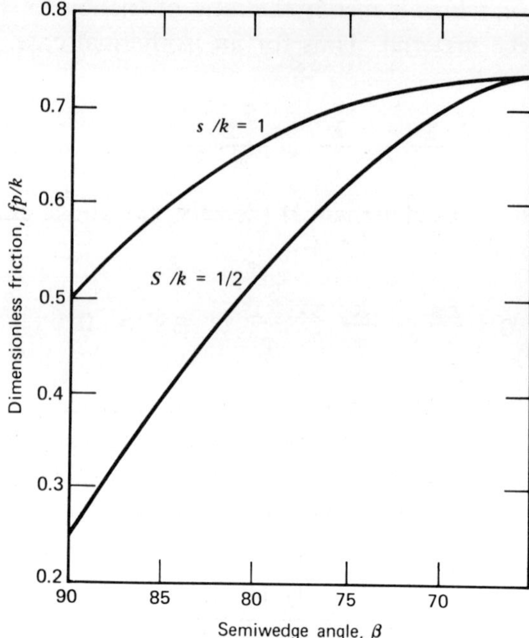

Figure 10.80

sponds to smoother surface. Figure 10.80 is to be compared qualitatively to Figure 10.79. It can be seen that the same trends are observed both theoretically and experimentally; that is, friction increases with increased roughness.

Secondly, data on k versus p shows that for high pressures k may be fitted as a straight line,[173] say

$$k = k_0 + \alpha p \quad ,$$

for $p_1 \leq p \leq p_2$, where p_1 and p_2 are some limiting values of p for which the above is a good approximation. With the above, [10.39] reads

$$f = B \frac{k_0 + \alpha p}{p} + \frac{S_i}{2p} \quad . \qquad [10.40]$$

For large values of p, k_0/p, and S_i/p are negligibly small; this leaves

$$f \cong B\alpha,$$

a constant for the extreme pressure regime described above. In almost all the experimental data obtained the friction has tended to be a constant value at extreme pressures.

Thirdly, for the deformation regime, with lubricant, the surface roughness is reduced so as to give an effective value of $\beta = 90°$. Thus [10.39] becomes

$$f \cong \frac{S_i}{2p} \, ,$$

with the first term vanishing.

Fourthly, for unlubricated or cases where lubricity is not as extreme as above, [10.39] generally gives results corresponding to the constant shear-stress theory. When k is independent of p, the line is straight; otherwise the f versus $1/p$ curve reflects the pressure dependence of k. Also, according to experimental findings, not only is k dependent on pressure but also the k at high pressure is not the same as that at atmospheric pressure. The last point explains, at least in part, the steepness of the slope of the f versus $1/p$ curve.

Equation 10.40 says that whereas $B\alpha$ gives the extreme pressure value of f, $(S_i/2 + Bk_0)$ gives the slope of the f versus $1/p$ curve. In short, the slope represents the sum of adhesion effects through S_i and B, the shear strength at atmospheric pressure through k_0 and surface roughness through B.

Z.   ON McCUTCHEN'S CONJECTURE ON HUMAN JOINT LUBRICATION

As indicated in Section 1.F, there are many diffusive processes, for example, [1.21], for the slow moving fluid in porous media. When this is coupled with the elasticity equation, they form the governing equations of poroelastic materials. This is applicable to soils and other newer applications.

Bone terminals forming animal joints are covered by a layer of porous material called hyaline cartilage. It is this relatively soft cartilage immersed in a bath of synovial fluid that serves as a lubricating system.

Recent investigations by McCutchen[174-176] suggest that the porous nature of cartilage plays a singular role in the joint behavior. Of particular interest is the observed low value of coefficient of friction which is in the range, 0.005 to 0.01. Teflon, the smoothest material available to man, has a coefficient of friction of 0.04. For friction lower than 0.04 or thereabouts, lubrication must be achieved by separating the mating surfaces.

McCutchen proposed that nature has ingeniously provided animals with hydrostatically lubricated joints. It was suggested that with low elastic modulus and permeability of the cartilage, it is the hydrostatic pressure of the fluid within the pores that supports a large portion of the external normal traction. The low permeability prevents rapid wiring-out of fluid; consequently transfer of stress to the matrix in porous material is frustrated. This hydrostatic bearing is the most plausible explanation of the low frictional coefficients in joints when the relative sliding velocity between the rubbing surfaces may be zero.

The governing equations consist of the elastic equations which describe the motion of the matrix and a diffusion equation which describes the motion of the fluid in the pores.[5] As such the solution of boundary value problems, once a model is adopted, is akin to those described in Chapter 5 on thermoelasticity with the heat equation playing the role of the Darcy equation for pore pressure.

Figure 10.81 is a model chosen for the analysis.[177] The elastic half-space simulates the bone end while the layer simulates the articulate cartilage. The figure shows two sets of Lamé constants with (λ, μ) for the matrix of the assumed porous layer, for which an additional constant is prescribed. That is, the permeability K which is simply related to Γ in [1.21]. In the theory the total stress tensor

$$\sigma_{\alpha\beta} = \sigma'_{\alpha\beta} + \sigma \delta_{\alpha\beta} \quad ,$$

where $\sigma'_{\alpha\beta}$ is the effective stress tensor which obeys the elastic law and σ is the pore stress.

In the exploratory work, without going into a contact problem, the following artifice is used to handle the important boundary conditions. Other details are not shown in view of Chapter 5. On the surface, the normal stress component of $\sigma_{\alpha\beta}$,

$$\sigma'_{22} = \begin{cases} -(1-\beta) \, f(x_1) & (|x_1| \le \ell, \; x_2 = 0) \\ 0 & (|x_1| > \ell, \; x_2 = 0) \end{cases} \quad ,$$

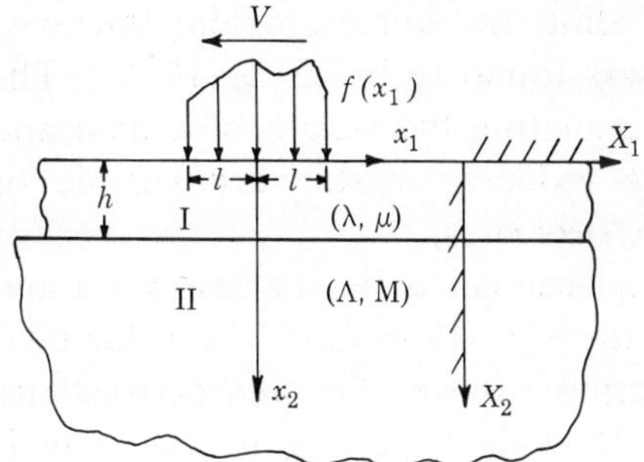

Figure 10.81

the shear stress

$$\sigma_{12} = 0 \qquad (|x_1| < \infty, \ x_2 = 0) ,$$

and the pore stress was

$$\sigma = \begin{cases} -\beta \, f(x_1) & (|x_1| \le \ell, \ x_2 = 0) \\ 0 & (|x_1| > \ell, \ x_2 = 0) \end{cases} .$$

In the above, see Figure 10.81, $f(x_1)$ stands for the total traction on the surface. The β is a constant parameter introduced to give a simple measure of what portion of $f(x_1)$ is carried by the fluid and the matrix, respectively. In general, of course, β is a function of x_1.

Table 15 shows the constants used for a numerical example following an analysis akin to that in Section 5.D. The geometrical parameters are $h = 0.02$ in. and $l = 0.5$ in. Two values of V are used. These are 5 and 15 in./sec.

In Section 10.E, elastohydrodynamic lubrication was referred to. This theory has been applied to the human joint problem.[181] Using an elasto-

Table 15.   Physical Constants

	Porous Layer	Substrate
Young's moduli (psi)	380	20,000
Poisson's ratio	0.4999	0.3
Permeability (in.^{4}/lb-sec)	6×10^{-9}	—

From references 175, 178, 179, and 180.

hydrodynamic theory for a bone joint, the minimum film thickness required for lubricating the joint was found to be 4.9 $\times$ 10^{-4}.[182] The above was for a load of 150 lb, simulating the weight of a man and walking speeds. When speeds are reduced, elastohydrodynamic or hydrodynamic action, by definition, decreases.

Figure 10.82 shows the surface displacement, ω, for the case $\beta = 1$ and parabolically distributed loading with $V = 15$ in./sec. The value of f_0 used corresponds to the case described above. The vertical distance between the peaks and the peak in the valley is 2.1 $\times$ 10^{-4} in. This is of the order found for the elastohydrodynamic case. For a cursory view of the friction due to the shear of fluid between load and the layer, a film thickness of 2.1 $\times$ 10^{-4} in. may be adopted. Assuming the velocity gradient across the fluid film to be linear and using the viscosity of synovial fluid, which ranges from 6 to 1200 centipoise, the friction coefficient has a range 4 $\times$ 10^{-1} to 8 $\times$ 10^{-2}. This cursory calculation merely attests to the feasibility of a poroelastic layer's ability to contain fluid and shear with a friction coefficient in a range that is found in actual experience. In this regard, this analysis shows a similarity to the

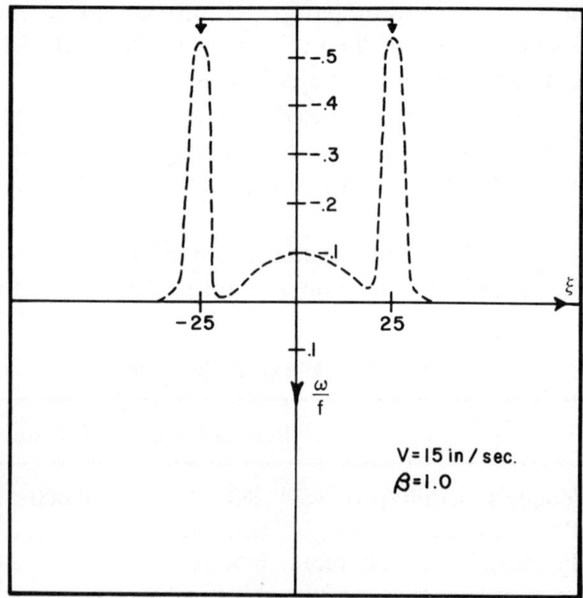

Figure 10.82

elastohydrodynamic theory in that lubrication is achieved through motion.

The pore stress distribution at the layer-substrate interface suggests that the hydrostatic bearing is indeed feasible. The normal traction is supported by the pore pressure developed within the layer. The variation along the depth is very small, with a slight increase as the depth increases for $\xi < 0$ and the opposite for $\xi > 0$, ξ being the dimensionless variable for (x_1, x_2). These results support the notion that the flow is primarily along the layer and away from $\xi = 0$. Interestingly, the flow at the leading edge of the load is directed toward the surface and nearly parallel to it. At the trailing edge, the flow is directed into the surface. This exudation and imbibition of fluid support the notion of weeping lubrication.[174]

Figure 10.83 is a plot of the interface pore stress. An interesting feature is the pore tension developed for $\xi > 0$. The tension increased as the load is increasingly supported by the effective stress. This phenomenon has been observed by workers in the field of soil mechanics. It indicates the tendency of the porous structure to expand when it is

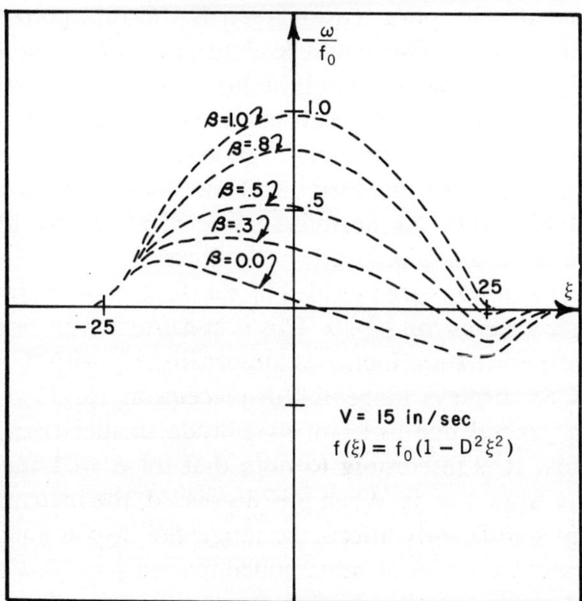

Figure 10.83

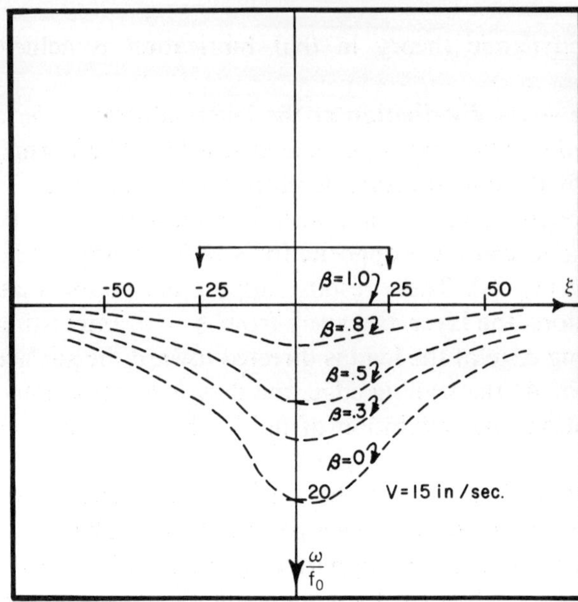

Figure 10.84

being unloaded. The pore tension retards the return to the original unload configuration. This could lead to cavitation. The data shown indicated that this possibility is minimized when β is restricted to be nearly 1.0. Velocity effect is small according to data not shown here. This confirms earlier experimental observations.[183] For higher velocity a greater pressure is generated for $\xi < 0$ and smaller pressures for $\xi > 0$.

Figure 10.84 shows the normal surface displacement at 15 in/sec. Comparison of results obtained for the slower velocity indicated small variations. The slower speed yielded larger displacements for $\xi < 0$ and smaller displacements for $\xi > 0$. This is consistent with results on pore tension. The deformation increases almost linearly with β.

Figure 10.85 displays tangential displacements for $V = 15$ in./sec. They are in general one order of magnitude smaller than the normal displacements. It is interesting to note that for $\beta = 1$ the material is pushed away from $\xi = 0$. When β is decreased, the increase in normal displacement significantly affects the tangential displacement. The case $\beta = 0$ indicates the material being pulled toward $\xi = 10$. Velocity effect is small and yielded smaller displacements.

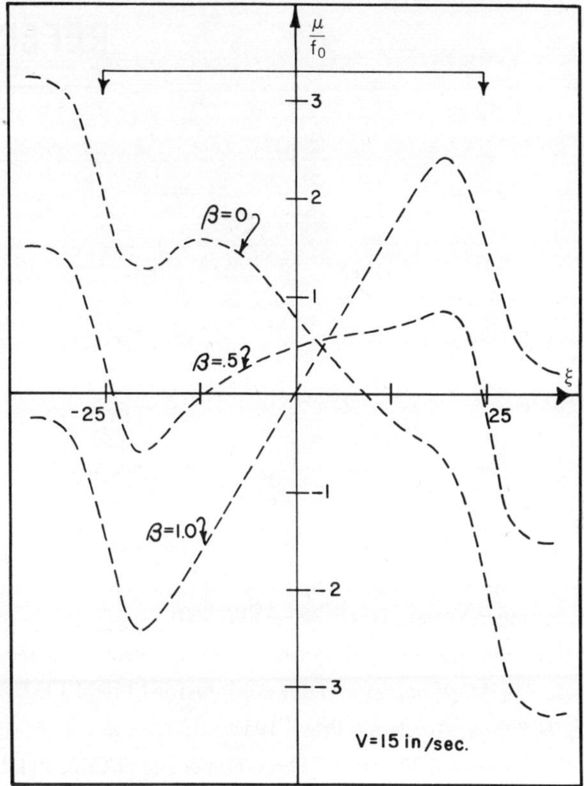

Figure 10.85

A final point of interest is the interface shearing stress. Computations indicate the shearing stress to have a maximum at the extremities of the load for both the parabolically and uniformly distributed cases. The shearing stress showed very little variation with velocity and β. It had a magnitude of about 10% of the normal traction f_0.

REFERENCES

1. C. Truesdell and R. Toupin, "The Classical Field Theories," *Encyclopedia of Physics*, Springer-Verlag, **III**/**1**, 226–850 (1960).

2. Y. C. Fung, *Foundations of Solid Mechanics*, Prentice-Hall, Inc., 1965.

3. J. Cranck, *The Mathematics of Diffusions*, Oxford Press, 1957.

4. W. Jost, *Diffusion*, Academic Press, 1952.

5. M. A. Biot, "General Theory of Three-Dimensional Consolidation," *Journal of Applied Physics*, **12**, 155–164 (1941).

6. H. Lamb, *Hydrodynamics*, Dover Publications, 1932.

7. W. Prager and P. G. Hodge, *Theory of Perfect Plastic Solids*, John Wiley and Sons, Inc., 1951.

8. R. Hill, *Mathematical Theory of Plasticity*, Clarendon Press, 1950.

9. G. Joos, *Theoretical Physics*, Hafner Publishing Company, 1950.

10. F. F. Ling, "A Quasi-Iterative Method for Computing Interface Temperature Distributions," *Zeitschrift für Angewandte Mathematik und Physik*, **X**, 461–474 (1959).

11. F. F. Ling and J. S. Rice, "Surface Temperature with Temperature Dependent Thermal Properties," *American Society of Lubrication Engineers Transactions*, **9**, 195–201 (1966).

12. F. F. Ling and C. W. Ng, "On Temperature at the Interfaces of Bodies in Sliding Contact," *Proceedings of the 4th U.S. National Congress of Applied Mechanics*, 1343–1349 (1962).

13. S. L. Pu, "Surface Temperature on a Truncated Right Circular Cone," *Proceedings of the 5th Midwestern Conference on Mechanics* (1966).

14. C. K. Youngdahl and E. Sternberg, "Transient Thermal Stresses in a Circular Cylinder," *Journal of Applied Mechanics*, **28**, 25–34 (1961).

15. P. M. Morse and H. Feshback, *Methods of Theoretical Physics*, McGraw-Hill Book Company, 1953.

16. J. C. Samuels, "On Random Heat Conduction in Solids," *International Journal of Heat Transfer*, **9**, 301–314 (1966).

17. Y. M. Chen and C. T. Tien, "Penetration of Temperature Waves in a Random Medium," *Journal of Mathematical Physics*, **46**, 188–194 (1967).

18. E. W. Gaylord, W. F. Hughes, F. C. Appl, and F. F. Ling, "On the Theoretical Analysis of a Dynamic Thermocouple," *Transactions of the American Society of Mechanical Engineers*, **80**, 307–310 (1958).

19. W. F. Hughes and E. W. Gaylord, "On the Theoretical Analysis of a Dynamic Thermocouple: 2—The Continuous Area Interface," *Journal of Applied Mechanics*, **27**, 259–262 (1960).

20. H. H. H. Shu, E. W. Gaylord, and W. F. Hughes, "The Relation Between the Rubbing Interface Temperature Distribution and Dynamic Thermocouple Temperature," *Transactions of the American Society of Mechanical Engineers*, **86**, 417–422 (1964).

21. H. Blok, "Theoretical Studies of Temperature Rise at Surfaces of Actual Contact Under Oiliness Lubricating Conditions," *Proceedings of General Discussion of Lubrication and Lubricants, The Institute of Mechanical Engineers*, 225–235 (1937).

22. F. F. Ling and E. Saibel, "Thermal Aspects of Galling of Dry Metallic Surfaces in Sliding Contact," *Wear*, **1**, 80–91 (1957).

23. W. B. Konwenhoven and J. H. Potter, "Thermal Resistance of Metal Contacts," *Welding Research Supplements*, **27**, 5155–5205 (1948).

24. H. Fenech and W. M. Resenow, "Prediction of Thermal Conductance of Metallic Surfaces in Contact," *Journal of Heat Transfer, Transactions of the American Society of Mechanical Engineers*, **55**, 15–24 (1963).

25. F. F. Ling and T. E. Simkins, "Measurement of Pointwise Juncture Condition of Temperature at the Interface of Two Bodies in Sliding Contact," *Journal of Basic Engineering, Transactions of the American Society of Mechanical Engineers*, **85**, 481–487 (1963).

26. F. F. Ling, "Heat and Mass Transfer Effects in Sliding Metal Systems Lubricated by Solid Interfacial Films," AFML-TR-66-150 (1966).

27. F. F. Ling and S. L. Pu, "Probable Interface Temperatures of Solids in Sliding Contact," *Wear*, **7**, 23–34 (1964).

28. F. F. Ling, "Deformational and Geometrical Aspects of Surfaces in Sliding Contact," *Proceedings of the Second Symposium on Fundamental Phenomena in Material Sciences*, Plenum Press, 57–71 (1966).

29. H. Shore, "Thermoelectric Measurement of Cutting Tool Temperature," *Journal of the Washington Academy of Science*, **15**, 85–88 (1925).

30. K. Gottwein, "Die Messung der Schneiden Temperatur beim Abdrehen von Flusseisen," *Maschinenbau*, **4**, 1129–1135 (1925).

31. E. G. Herbert, "The Measurement of Cutting Temperature," *Proceedings of the Institute of Mechanical Engineers (London)*, **1**, 289–329 (1926).

32. G. Nieman and G. Lechuer, "The Measurement of Surface Temperatures on Gear Teeth," *Journal of Basic Engineering, Transactions of the American Society of Mechanical Engineers*, **87**, 641–654 (1965).

33. M. J. Furey, "Surface Temperature in Sliding Contact," *American Society of Lubrication Engineers Transactions*, **7**, 133–146 (1964).

34. F. Schwerd, "Uber die Bestimmung des Temperatur Felds beim Spanablauf," *Zeitschrift des Vereines Deutscher Ingenieure*, **71**, 211–216 (1933).

35. F. P. Bowden and P. H. Thomas, "The Surface Temperature of Sliding Solids," *Proceedings of the Royal Society of London*, **A223**, 29–40 (1954).

36. F. F. Ling, "On Temperature Transients in Sliding Interface," *Journal of Lubrication Technology, Transactions of the American Society of Mechanical Engineers*, **91**, 397–405 (1969).

37. M. Flamant, "Sur la répartition des pressions dans un solide rectangulaire chargé transversalement," *Comptes Rendus*, **114**, 1465–1468 (1892).

38. J. Boussinesq, *Application des potentials à l'étude de l'équilibre et du mouvement de solides elastiques*, Gauthier-Villars, 1885.

39. H. D. Conway, K. A. Farnham, and T. C. Ku, "The Indentation of a Transversely Isotropic Half Space by a Rigid Sphere," *Journal of Applied Mechanics*, **34**, 491–492 (1967).

40. V. Cerruti, "Ricerche intorno all'equilibrio de corpi elastici isotropi," *Atli academi nazl. Lincei, Mem. classe sci. fis., mat. c nat.*, **13**, 81 (1892).

41. C. W. Ng, "Green's Function of Radial Displacement in a Circular Disc Due to Unit Normal and Tangential Loads," *Wear*, **7**, 344–353 (1964).

42. S. D. Beck and F. F. Ling, "Stresses in a Layered Disc," *Proceedings of the Fifth U.S. National Congress of Applied Mechanics*, 244 (1966).

43. Y. C. Hsu and F. F. Ling, "Shear Stresses in a Layered Elastic System Under a Moving Load," *Recent Advances in Engineering Science*, **II**, Gordon and Breach, 323–351 (1965).

44. E. and F. Cosserat, *Théorie des Corps Déformables*, A. Hermann et Fils, 1909.

45. R. D. Mindlin and H. F. Tiersten, "Effects of Couple-Stresses in Linear Elasticity," *Archive of Rational Mechanics and Analysis*, **11**, 26–28 (1962).

46. G. Grioli, "Elasticità Asimmetrica," *Ann. di Mat. Pura ed Appl.*, **IV**, 389–417 (1960).

47. R. A. Toupin, "Elastic Material with Couple-Stresses," *Archive of Rational Mechanics and Analysis*, **11**, 385–414 (1962).

48. E. L. Aero and E. V. Kuvshinskii, "Fundamental Equations of the Theory of Elastic Media with Rotationally Interacting Particles," *Fizika Tverdogo Tela*, **II**, 1399–1409 (1960); *Trans. Soviet Physics Solid State*, **II**, 1272–1281 (1961).

49. R. F. Maye and F. F. Ling, "On the Solution of the Ring for Cosserat Materials," *Developments in Mechanics*, **4,** 151–169 (1967).

50. R. F. Maye and F. F. Ling, "General Solution of the Ring for Cosserat Materials," *Proceedings of the 5th U.S. National Congress of Applied Mechanics*, 304 (1966).

51. M. Hetenyi, "A Method of Solution for the Elastic Quarter-Plane," *Journal of Applied Mechanics*, **27,** 289–296 (1960).

52. M. Hetenyi, "A General Solution for the Elastic Quarter-Space," *Proceedings of the 5th U.S. National Congress of Applied Mechanics*, 278 (1966).

53. M. Hetenyi, "A General Solution for the Elastic Quarter-Space," *Journal of Applied Mechanics*, **37,** 70–76 (1970).

54. P. P. Teodorescu, "Sur le problème du quart d'espace élastique. I. Charges normales; II. Charges tangentielles," *Mécanique Appliquée*, **XIV,** 487–504 (1966).

55. R. A. Eubanks and E. Sternberg, "On the Completeness of the Boussinesq-Papkovich Stress Functions," *Journal of Rational Mechanics and Analysis*, 735–746 (1956).

56. A. I. Luré, *Three Dimensional Problems of the Theory of Elasticity* (English translation), Interscience Publishers, 1964.

57. E. A. Wilson and F. F. Ling, "Surface Displacements on the End of an Elastic Cylinder," AFML-TR-68-37 (1968).

58. E. Melan and H. Parkus, *Wärmespannungen infolge Stationärer Temperaturfelder*, Springer-Verlag, 1953.

59. H. Parkus, *Instationäre Wärmespannungen*, Springer-Verlag, 1959.

60. B. A. Boley and J. Weiner, *Theory of Thermal Stresses*, John Wiley and Sons, Inc., 1960.

61. W. Nowacki, *Thermoelasticity*, Addison-Wesley Publishing Company, 1962.

62. W. Sternberg and J. G. Chakravorty, "On Inertia Effects in a Transient Thermoelastic Problem," *Journal of Applied Mechanics*, **26,** 503–509 (1959).

63. V. I. Danilovskaya, "Thermal Stresses in an Elastic Half-Space Due to a Sudden Heating of Its Boundary," *Prikladnaya Matematika i Mekhanika*, **14,** 316–318 (1950).

64. H. Deresiewicz, "Planc Waves in a Thermoelastic Solid," *Journal of the Acoustical Society of America*, **29,** 204–209 (1957).

65. F. F. Ling and V. C. Mow, "Surface Displacement of a Convective Elastic Half-Space Under an Arbitrarily-Distributed Fast-Moving Heat Source," *Proceedings of the International Conference on Lubrication* (1964); also *Transactions of the American Society of Mechanical Engineers*, **87,** 729–734 (1965).

66. A. Jahanshahi, "Quasi-Static Stresses Due to Moving Temperature Discontinuity on a Plane Boundary," *Journal of Applied Mechanics*, **33,** 814–816 (1966).

67. M. A. Biot, "Theory of Stress-Strain Relations in Anisotropic Viscoelasticity and Relaxation Phenomena," *Journal of Applied Physics*, **25,** 1385–1391 (1954).

68. T. Alfrey, *Mechanical Behavior of High Polymers*, Interscience Publishers, 1948.

69. H. S. Tsien, "A Generalization of Alfrey's Theorem for Viscoelastic Media," *Quarterly of Applied Mathematics*, **8,** 104–106 (1950).

70. E. H. Lee, "Stress Analysis in Viscoelastic Bodies," *Quarterly of Applied Mathematics*, **13,** 183–190 (1955).

71. E. Sternberg, "On Transient Thermal Stresses in Linear Viscoelasticity," *Proceedings of the Third U.S. National Congress of Applied Mechanics*, 673–683 (1958).

72. S. C. Hunter, "The Rolling Contact of Rigid Cylinder with Viscoelastic Half Space," *Journal of Applied Mechanics*, **28,** 611–617 (1961).

73. D. G. Flom, "Rolling Friction of Polymeric Materials-Elastomers," *Journal of Applied Physics*, **31,** 306–314 (1960).

74. W. D. May, E. L. Morris, and D. Atack, "Rolling Friction of a Hard Cylinder Over a Viscoelastic Material," *Journal of Applied Physics*, **30,** 1713–1724 (1959).

75. L. W. Morland, "A Plane Problem of Rolling Contact in Linear Theory," *Journal of Applied Mechanics*, **29,** 345–352 (1962).

76. G. R. Abrahamson and J. N. Goodier, "The Hump Deformation Preceding a Moving Load on a Layer of Soft Material," *Journal of Applied Mechanics*, **28,** 608–610 (1961).

77. S. K. Batra and F. F. Ling, "Deformational Friction of a Viscoelastic Layered System Under Moving Load," *American Society of Lubrication Engineers Transactions*, **10,** 294–301 (1967).

78. W. Prager, *An Introduction to Plasticity*, Addison-Wesley Publishing Company, 1959.

79. R. Hill, "Some Special Problems of Indentation and Compression in Plasticity," *Proceedings of the 7th International Congress of Applied Mechanics*, **1,** 365–377 (1949).

80. H. Hencky, "Ueber einige statisch bestimemte Faelle des Gleichgewichts in plastischen Koerpern," *Zeitschrift für Angewandte Mathematik und Mechanik*, **3,** 241–251(1923).

81. L. Prandtl, "Anwendungsbeispiele zu einem Henckyschen Satz ueber das plastische Gleichgewicht," *Zeitschrift für Angewandte Mathematik und Mechanik*, **3,** 401–407 (1923).

82. C. Carathéodory and E. Schmidt, "Ueber die Hencky-Prandtlschen Kurvenscharen," *Zeitschrift für Angewandte Mathematik und Mechanik*, **3,** 468–475 (1923).

83. A. Haar and Th. v. Kármán, "Zur Theorie der Spannungszustaende in plastischen und sandartigen Medien," *Goettinger Nachricht, Mathematik-Physik*, **1909,** 204–218 (1909).

84. R. T. Shield, "On the Plastic Flow of Metals Under Conditions of Axial Symmetry," *Proceedings of the Royal Society*, **A233**, 267–286 (1955).

85. H. Lippmann, "Principal Line Theory of Axially-Symmetric Plastic Deformation," *Journal of Mechanics and Physics of Solids*, **10**, 111–122 (1962).

86. L. Prandtl, "Ueber die Haerte plastisher Koerper," *Goettinger Nachricht, Mathematik-Physik*, **1920**, 74–85 (1920).

87. R. Hill, "The Plastic Yielding of Notched Bars under Tension," *Quarterly Journal of Mechanics and Applied Mathematics*, **2**, 40–52 (1949).

88. A. P. Green, "The Use of Plasticine Models to Simulate the Plastic Flow of Metals," *Philosophical Magazine*, **42**, 365–373 (1951).

89. F. F. Ling, "Some Factors Influencing the Area-Load Characteristics of Semi-Smooth Surfaces under 'Static' Loading," *Transactions of the American Society of Mechanical Engineers*, **80**, 1113–1120 (1958).

90. C. H. Popelar, "On the Basic Equation of Junction Growth," *Journal of Applied Mechanics*, **36**, 132–133 (1969).

91. R. Hill, E. H. Lee, and S. J. Tupper, "The Theory of Wedge Indentation of Ductile Materials," *Proceedings of the Royal Society*, **A188**, 273–289 (1945).

92. V. V. Sokolovskii, *Teoriia plastichnosti*, Moscow, Izd-vo AN SSSR (1949).

93. A. In. Ishlinskii, "Osesimmetrichnaia zadacha teorii plastichnosti i proba Brinella," *Prikladu. matem. i mechanika*, **8**, Moscow, Izd-vo AN SSSR (1944).

94. V. N. Marochkin, "The Limiting Plastic State in Indenting and Compressing a Truncated Cone," *Friction and Wear in Machinery* (translated from Russian by the American Society of Mechanical Engineers), **13**, 79–131 (1959).

95. S. Ross and J. Olivier, *On Physical Adsorption*, John Wiley and Sons, Inc., 1964.

96. P. Beckmann and A. Spizzichino, *The Scattering of Electromagnetic Radiation from Rough Surfaces*, Pergamon Press, 1963.

97. R. Holm, *Electric Contacts*, Hugo Gebers, 1946.

98. F. P. Bowden and D. Tabor, *The Friction and Lubrication of Solids*, Vols. **I,** and **II,** Oxford Press, 1950, 1964.

99. R. A. Burton, "Effect of Two-dimensional, Sinusoidal Roughness on the Load Support Characteristics of a Lubricant Film," *Journal of Basic Engineering, Transactions of the American Society of Mechanical Engineers*, **85**, 258–262 (1963).

100. S. T. Tzeng and E. Saibel, "Surface Roughness Effect on Slider Bearing Lubrication," *American Society of Lubrication Engineers Transactions*, **10**, 334–338 (1967).

101. J. B. P. Williamson, J. Pullen, and R. T. Hunt, "The Shape of Solid Surfaces," *Surface Mechanics, a symposium volume*, American Society of Mechanical Engineers, 24–35 (1969).

102. D. J. Whitehouse and J. F. Archard, "The Properties of Random Surfaces in Contact," *Surface Mechanics, a symposium volume*, American Society of Mechanical Engineers, 36–57 (1969).

103. T. Nakamura, "The Analysis of Surface Roughness Curves, Optical Correlators and Their Utilizations," *Journal of the Society of Precision Engineering (Japan)*, **25**, 110–117 (1959).

104. Y. Onishi, "New Theory Evaluating Surface Roughness and Its Applications," Dissertation, Ehime University, (Japan), 1969.

105. P. E. D'yachenko, N. N. Tolkacheva, G. A. Andreev and T. M. Karpova, *The Actual Contact Area Between Touching Surfaces* (the original in Russian was published by the Institute of Mechanical Engineering, Academy of Sciences Press, Moscow, 1963), translated into English by Consultants Bureau, New York, 1964.

106. E. J. Abbott and F. A. Firestone, "Specifying Surface Quality," *Mechanical Engineering*, **55**, 569–572 (1933).

107. I. V. Kragelsky and N. B. Demkin, "Contact Area of Rough Surfaces," *Wear*, **3**, 170–187 (1960).

108. J. Wallach, "Surface Topography Description and Measurement," *Surface Mechanics, a symposium volume*, American Society of Mechanical Engineers. 1–23 (1969).

109. T. Nakamura, "The Analysis of Surface Roughness Curves," *Journal of the Society of Precision Engineering (Japan)*, **25**, 56–62 (1959).

110. V. Odvody and B. Pardubsky, "Festlegung der zweckmäsigen Bezugs- und Prüflangen für die zahlenmässige Bestimmung der Charakteristik der Oberflächenrauheit R_a im 'M-System' mittels elektrischer Profilometer," *C.I.R.P. Annalen*, **X**, 279–286 (1962).

111. W. O. Myers, "Characterization of Surface Roughness," *Wear*, **5**, 182–189 (1962).

112. J. A. Greenwood, "The Area of Contact between Rough Surfaces and Flats," *Transactions of the American Society of Mechanical Engineers*, **89**, 81–91 (1967).

113. J. Peklenik, "Contribution to the Theory of Surface Characterization," *C.I.R.P.-Annalen*, **XII**, 173–178 (1964).

114. J. F. Archard, "Single Contacts and Multiple Encounters," *Journal of Applied Physics*, **32**, 1420–1425 (1961).

115. A. Iwaki and M. Mori, "On the Distribution of Surface Roughness When Two Surfaces Are Pressed Together," *Bulletin of the Japanese Society of Mechanical Engineers*, **1**, 329–337 (1958).

116. T. Tsukizoe and T. Hisakado, "On the Mechanism of Contact Between Metal Surfaces—The Penetrating Depth and the Average Clearance," *Journal of Lubrication Technology, Transactions of the American Society of Mechanical Engineers*, **87**, 666–674 (1965).

117. F. F. Ling, "On Asperity Distributions of Metallic Surfaces," *Journal of Applied Physics*, **29**, 1168–1174 (1958).

118. M. C. Shaw, "A Yield Criterion for Ductile Metals Based Upon Atomic Structure," *Journal of the Franklin Institute*, **254**, 109–126 (1952).

119. E. A. Wilson, "The Area of Contact Between Rough Surfaces and Flow," *Journal of Lubrication Technology, Transactions of the American Society of Mechanical Engineers*, **90**, 521–522 (1968).

120. *The Science of Engineering Materials*, J. E. Goldman (Ed.), John Wiley and Sons, Inc., 1957.

121. V. I. Likhtman, E. D. Shchukin, and P. A. Rehbinder, *Physico-chemical Mechanics of Metals*, Academy of Sciences of the U.S.S.R. Institute of Physical Chemistry (translated by Israel Program for Scientific Translations, Jerusalem, 1962).

122. H. Blok, "The Dissipation of Frictional Heat," *Applied Scientific Research*, **A5**, 151–181 (1954).

123. J. Bandry, "Conditions de stabilité de l'équilibre thermique d'un palier lisse fonctionnant en régime de graissage onctueux," *C. R. Acad. Sci. Paris*, **227**, (1948).

124. H. Hertz, "Über die Berührung fester elastischer Körper," *Journal für die Reine und Angewandte Mathematik*, **92**, 156–171 (1881).

125. A. N. Grubin and I. E. Vinogradova, "Investigation of the Contact of Machine Components," *Central Scientific Research Institute for Technology and Mechanical Engineering, Book No.* 30 (*Moscow*), D.S.I.R. Translation No. 337 (1949).

126. D. Dowson and G. R. Higginson, "A Numerical Solution to the Elasto-hydrodynamic Problem," *Journal of Mechanical Engineering Science*, **1**, 6–15 (1959).

127. D. Dowson, "Elastohydrodynamic Lubrication: An Introduction and a Review of Theoretical Studies," *Proceedings of the Symposium on Elastohydro-dynamic Lubrication*, The Institute of Mechanical Engineers, 7–16 (1966).

128. H. S. Cheng, "A Refined Solution to the Thermal-Elastohydrodynamic Lubrication of Rolling and Sliding Cylinders," *American Society of Lubrication Engineers Transactions*, **8**, 397–410 (1965).

129. B. Jacobson, "On the Lubrication of Heavily Loaded Spherical Surfaces Considering Surface Deformations and Solidification of the Lubricant," *Acta Polytechnica Scandinavica, Me 54* (1970).

130. M. D. Hersey and V. Hopkins, "Viscosity of Lubricants under Pressure," *American Society of Mechanical Engineers Research Communication on Lubrication*, New York (1954).

131. P. W. Bridgman, "The Compression of Twenty-One Halogen Compounds and Eleven Other Simple Substances to 100,000 Kg/cm²," *Proceedings of the American Academy of Arts and Sciences*, **76**, 9–24 (1945).

132. J. W. Jackson and M. Waxman, "An Analysis of Pressure and Stress Distribution Under Rigid Bridgman-Type Anvils," *High Pressure Measurement*, Giadini and Lloyd (Eds.), Butterworth, 1963.

133. E. G. Thomsen, C. T. Yang, and S. Kobayashi, *Plastic Deformation in Metal Processing*, Macmillan Company, 1965.

134. A. T. Male and M. G. Cockcroft, "A Method for Determination of the Coefficient of Friction of Metals Under Conditions of Bulk Plastic Deformation," *Journal of the Institute of Metals*, **93**, 38–46 (1964).

135. E. A. Wilson, "Compression of a Thin Plastic Mass Between Two Elastic Cylinders," *Journal of Lubrication Technology, Transactions of the American Society of Mechanical Engineers*, **91**, 342–350 (1969).

136. J. J. Kalker, "Transient Phenomena in Two Elastic Cylinders Rolling over Each Other with Dry Friction," *Journal of Applied Mechanics*, **37**, 677–688 (1970).

137. H. Butler, "Zur Theorie der rollenden Reibung," *Ingenieur Archiv*, **XXVII**, 137–152 (1959).

138. C. Cattaneo, "Sul contatto di due corpi elastici; distribuzione locale degli sforzi," *Rend. Acad. Lincei*, **XXVII**, 342–478 (1938).

139. F. W. Carter, "On the Action of a Locomotive Driving Wheel," *Proceedings of the Royal Society*, **A112**, 151–157 (1926).

140. K. L. Johnson, "Tangential Traction and Microslip in Rolling Contact," *Proceedings of the Symposium of Rolling Contact Phenomena*, J. B. Bidwell, (Ed.), Elsevier, 6–26 (1962).

141. Y. C. Pao, T. S. Wu, and Y. P. Chiu, "Bounds on the Maximum Contact Stress of an Indented Elastic Layer," *Journal of Applied Mechanics*, **38**, 608–614 (1971).

142. D. C. Gakenheimer, "Response of an Elastic Half Space to Expanding Surface Loads," *Journal of Applied Mechanics*, **38**, 99–110 (1971).

143. K. J. Tong, "Dynamic Response of a Homogeneous Isotropic Elastic Half-Space to a Spreading Blast Wave," *Ph.D. Thesis*, Stanford University, 1968.

144. I. N. Sneddon, *Fourier Transforms*, McGraw-Hill Book Company, 1951.

145. Y. O. Tu and D. G. Gazis, "The Contact Problem of a Plate Pressed Between Two Spheres," *Journal of Applied Mechanics*, **31**, 659–666 (1964).

146. S. L. Pu, M. A. Hussain, and G. Anderson, "Lifting of a Plate from the Foundation Due to Axisymmetric Pressure," *Developments in Mechanics*, **5**, 577–590 (1970).

147. S. L. Pu and M. A. Hussain, "Note on the Unbonded Contact Between Plates and an Elastic Half Space," *Journal of Applied Mechanics*, **37**, 859–861 (1970).

148. G. B. Warburton, J. D. Richardson, and J. J. Webster, "Forced Vibrations of Two Masses on an Elastic Half Space," *Journal of Applied Mechanics*, **38**, 148–156 (1971).

149. W. T. Chen, "Stresses in Some Anisotropic Materials Due to Indentation and Sliding," *International Journal of Solids and Structures*, **5**, 191–214 (1969).

150. H. C. Hu, "On the Equilibrium of a Transversely Isotropic Elastic Half Space," *Acta Scientia Sinica*, **3**, 463–479 (1954).

151. H. B. Huntington, "The Elastic Constants of Crystals," *Advances in Research and Development*, F. Seitz and D. Turnbull (Eds.), **7**, 213 Academic Press, 1957.

152. W. T. Chen and P. A. Engel, "Impact and Contact Stress Analysis in Multi-layer Media, *IBM SDD Technical Report TR* 01.1484 (1971).

153. H. D. Conway, H. C. Lee and R. G. Bayer, "The Impact Between a Rigid Sphere and a Thin Layer," *Journal of Applied Mechanics*, **37**, 159–162 (1970).

154. R. M. Carson and K. L. Johnson, "Surface Corrugations Spontaneously Generated in a Rolling Contact Disc Machine," *Wear*, **17**, 59–72 (1971).

155. F. F. Ling, "Mechanics of Sliding Surfaces," *Metals Engineering Quarterly*, **10**, 1–4 (1967).

156. K. Y. Li and F. F. Ling, "The Sliding of Copper-Based Sintered Material Against Steel in Paraffinic Mineral Oil," *Wear*, **15**, 249–256 (1970).

157. E. F. Finkin, "An Annular Contact Friction and Boundary Lubrication Tester," *The Review of Scientific Instruments*, **39**, 690–696 (1968).

158. M. D. Hersey, *Theory and Research in Lubrication*, John Wiley and Sons, Inc., 1966.

159. F. F. Ling and C. F. Yang, "Temperature Distribution in a Semi-Infinite Solid Under a Fast-Moving Arbitrary Heat Source," *International Journal of Heat and Mass Transfer*, **14**, 199–206 (1970).

160. V. C. Mow and H. S. Cheng, "Thermal Stresses in an Elastic Half-space Associated with an Arbitrarily Distributed Moving Heat Source," *Zeitschrift für Angewandte Mathematik und Physik*, **18**, 500–507 (1967).

161. J. J. Wu and F. F. Ling, "A Method for Micro-hardness Analysis of an Elastoplastic Material," *Developments in Mechanics*, **6**, 359–372 (1971).

162. Y. Yamada, N. Yoshimura and T. Sakurai, "Plastic Stress-Strain Matrix and Its Application for the Solution of Elastic-Plastic Problems by the Finite Element Method," *International Journal of Mechanical Science*, **10**, 343 (1968).

163. A. W. Crook, "Simulated Gear-Tooth Contacts: Some Experiments upon Their Lubrication and Subsurface Deformation," *Proceedings of the Institute of Mechanical Engineers*, **171**, 187 (1957).

164. J. E. Merwin and K. L. Johnson, "An Analysis of Plastic Deformation in Rolling Contact," *Proceedings of the Institute of Mechanical Engineers*, **177**, 676–685 (1963).

165. K. L. Johnson, "A Shakedown Limit in Rolling Contact," *Proceedings of the Fourth U.S National Congress of Applied Mechanics*, **2**, 971–975 (1962).

166. C. H. Yew and W. Goldsmith, "Stress Distributions in Soft Metals Due to Static and Dynamic Loading by a Steel Sphere," *Journal of Applied Mechanics*, **31**, 635–646 (1964).

167. E. G. Thomsen, C. T. Yang, and S. Kobayashi, *Mechanics of Plastic Deformation in Metal Processing*, The Macmillan Company, 1965.

168. D. Tabor, *The Hardness of Metals*, Oxford, 1951.

169. D. Tabor, "The Hardness of Solids," *Review of Physics in Technology*, **1,** 145–179 (1970).

170. G. Amontons, "De la Résistance Causée dans les Machines," *Mémoires Académie Royal Science*, 206 (1699); 96 (1704).

171. F. F. Ling and M. B. Peterson, "Friction and Lubrication in Metalworking Processes," *Proceedings of the International Conference on Manufacturing Technology, American Society of Tool and Manufacturing Engineers*, 1181–1192 (1967).

172. M. B. Peterson and F. F. Ling, "Frictional Behavior in Metalworking Processes," *Journal of Lubrication Technology, Transactions of the American Society of Mechanical Engineers*, **92,** 535–542 (1970).

173. F. F. Ling and J. D. Vasilakis, "Pressure Dependence of Shear Strength and Sliding Friction," *Journal of Applied Mechanics*, **37,** 192–195 (1970).

174. C. W. McCutchen, "Animal Joints and Weeping Lubrication," *New Scientist*, **23,** 412–415 (1961).

175. C. W. McCutchen, "The Frictional Properties of Animal Joints," *Wear*, **5,** 1–17 (1962).

176. C. W. McCutchen, "Boundary Lubrication by Synovial Fluid: Demonstration and Possible Osmotic Explanation," *Federation Proceedings*, **25,** 1061–1068 (1966).

177. M. C. Mow and F. F. Ling, "On Weeping Lubrication Theory," *Zeitschrift für Angewandte Mathematik und Physik*, **20,** 156–166 (1969).

178. F. G. Evans and A. I. King, "Regional Differences in Some Physical Properties of Human Spongy Bone," *Biomedical Studies of the Musculo-skeletal System*, F. G. Evans (ed.), Springfield, Illinois, Charles C Thomas, 49–67 (1961).

179. L. Sokoloff, "Elasticity of Aging Cartilage," *Federation Proceedings*, **25,** 1089–1095 (1966).

180. S. M. Elmore, L. Sokoloff, G. Norris, and P. Carmeco, "Nature of Imperfect Elasticity of Articular Cartilage," *Journal of Applied Physics*, **18,** 393–396 (1963).

181. D. Dowson, "Modes of Lubrication in Human Joints," *Symposium on Lubrication and Wear in Living and Artificial Human Joints: Proceedings of the Institute of Mechanical Engineers*, **181** (1967).

182. V. C. Mow, "The Role of Lubrication in Biomechanical Joints," *Journal of Lubrication Technology, Transactions of the American Society of Mechanical Engineers*, **91,** 320–328 (1969).

183. F. C. Linn, "Lubrication of Animal Joints," *Journal of Lubrication Technology, Transactions of the American Society of Mechanical Engineers*, **91,** 329–341 (1969).

APPENDIX
———————

SINGULAR INTEGRAL EQUATIONS

In many of the problems discussed in this book, results have been put in the form of finite singular integral equations of the first kind

$$f(s) = \int g(t)\, K(s,t)dt \quad , \qquad [A.1]$$

where $K(s, t)$ is singular, in that K becomes unbounded as $s \to t$ and that the integral often exists only in the Cauchy principle value sense. Moreover $g(t)$ in [A.1] is sought while $f(s)$ is given. In this case, it is desired to have the singular integral equation [A.1] inverted. Several known singular equations, their inverses and examples of their application are given in this appendix.

A.   ABEL'S INTEGRAL EQUATION OF THE FIRST KIND

This is a singular integral equation of the first kind which is of the Volterra type,[A1] that is,

$$f(s) = k \int_{-1}^{s} \frac{g(t)dt}{\sqrt{s-t}} \quad . \qquad [A.2]$$

[301]

The inverse formula of [A.2] is

$$g(s) = (k\pi)^{-1} \int_{-1}^{s} \frac{f'(t)dt}{\sqrt{s-t}} \quad , \qquad [A.3]$$

where the prime refers to differentiation.

For the problem shown in Figure 2.1, without convection, the fundamental solution for $R \gg 1$ is given by [2.16]. The general solution, as given by [2.15] with [2.16], applies with $H = 0$ and is of the form [A.2].

As an example, how should $P(\xi)$ be distributed in $-1 \leq \xi \leq 1$ so that $\phi(\xi)$ would be linearly distributed? More specifically $\phi(\xi) = \sqrt{(\pi/R)}$ ξ in $-1 \leq \xi \leq 1$. Then $\phi'(\xi) = \sqrt{\pi/R}$ and the application of the inverse formula leads immediately to

$$P(\xi) = \sqrt{\xi + 1} \quad .$$

In other words, if the heat source $P(\xi)$ is parabolically distributed, then the temperature $\phi(\xi)$ would be linearly distributed.

B.   ABEL'S INTEGRAL EQUATION OF THE SECOND KIND

This is a singular equation of the form

$$f(s) = \int_{0}^{1} \frac{g(t)\,dt}{\sqrt{s^2 - t^2}} \quad . \qquad [A.4]$$

Equation A.4 possesses an inverse[A2] of the form

$$g(s) = \pi^{-2} \int_{0}^{1} \frac{f(t)\,dt}{\sqrt{s^2 - t^2}} \quad . \qquad [A.5]$$

This was used in deriving [3.12].[A3] Also, [A.5] and [A.3] apply to [4.19] on the question of the pressure distribution between a rigid flat punch and a large elastic body.[A1]

First, [4.19] may be alternatively expressed as

$$w(\rho,0) = \pi^{-1}(\kappa + 2)(\kappa + 1)^{-1} \int_{0}^{\rho} \frac{\Phi(\xi)\,d\xi}{\sqrt{\rho^2 - \xi^2}} \quad ,$$

where

$$\Phi(\xi) = \int_{\xi}^{1} \frac{q(\zeta)\,\zeta\,d\zeta}{\sqrt{\zeta^2 - \xi^2}} \quad .$$

Given $w = $ constant, $\Phi(\xi)$ may be found by means of [A.5]. Given $\Phi(\xi)$, thus found, $q(\zeta)$ may be found by means of [A.3] after a simple transformation,

$$q(\rho) = \frac{q_0}{2\sqrt{1 - \rho^2}} \quad , \qquad [A.6]$$

where q_0 is the average value of q.

For an axisymmetric parabolic punch, which is expressible as $\alpha - \beta\rho^2$, $q(\rho)$ may be found by the same procedure as above,

$$q(\rho) = \frac{3}{2}\,q_0\,\sqrt{1 - \rho^2} \quad . \qquad [A.7]$$

Equation A.7 is the Hertz pressure.[A4]

C. CARLEMAN'S INTEGRAL EQUATION

This is a singular equation of the first kind

$$f(s) = \int_{-1}^{1} g(t)\,\ell n\,|s - t|\,dt \quad . \qquad [A.8]$$

The inverse[A5] of [1.8] is

$$g(s) = \frac{1}{\pi\sqrt{1 - s^2}}\,[A^*(f) + L(f')] \quad , \qquad [A.9]$$

where

$$A^*(z) = \frac{1}{\pi\,\ell n\,2}\int_{-1}^{1}\frac{z(t)dt}{\sqrt{1 - t^2}} \quad ,$$

$$L(z) = -\frac{1}{\pi}\int_{-1}^{1}\frac{\sqrt{1 - t^2}\,z(t)dt}{t - s} \quad . \qquad [A.10]$$

This formula can be generalized; the one used in reference A6 is

$$f(s) = \int_{-1}^{1} g(t)[\ln 2\beta - \ln |s - t|]\, dt \quad , \qquad [A.11]$$

whose inverse is shown to be

$$g(s) = \frac{1}{\pi\sqrt{1 - s^2}}\left[\frac{A^*(f)}{1 + a} - \frac{aL^*(f')}{1 + a} + L(f')\right] \quad , \qquad [A.12]$$

where in addition to [A.10],

$$a = \frac{\ln 2\beta}{\ln 2} \quad , \qquad L^*(z) = \int_{-1}^{1} \frac{L\, ds}{\pi\sqrt{1 - s^2}} \quad .$$

In fact it is possible to work out the inverse of

$$f(s) = \int_{-1}^{1} g(t)[\ln 2\beta - h(s,t)\ln |s - t|]\, dt \quad ,$$

where β is any constant and so long as $h(s, t)$ is a power series of $(s - t)$.

As examples in the use of [A.9], reference is made of [4.13]. The pressure distribution between a rigid, flat punch and a large elastic body is then

$$q(\xi) = \frac{q_0}{\pi\sqrt{1 - \xi^2}} \quad . \qquad [A.13]$$

This is the Sadowsky[A7] solution, although it is obtained by another method.

For a punch, parabolic in cross section, [4.13] leads to

$$q(\xi) = \frac{2q_0}{\pi}\sqrt{1 - \xi^2} \quad , \qquad [A.14]$$

which is, of course, also Hertz pressure in two-dimensions.[A4]

D. OTHER SINGULAR INTEGRAL EQUATIONS

In addition, the generalized Abel equation[A1,A8]

$$f(s) = \int_0^s \frac{g(t)\,dt}{(s-t)^\alpha} \quad (0 < \alpha < 1) \quad , \quad [A.15]$$

has an inverse

$$g(s) = \frac{\sin(\alpha\pi)}{\pi} \left[\frac{f(0)}{s^{1-\alpha}} + \int_0^s \frac{f'(t)\,dt}{(s-t)^{1-\alpha}} \right] \quad . \quad [A.16]$$

The finite Hilbert transformation integral[A9]

$$f(s) = \frac{1}{\pi} \int_{-1}^1 \frac{g(t)\,dt}{s-t} \quad , \quad\quad [A.17]$$

has the inverse, if a solution exists,

$$g(s) = -\frac{1}{\pi} \int_{-1}^1 \sqrt{\frac{1-t^2}{1-s^2}} \cdot \frac{f(t)\,dt}{s-t} + \frac{C}{\sqrt{1-s^2}} \quad , [A.18]$$

where

$$C = \frac{1}{\pi} \int_{-1}^1 g(t)\,dt \quad .$$

The Hilbert-Betz integral

$$f(s) = \int_0^\ell g(t)\,\mathrm{csch}(s-t)\,dt \quad\quad [A.19]$$

has the inverse

$$\frac{g(s)}{\sqrt{1-s^2}} = -\frac{1}{\pi\sqrt{s(\tanh \ell - s)}} \left\{ \int_0^{\tanh \ell} \frac{\sqrt{t(\tanh \ell - t)}}{s-t} f(t)\,dt + \mathrm{constant} \right\} \quad .$$

$$[A.20]$$

Also

$$f(s) = \int_{-c}^{c} g(t)\cot(s-t)dt \quad , \qquad [A.21]$$

has an inverse

$$g(s) = \frac{kc\,\cos s}{\pi\sqrt{\sin^2 c - \sin^2 s}} + \frac{2}{\pi\sqrt{\sin^2 c - \sin^2 s}} \int_{-c}^{c} \frac{f(t)\sqrt{\sin^2 c - \sin^2 t}\,dt}{\sin(s-t)} \quad .$$

$$[A.22]$$

It should be noted that the above does not include many variations of the same basic integrals. Moreover it should be of interest to note that a general method exists for investing integral equations of the first kind,[A10,A11] which is of the form,

$$\phi(x) = \int_{a}^{b} K(x-t)f(t)dt \qquad (a < x < b) \quad ,$$

$$[A.23]$$

where the kernel $K(x)$ has the form

$$K(x) = p(x)j(x) + q(x) \quad . \qquad [A.24]$$

According to the method of Latta, $j(x)$ satisfies a differential equation, while p and q are polynomials.

The method will be outlined as follows: define the operational notation

$$\int_{a}^{b} K(x-t)f(t)dt = \Lambda f \quad , \qquad [A.25]$$

$$\int_{a}^{b} j(x-t)f(t)dt = \Gamma f \quad , \qquad [A.26]$$

where the kernel of Γf is j of [A.24].

The polynomial q of [A.24] is immaterial (i.e., it can be easily incorporated into $\phi(x)$ of [A.23]); thus without loss of generality,

$$K(x) = p(x) j(x) \quad ,$$

in lieu of [A.24].

Letting

$$p(x) = \sum_n p_n x^n \quad ,$$

then

$$\Lambda f = \sum_n p_n \sum_\nu (-1)^\nu \binom{n}{\nu} x^{n-\nu} \int_a^b j(x-t) t^\nu f(t) dt$$

$$= \sum_n p_n \sum_\nu (-1)^\nu \binom{n}{\nu} x^{n-\nu} \Gamma x^\nu f \quad . \qquad [A.27]$$

Latta's condition that $j(x)$ must satisfy a differential equation allows one to relate Γxf (and as a consequence, $\Gamma x^\nu f$) to Γf by a linear differential equation. Using this fact, one can derive from [A.27] a differential equation relating Γf and Λf. The differential equation between Γf and Λf yields as a result

$$\psi(x) = \Gamma f \quad . \qquad [A.28]$$

As a first example, consider

$$K(x) = x \log |x| \quad . \qquad [A.29]$$

Now $j(x) = \log |x|$, satisfying a differential equation of the type $xj'(x) - 1 = 0.$

From the operator Γf defined in [A.26],

$$\Gamma f = \int_a^b \ell n |x-t| f(t) dt$$

$$\Gamma xf = \int_a^b t \ln |x-t| f(t)dt$$

$$= -\int_a^b (x-t) \ln |x-t| f(t)dt + x\Gamma f \quad .$$

Thus

$$\frac{d}{dx}(\Gamma xf) = \Gamma' xf = x\Gamma' f - \int_a^b f(t)dt \quad .$$

The moment is denoted by $\mu_i = \int_a^b t^i f(t)\,dt$, where $i = 0, 1, 2, \ldots, n$. Thus

$$\Gamma' xf = x\Gamma' f - \mu_0 \quad . \qquad\qquad [A.30]$$

Equations A.25 and A.29 lead to

$$\Lambda f = x\Gamma f - \Gamma xf$$

or

$$\Lambda' f = x\Gamma' f + \Gamma f - \Gamma' xf \quad .$$

With the aid of [A.30],

$$\Lambda' f = \Gamma f + \mu_0 \quad .$$

Using [A.28] and $\Lambda' f = \phi'$,

$$\psi = \phi' - \mu_0 \quad , \qquad\qquad [A.31]$$

where ϕ, in general, is restricted to polynomials or an exponential polynomial. In particular, if $\phi(x) = x$, $b = -a = 1$, [A.28] and [A.29] lead to

$$\Gamma f = 1 - \mu_0 \quad .$$

According to Carleman's solution with the right-hand side being constant,

$$f = \frac{\mu_0 - 1}{\pi \, \log 2} (1 - x^2)^{-1/2} \quad .$$

Integrating

$$\mu_0 = \int_{-1}^{1} f(t)dt = \frac{\mu_0 - 1}{\ln 2}$$

or

$$\mu_0 = \frac{1}{1 - \ln 2}$$

and the solution is

$$f = \frac{(1 - x^2)^{-1/2}}{\pi (1 - \ln 2)} \quad . \qquad [A.32]$$

As second example, consider the kernel of the Pearson type,

$$K(x) = (1 + \alpha x) \log |x| \quad . \qquad [A.33]$$

Equations A.33 and A.25 yield

$$\Lambda f = (1 + \alpha x)\Gamma f - \alpha \Gamma x f \quad ,$$

so that

$$\Lambda' f = (1 + \alpha x)\Gamma' f + \alpha \Gamma f - \alpha \Gamma' x f \quad .$$

And, with the aid of [A.30],

$$\Lambda' f = \Gamma' f + \alpha \Gamma f + \alpha \mu_0 \quad .$$

Now [A.28] and [A.23] lead to

$$\psi' + \alpha \psi = \phi - \alpha \mu_0 \quad . \qquad [A.34]$$

What remains to be solved is [A.28], where Ψ is some solution of [A.34].
Setting $\phi(x) = 1$, $b = -a = 1$ in [A.24], and solving

$$\psi(x) = -\mu_0 + Ae^{-\alpha x} \quad .$$

For α small,

$$\psi(x) = (A - \mu_0) - A\alpha x + 0(\alpha^2) \quad , \qquad [A.35]$$

Where A is a constant.

Defining $f_0(x)$ by the equation $\Gamma f_0 = 1$,

$$f_0(x) = -\frac{1}{\pi\ \ell n\ 2}\ (1 - x^2)^{-1/2} \quad . \qquad [A.36]$$

From [A.30]

$$\Gamma' x f_0 = x\Gamma' f_0 - \mu_0^{*}$$

$$= -\mu_0^{*} = -\int_{-1}^{1} f_0(x)dx = \frac{1}{\ell n\ 2} \quad .$$

Integrating,

$$\Gamma x f_0 = \frac{x}{\ell n\ 2} \quad . \qquad [A.37]$$

Considering [A.28] with the value in [A.35],

$$\Gamma f = \psi = (A - \mu_0) - A\alpha x + 0(\alpha^2) \quad .$$

With the aid of [A.36] and [A.37],

$$\Gamma f = (A - \mu_0)\Gamma f_0 - (A\alpha\ \ell n\ 2)\Gamma x f_0 + 0(\alpha^2) \quad . \quad [A.38]$$

Equation A.38 has the form of Carleman type, the solution of which has the singularities of the type $(1 - x^2)^{-1/2}$. It follows from [A.38]

$$f = (A - \mu_0)f_0 - (A\alpha\ \ell n\ 2)x f_0 + 0[\alpha^2(1 - x^2)^{1/2}] \quad .$$

Substitution f_0 from [A.36],

$$f = (1 - x^2)^{-1/2}\ [\ \frac{\mu_0 - A}{\pi\ \ell n\ 2} + \frac{A\alpha}{\pi}\ x + 0(\alpha^2)] \quad . \quad [A.39]$$

The constant A can be evaluated by integrating [A.39]

$$\mu_0 = \int_{-1}^{1} f\,dx = \frac{\mu_0 - A}{\ell n\ 2} + 0(\alpha^2) \quad .$$

Thus

$$A = \mu_0(1 - \ln 2) + O(\alpha^2) \quad ,$$

and

$$f = (1 - x^2)^{1/2} \left[\frac{\mu_0}{\pi} + \frac{\mu_0 \alpha}{\pi} (1 - \ln 2)x + O(\alpha^2) \right] \quad . \quad [\text{A}.40]$$

To evaluate μ_0, recall that

$$\Lambda f = 1 = (1 + \alpha x)\Gamma f - \alpha \Gamma x f \quad . \qquad [\text{A}.41]$$

Now putting [A.40] into f_0 with the aid of [A.36] and [A.37],

$$f = -(\mu_0 \ln 2)f_0 - (\mu_0 \alpha \ln 2)(1 - \ln 2)xf_0 + O[\alpha^2(1 - x^2)^{-1/2}] \quad .$$

$$[\text{A}.42]$$

Hence

$$\Gamma f = -\mu_0 \ln 2 - \mu_0 \alpha(1 - \ln 2)x + O[\alpha^2] \quad . \quad [\text{A}.43]$$

Multiplying [A.42] by x and the operator by Γ,

$$\Gamma x f = -\mu_0 x + O(\alpha) \quad . \qquad [\text{A}.44]$$

Substituting [A.43] and [A.44] into [A.41] results in

$$1 = (1 + \alpha x)[-\mu_0 \ln 2 - \mu_0 \alpha (1 - \ln 2)x] + \mu_0 \alpha x + O(\alpha^2)$$

$$= -\mu_0 \ln 2 + O(\alpha^2) \quad .$$

Thus

$$\mu_0 = -\frac{1}{\ln 2} + O(\alpha^2) \quad ,$$

and

$$f = (1 - x^2)^{-1/2} \left[-\frac{1}{\pi \ln 2} - \frac{\alpha}{\pi} \left(\frac{1}{\ln 2} - 1 \right)x + O(\alpha^2) \right] \quad ,$$

which check with Pearson's solution.

REFERENCES

A1. N. H. Abel, "Solution de quelques problième à láide d'integrales definies," *Magazin for Naturvidenskaberne*, **1** (1823); also collected works, *Christiania*, **1**, 11–27 (1881).

A2. C. E. Pearson, "On the Finite Strip Problem," *Quarterly of Applied Mathematics*, **XV**, 203–211 (1957).

A3. J. C. Samuels, "On Random Heat Conduction in Solids," *International Journal of Heat Transfer*, **9**, 301–314 (1966).

A4. H. Hertz, "Über die Beruhrung fester elastischer Körper," *Journal für die Reine und Angewandte Mathematik*, **92**, 156–171 (1881).

A5. T. Carleman, "Über die Abelsche Integralgleichung mit Konstanten Integrations Grenzen," *Mathematik Zeitschrift*, **15**, 111–120 (1922).

A6. F. F. Ling, "A Quasi-Iterative Method for Computing Interface Temperature Distributions," *Zeitschrift für Angewandte Mathematik und Physik*," **X**, 461–474 (1959).

A7. M. A. Sadowsky, "Zweidimensionale Probleme der Elastizitätstheorie," *Zeitschrift für Angewandte Mathematik und Merchanik*, **9**, 107–121 (1928).

A8. F. G. Tricomi, *Integral Equations*, Interscience Publishers, 1957.

A9. F. G. Tricomi, "On the Finite Hilbert Transformation," *Quarterly Journal of Mathematics*, **2**, 199–211 (1951).

A10. A. Shinbrot, "A Generalization of Latta's Method for the Solution of Integral Equations," *Quarterly of Applied Mathematics*, **XVII**, 415–421 (1959).

A11. G. E. Latta, "The Solution of a Class of Integral Equations." *Journal of Rational Mechanics and Analysis*, **5**, 821–833 (1965).

INDEX